POLISH
ACADEMY
OF SCIENCES
INSTITUTE
OF PHYSICS

QUANTUM OPTICS

Proceedings
of the VI International School
of Coherent Optics
September 19–26, 1985, Ustroń (Poland)

Edited by

ADAM KUJAWSKI, MACIEJ LEWENSTEIN

D. Reidel Publishing Company

A MEMBER OF THE KLUWER ACADEMIC PUBLISHERS GROUP

Dordrecht / Boston / Lancaster / Tokyo

Library of Congress Cataloging-in-Publication Data

International School of Coherent Optics (6th: 1985: Ustroń, Poland)

Quantum optics.

(Proceedings of conferences in physics / Polish Academy of Sciences, Institute of Physics; v. 7)

1. Quantum optics — Congresses. I. Kujawski, Adam, 1933 —   . II. Lewenstein, Maciej, 1955 —   III. Title. IV. Series: Proceedings of conferences in physics; v. 7.

QC446.15.I63 1985   535'.2   86-13112

ISBN 90-277-2281-1

Published by Ossolineum Publishing House, Wrocław, Poland
in co-edition with
D. Reidel Publishing Company, Dordrecht, Holland

Sold and distributed in the U.S.A. and Canada
by Kluwer Academic Publishers,
101 Philip Drive, Assinippi Park, Norwell, MA 02061, U.S.A.

Sold and distributed in Albania, Bulgaria, China, Czechoslovakia, Cuba, German Democratic Republic, Hungary, Mongolia, Northern Korea, Poland, Rumania, U.S.S.R., Vietnam, and Yugoslavia by Foreign Trade Enterprise, Ars Polona, Krakowskie Przedmieście. 7, 00-068 Warszawa, Poland

Sold and distributed in all remaining countries
by Kluwer Academic Publishers Group,
P.O. Box 322, 3300 AH Dordrecht, Holland

All rights reserved

© 1986 by Ossolineum, Publishing House of the Polish Academy of Sciences, Wrocław, Poland. No part of the material protected by this copyright notice may be reproduced or utilised in any form or by any means, electronic or mechanical, including photocopying, recording or by any information storage and retrieval system, without written permission from the copyright owner.

Printed in Poland

Wrocławska Drukarnia Naukowa

The VI International School of Coherent Optics "Quantum Optics" was organized by the Institute of Physics and Institute for Theoretical Physics of the Polish Academy of Sciences.

Programme and organizing Committee:

I. Białynicki-Birula (Chairman of the Programme Committee)
Z. Białynicka-Birula
L. Borg
M. Gajda
M. Głódź
B. Jakubczak
W. Jastrzębski
K. Kolwas
M. Kolwas
A. Kujawski (Chairman of the Organizing Committee)
M. Lewenstein
M. Łukaszewski
J. Mostowski
K. Rosiński
K. Rząźewski
K. Wódkiewicz
W. Żakowicz

Editors:
A. Kujawski
M. Lewenstein

LIST OF INVITED SPEAKERS:

I. J. Bersons
M. Bertolotti
Z. Białynicka-Birula
R. G. Brewer
W. Brunner
J. H. Eberly
Gy. Farkas
M. V. Fedorov
R. J. Glauber
R. Graham
F. Haake
A. P. Kazantsev
S. Kielich
P. L. Knight
A. V. Masalov
P. Meystre
P. W. Milonni
J. Peřina
K. Rząźewski
M. Schubert
B. W. Shore
S. J. Smith
S. Stenholm
G. I. Surdutovich
H. Walther
K. Wódkiewicz
P. Zoller

CONTENTS

1. A. KUJAWSKI, Introduction 7
2. R. J. GLAUBER, Simple Models for Linear Damping and Amplification 9
3. R. GRAHAM, Chaos in Quantum Optics 31
4. M. V. FEDOROV, Resonances and Saturation in the Continuum 34
5. P. FILIPOWICZ, J. JAVANAINEN, P. MEYSTRE, The Ultimate Maser 47
6. A. P. KAZANTSEV, Kinetic Phenomena of Atomic Motion in a Light Field 55
7. M. SCHUBERT, New Photon States and Quantum-Statistical Analysis of Multiphoton Processes 56
8. S. STENHOLM, Theory of Laser Cooling 74
9. P. W. MILONNI, Chaos in Quantum Optics 83
10. J. PERINA, Sub-Poisson Light 87
11. K. WÓDKIEWICZ, Noise in Strong Laser-Atom Interactions 97
12. W. BRUNNER, R. FISCHER, H. PAUL, Regular and Irregular Behaviour of Multimode Laser Radiation 107
13. Gy. FARKAS, Experimental Investigations of the Laser Induced Multiphoton and Tunneling Processes in Atoms and Solids 116
14. J. H. EBERLY, Essential States in Multiphoton Ionisation and Electron Scattering 125
15. S. M. BARNETT, P. L. KNIGHT, P. M. RADMORE, Dephasing and Decay of Quantum Optical Systems 133
16. G. REMPE, H. WALTHER, P. DOBIASCH, The One-Atom Maser – A Test System for Simple Quantumelectrodynamical Effects 144

17. Z. BIAŁYNICKA-BIRULA, Ionisation of Atoms by High-Intensity Lasers: Collective Effects 165
18. E. G. PESTOV, Relaxation of Quantum Systems in Strong Electromagnetic Field - New Nonlinear Effects 173
19. S. Ya. KILIN, The Quantum Statistics of Light Scattering and Propagation Effects 185
20. F. HAAKE, Quantum Pencils and a Kicked Spinning Top: Two Different Types of Large Fluctuations 196
21. A. V. MASALOV, Spectral and Temporal Fluctuations of Broad-Band Laser Radiation 207
22. P. ZOLLER, Rydberg Thresholds in Laser Fields: Applying Multichannel Quantum Defect Theory to Quantum Optics Problems 219
23. B. W. SHORE, Stochastic Processes in Quantum Optics ... 230
24. D. L. SHEPELYANSKY, Intrisic Chaos in Quantum Systems 240
25. J. W. HAUS, Quantum Beat Superfluorescence 241
26. I. J. BERSONS, Semiclassical Theory of Multiphoton Processes in Rydberg Atoms 252
27. P. R. BERMAN, R. G. BREWER, Modified Bloch Equations for Solids 253
28. R. G. DE VOE, C. FABRE, R. G. BREWER, Laser Frequency Division and Stabilisation 259
29. S. J. SMITH, Experimental Investigations of the Role of Laser Field Fluctuations in Non-linear Absorption Processes 265
30. S. KIELICH, R. TANAŚ, Self-Squeezing as a Novel Potent Source of Quantum Field 275
31. G. I. SURDUTOVICH, Light Pressure and Bistability Phenomenon 288
32. K. RZĄŻEWSKI, Spectra Generated by Short Laser Pulses 290
33. M. BERTOLOTTI, Historical Paths in Lasers and Quantum Optics 298
34. B. SHORE, Concluding Remarks 316
35. List of Presented Posters 320

INTRODUCTION

This volume contains the proceedings of the Sixth International School of Coherent Optics (ISCO-6) held in Ustroń, in the low mountain region of southern Poland, in September 19-26, 1985. The Schools of Coherent Optics were initiated in the Academy of Sciences of the USSR under the rectorship of Academician A. M. Prokhorov. ISCO-6, devoted to quantum optics, was organized by the Institute of Physics of the Polish Academy of Sciences. The two previous schools were held at Bechyne, Czechoslovakia (1983) and at Jena, German Democratic Republic (1984).

The contributions printed here are based on lectures presented by leading scientists in quantum optics and represent both the invited papers, seminars and the titles of posters. The invited lectures covered fundamental aspects of quantum optics including particularly multiphoton processes, statistical properties of light, stochastic methods and chaos, single-atom QED in cavity, light pressure and atomic coherence. To mark the 25th anniversary of the invention of laser Professor M. Bertolotti gave an after-dinner lecture on "Historical paths in lasers and quantum optics", which is reproduced at the end of this volume.

Our School enjoyed the participation of people from 15 nations. The interchange of scientific results was possible not only through the lectures, seminars and posters but also through personal contacts and lively discussions. We are grateful to Dr R. Sinclair for enjoyable evening when he showed the film "Sun Dagger" and discussed his participation in recent discoveries concerning the astronomy of the pre--Columbian Pueblo Indians of North America - ca 1000 A. D.

Many people contributed to work on the program under the leadership of Iwo Białynicki-Birula, chairman of the program committee. All of the participants appreciated the ready assistance of J. Kosiński, director of the recreation center Muflon, when the meeting was held. My heartiest thanks are addressed to Z. Białynicka-Birula, L. Borg, B. Jakubczak, K. Kolwas, M. Kolwas, H. Lewandowska, M. Lewenstein, M. Łukaszewski and J. Mostowski, and others from the program and organizing committee for their effort and efficient work on the preparation and smooth running of the School.

Attendance and participation at the meeting of delegates from the USSR and the USA was materially assisted by the advice and support given by Dr V. D. Novikov, Scientific Secretary of the Scientific Council of Coherent and Nonlinear Optics of the USSR Academy of Sciences and Dr B. W. Shore from the Lawrence Livermore National Laboratory.

Chairman of the Organizing Committee

Adam Kujawski

R. J. GLAUBER

Lyman Laboratory,
Harvard University,
Cambridge, Mass. 02138

SIMPLE MODELS FOR LINEAR DAMPING AND AMPLIFICATION[*]

I. Introduction

The subject I shall discuss is the way in which signals at the quantum level can be increased or decreased in magnitude. That they can be diminished in magnitude is an old story. Attenuation by damping is part of nature and often unavoidable. A much more interesting assignment is to amplify quantum signals and, if possible, to do it coherently, that is to preserve as much of the original signal as possible. Our problem then is to construct mathematical models of devices which accomplish that.

These amplifiers are intrinsically irreversible in principle. They can take an input signal from a state defined at the quantum level to a state which involves field strengths that are arbitrarily great and classical in nature. In doing that they inevitably introduce noise into the signal. There is no such thing as an amplifier that functions without noise and the reasons are intrinsically quantum mechanical. These devices are inevitably noise generators too and, to the extent that they always corrupt at least a little our knowledge of the original state, their action is irreversible.

There is a closely related mechanism which is far more widespread than amplification, although, I must say, it seems that there are ampli-

---

[*]This notes were prepared from a tape recording of Professor Glauber's lecture with a minimum of editing by J. Haus and M. Lewenstein.

fiers everywhere these days. That is the mechanism that attenuates signals and such damping or dissipation is present in just about everything. It has been a long standing problem how to represent dissipative processes quantum mechanically. One of the first place we encounter the problem is in dealing with the harmonic oscillator. In fact we encounter that one so often you might question seriously whether we have learned to solve any other problem. As you know, an amazing number of problems can be reduced to harmonic oscillator problem.

There is one version of the harmonic oscillator problem that we learn about very early in our careers as physicists. We learn to account for dissipation in the harmonic oscillator, so that when we talk about resonant phenomena like the effects of Napoleon's army marching over a bridge the vibration amplitude need not always diverge; the bridge need not always collapse. The resonant amplitude remains finite because of the existence of dissipation.

In classical physics accounting for dissipation is relatively easy. A simple dissipative term is added to the equation of motion for the harmonic oscillator; an acceleration is added proportional to the velocity of the oscillation and directed oppositely to it.

Why can't we do the same thing quantum mechanically? We have a Heisenberg equation of motion so why not just add a dissipative term which has the same form as its classical counterpart? What results from this prescription, of course, is considerable nonsense. The theory is not unitary; it does not preserve probabilities and causes the uncertainty principle to be violated as well. The commutator of position and momentum which results decays exponentially.

One of the basic ideas that was understood twenty-five years or so ago was that dissipation and fluctuation are always related to one another. This relationship must be preserved in the quantum mechanical description. The existence of a dissipative terms means that there must be an accompanying quantum fluctuation term. It is this fluctuation that maintains, for example, the commutation relation in dealing with the damped harmonic oscillator [1].

The first problem I shall discuss in this lecture is the damped harmonic oscillator, and then I will say a few things about a closely related amplifier and some of its properties.

Senitzky, who had been working with Schwinger, may have been the first to take seriously the coupling of the harmonic oscillator to its thermal environment [2]. This environment is the physical agent responsible not only for damping by absorbing energy from the oscillator, but responsible too for fluctuation, by returning energy randomly to the oscillator.

That fundamental idea was rediscovered a great many times in the early 1960's. It illustrates a phenomenon we often see in our field; the means by which people gain understanding is not to read but to write their own paper on it. There is strikingly little evidence that these dozens of papers have been read, although a long litany of references to them did accumulate. All had at their heart the idea of somehow coupling one harmonic oscillator of particular interest to an environment consisting of other oscillators and then solving the coupled oscillator problem by any of several techniques. The choice of oscillator for the environment is not a completely realistic element of these calculations, but it has the important characteristic of keeping the Heisenberg equations of motion linear.

The unique advantage of such linear models in contrast to some of the nonlinear ones which will be explored in this school is that they can furnish you with deep analytic insight. Of course, I may be the guiltiest among all the parties I have mentioned for not having read the papers of others and having instead constructed a rather special model of my own. It was a late entry however, and so I published it somewhat obscurely in the lecture notes [3] of the Varenna School of 1967.

The subsequent years have shown that the model has a certain mathematical elegance. It is furthermore easily converted to a model of an amplifier, and can be applied in a number of other contexts as well. So it may not have deserved quite all of the obscurity which it has enjoyed to date.

This is a subject I can not return to however, without recalling the loss during the past year of my good friend Mark Kac. He was, as you surely know, Poland's great gift to statistical physics. My own interest in the area was indeed inspired by his wonderful lectures. We discussed statistical models, including coupled oscillator models, on many occasions, and he always needled me to be more specific about the role of irreversibility. The work I shall describe in fact represents in part an attempt to answer many of those provocative and precisely stated questions.

II. Models of Attenuation

The model contains one harmonic oscillator of central interest which is described by the creation and annihilation operators a^+ and a, respectively. This oscillator is coupled to many other oscillators which form, in effect, a heat bath. The creation and annihilation operators for each mode of the bath are b_k^+ and b_k where the subscript k labels the mode. The simplest imaginable Hamiltonian coupling of the central oscillator, the one that interests us most, to the bath is

$$H = \hbar\omega a^+ a + \sum_k \hbar\omega_k b_k^+ b_k + \hbar \sum_k (\lambda_k a^+ b_k + \lambda_k^* b_k^+ a). \tag{1}$$

The frequency of the central oscillator is ω and the bath oscillators have many different frequencies $\{\omega_k\}$ labeled by the modes to which they correspond. The are coupled together by a sum of rotating wave terms with coupling constants λ_k. This Hamiltonian can either be used as the basis of a model or it can be obtained from more general systems by neglecting antiresonant terms.

This Hamiltonian has an invariance law built into it. When the phases of the operators a and b are changed in the same way, $a \rightarrow a e^{i\theta}$ and $b_k \rightarrow b_k e^{i\theta}$ then the Hamiltonian is invariant. That implies that there exists a conservation law; for this system the total number of quanta is fixed

$$a^+ a + \sum_k b_k^+ b_k = \text{const.} \tag{2}$$

Let us consider now the states of the harmonic oscillator. The n-quantum states are those that we secure from the ground state $|0\rangle$ by applying the creation operator n times and then normalizing

$$|n\rangle = \frac{(a^+)^n}{\sqrt{n!}} |0\rangle. \tag{3}$$

The annihilation operator on the other hand takes us from the n--quantum state the n-1-quantum state

$$a|n\rangle = \sqrt{n}\,|n-1\rangle. \tag{4}$$

There is another specially useful set of states which did not receive much attention until we started considering multiquantum problems, usually in connection with laser beams. They are the coherent states, which are eigenstates of the annihilation operator a:

$$a |\alpha\rangle = \alpha |\alpha\rangle. \tag{5}$$

There is an infinite two-dimensional set of such states, one for every point in the plane of complex eigenvalues α. These states are easily expanded in terms of the n-quantum states as

$$|\alpha\rangle = e^{-\frac{1}{2}|\alpha|^2} \sum_{n=0}^{\infty} \frac{\alpha^n}{n!} |n\rangle. \tag{6}$$

This expansion yields a Poisson distribution for the number of quanta present in each state

$$|\langle n|\alpha\rangle|^2 = \frac{|\alpha|^{2n}}{n!} \exp(-|\alpha|^2). \tag{7}$$

The mean number of quanta present is the square of the modulus of the complex eigenvalue

$$\langle n \rangle = |\alpha|^2. \tag{8}$$

The completeness of the states is expressed via the resolution of the unit operator

$$\frac{1}{\pi} \int d_2\alpha \, |\alpha\rangle\langle\alpha| = \mathbb{1}, \tag{9}$$

in which the element of area of the complex plane is simply

$$d_2\alpha = d(\text{Re}\,\alpha)\, d(\text{Im}\,\alpha).$$

The equations of motion for amplitude of the central oscillator and those of the coupled heat bath oscillators are:

$$\dot{a} = -i\omega a - i\sum_k \lambda_k b_k,$$
$$\dot{b}_k = -i\omega_k b_k - i\lambda_k^* a. \tag{10}$$

They have the property that the a operators are connected to the $\{b_k\}$ operators and the $\{b_k\}$ operators are connected to the a operator, but there are no terms connecting a to the set $\{b_k^+\}$ or the $\{b_k\}$ to a^+. That will give rise to an interesting simplification in finding the quantum state of the system.

The time-dependent solution for the amplitude of the central oscillator can be cast in the general form

$$a(t) = a(0) U(t) + \sum_k b_k(0) \vartheta_k(t), \qquad (11)$$

in which the functions $U(t)$ and $\vartheta_k(t)$ remain unknown but obey the initial conditions $U(0) = 1$ and $\vartheta_k(0) = 0$, for all k. Of course, the functions $U(t)$ and $\vartheta_k(t)$ must eventually be calculated by solving the appropriate differential equations, but there are some very general and precise statements that we can make before evaluating these functions.

Let us use the notation $|\alpha, \{\beta_k\}\rangle$ for a coherent state of the entire system of oscillators. It corresponds to the eigenvalue α for the central oscillator:

$$a|\alpha, \{\beta_k\}\rangle = \alpha |\alpha, \{\beta_k\}\rangle \qquad (12)$$

and to a whole set of eigenvalues $\{\beta_k\}$ for the bath oscillators:

$$b_k |\alpha, \{\beta_k\}\rangle = \beta_k |\alpha, \{\beta_k\}\rangle. \qquad (13)$$

There is an interesting theorem [4] which applies to just the sort of linear system we are considering here. The theorem is that if we begin with a coherent state for such a system it always remains in a coherent state. In other words, since coherent states have a certain classical or minimum uncertainty quality, that quality once it is present initially, is preserved as a function of time. For this particular coupling then the problem reduces to that of the evolution of a classical system as a function of time. The minimum uncertainty packet is simply transported along a classical trajectory.

The fixed Heisenberg state of the system is defined by the coherent state relations:

$$a(0)|\alpha, \{\beta_k\}\rangle = \alpha |\alpha, \{\beta_k\}\rangle, \qquad (14)$$

and

$$b_k(0) | \alpha, \{\beta_k\} > = \beta_k | \alpha, \{\beta_k\} >. \tag{15}$$

By forming the linear combination of operators given by Eq. (11) and applying it to the coherent state we can define a time dependent eigenvalue

$$\alpha(t) = \alpha U(t) + \sum_k \beta_k \vartheta_k(t). \tag{16}$$

We can likewise form a set of time-dependent eigenvalues $\{\beta_k\}$ from the set of solutions for the operators $\{b_k\}$ which we need not write out explicitly. Our coherent state then obeys the relations

$$a(t) | \alpha, \{\beta_k\} > = \alpha(t) | \alpha, \{\beta_k\} > \tag{17}$$

and

$$b_k(t) | \alpha, \{\beta_k\} > = \beta_k(t) | \alpha, \{\beta_k\} >. \tag{18}$$

Note that the state vector is the Heisenberg state or the initial state and not the time-dependent state of the system. To talk about the time-dependent state we must transform from the Heisenberg picture, where the operators change with time:

$$a(t) = U^{-1}(t) a(0) U(t) \tag{19}$$

and

$$b_k(t) = U^{-1}(t) b_k(0) U(t) \tag{20}$$

to the Schrödinger picture, where the state evolves in time

$$| t > = U(t) | \alpha, \{\beta_k\} >. \tag{21}$$

The transformation operator $U(t)$ from the Heisenberg to the Schrödinger picture is unitary. We can now show that the transformation just carries the system to a new coherent state from the initial one. We can do that by referring back to Eq. (17) substituting $a(t)$ from Eq. (19) in it and multiplying through by the unitary operator $U(t)$. We then have

$$a(0)|t\rangle = \alpha(t)|t\rangle \qquad (22)$$

and a similar manipulation for the heat-bath amplitudes gives

$$b_k(0)|t\rangle = \beta_k(t)|t\rangle. \qquad (23)$$

These equations for the time-dependent state $|t\rangle$ have a solution which is unique to within a phase factor. It is just the coherent state

$$|t\rangle = (c - \text{number phase factor}) |\alpha(t), \{\beta_k(t)\}\rangle. \qquad (24)$$

Undetermined phase factor cancels out of the density operator $\rho = |t\rangle\langle t|$, and so we need not to determine it. The time dependent state is determined in effect by just the time dependent eigenvalues, $\alpha(t)$ given by Eq. (16) and the analogous heat-bath amplitudes. Thus, for the simple model we are discussing here, all you ever need are the time-dependent eigenvalues; quantum mechanics very nearly reduces in other words to classical mechanics for this particular model.

To summarize, if we let the density operator correspond to a pure coherent state at the initial time

$$\rho(0) = |\alpha, \{\beta_k\}\rangle\langle\alpha, \{\beta_k\}|, \qquad (25)$$

then at all later times it remains a coherent state

$$\rho(t) = |\alpha(t), \{\beta_k(t)\}\rangle\langle\alpha(t), \{\beta_k(t)\}|, \qquad (26)$$

where the eigenvalues are the appropriately defined time-dependent eigenvalues corresponding to Eq. (16) etc. These c-number quantities satisfy the same equations of motion as the operators $a(t)$ and $\{b_k(t)\}$, so it is indeed just as though the operators were replaced by classical quantities.

All that really interests us at present is behavior of the central oscillator. To find the statistical properties of this oscillator we define a partial or reduced density operator by taking a trace over all the states or variables of the heat bath:

$$\rho_A(t) = \text{Trace}_B \, \rho(t) = |\alpha(t)\rangle\langle\alpha(t)|. \qquad (27)$$

All that is left in this simple example is a pure coherent state for the central oscillator.

Now, of course the system does not always begin in a coherent state; there might be a mixture of some sort. To treat such more general initial states, we can introduce a quasiprobability weighting $P(\{\beta_k\})$ for the initial amplitudes $\{\beta_k\}$ and integrate over those variables. Such an initial mixture evolves into the state given by

$$\rho_A(t) = \int |\alpha(t)\rangle\langle\alpha(t)| P(\{\beta_k\}) \prod_k d^2\beta_k. \qquad (28)$$

We must remember here that $\alpha(t)$ is a linear combination of the parameters $\{\beta_k\}$, so that this may be a nontrivial integration to carry out.

Let us assume for example that the heat bath is initially in a chaotic state, i.e. that the parameters $\{\beta_k\}$ that describe the heat bath all have Gaussian distributions with zero mean value, and that the average number of quanta in the k-th mode of the heat bath is $\langle n_k \rangle$. We write in other words

$$P(\{\beta_k\}) = \prod_k e^{-|\beta_k|^2/\langle n_k\rangle} / (\prod_k \langle n_k \rangle). \qquad (29)$$

An easy way of expressing the reduced density operator is then

$$\rho_A(t) = \prod_k \frac{1}{\langle n_k\rangle} \int d^2\beta_k\, e^{-\frac{|\beta_k|^2}{\langle n_k\rangle}} \int d^2\gamma\, \delta^{(2)}(\gamma - \alpha(t)) |\gamma\rangle\langle\gamma|, \qquad (30)$$

where the auxiliary variable γ has been introduced as a device to separate the integration into two stages. There is an integration to carry out over all of the variables $\{\beta_k\}$. This is an easy integration to perform since it is Gaussian in nature:

$$I = \prod_k \frac{1}{\langle n_k\rangle} \int d^2\beta_k\, e^{-\frac{|\beta_k|^2}{\langle n_k\rangle}} \delta^{(2)}(\gamma - \alpha u(t) - \sum_k \beta_k \vartheta_k(t)). \qquad (31)$$

The result when put together is again Gaussian in form:

$$I = \frac{1}{\pi(\sum_k \langle n_k\rangle |\vartheta_k(t)|^2)} \exp\left[\frac{-|\gamma - \alpha u(t)|^2}{\sum_k \langle n_k\rangle |\vartheta_k(t)|^2}\right]. \qquad (32)$$

If the heat bath oscillators are initially in a chaotic state, the reduced density operator is then a mixture of coherent states $|\gamma\rangle$,

$$\rho_A(t) = \int P(\alpha 0 | \gamma t) | \gamma \rangle \langle \gamma | d^2\gamma \qquad (33)$$

weighted by a function $P(\alpha 0 | \gamma t)$ given by the Gaussian function

$$P(\alpha 0 | \gamma t) = \frac{1}{\pi D(t)} \exp\left[\frac{-|\gamma - \alpha u(t)|^2}{D(t)} \right] \qquad (34)$$

in which the dispersion is:

$$D(t) = \sum_k \langle n_k \rangle | \vartheta_k(t) |^2. \qquad (35)$$

If this were a classical theory the weight function P would be called a conditioned probability distribution. In the quantum context distributions which play an analogous role in classical terms and are usually called quasi-probability distributions: this is a conditioned quasi-probability distribution for the amplitude γ at time t given the amplitude α at time zero.

A simple illustration is given by setting the heat bath temperature equal to zero, i.e. taking all $\langle n_k \rangle = 0$. Then the dispersion $D(t)$ vanishes and the quasi-probability function reduces to a delta function

$$P(\alpha 0 | \gamma t) = \delta^{(2)}(\gamma - \alpha u(t)). \qquad (36)$$

This delta function represents a pure quantum-mechanical state. We have begun with both the central oscillator and the zero temperature heat bath in a pure coherent state. The dynamical behavior of this system is perfectly reversible. If α is the excitation amplitude of the central oscillator, the oscillator will begin by losing that excitation to the bath; but in fact, because the state is a pure one described by the classical equations of motion for the amplitudes there will always be Poincaré recurrences. The function $u(t)$ (which we have not yet solved for) will be a complicated function of time; it is only when we look over a relatively short period of time that we will have the illusion of dissipation of the energy into the heat bath; sooner or later the energy will return to the central oscillator to any degree we like, and that remains just as true quantum mechanically as it is classically.

How do we find the functions $u(t)$ and $\{\dot{\vartheta}_k(t)\}$? In practice we must resort to approximations unless there happens to be only some small number of heat-bath modes.

For the useful case in which there are so many heat-bath modes, they form a continuum, and that continuum has the frequency of the central oscillator embedded within it, one can approximate the functions $u(t)$ and $\{\vartheta_k(t)\}$ by using a method developed many years ago by Weiskopf and Wigner [5]. It is an approximation which has many simple consequences, one of which is to throw away the Poincaré recurrences entirely. It is only correct when the coupling is not terribly strong, but it is in no sense a perturbation expansion, and seems to be remarkably accurate where it does hold. In this approximation the function $u(t)$ takes the form

$$u(t) = e^{-i\omega' t - \kappa t}, \qquad (37)$$

where the frequency has been slightly shifted from ω: i.e. $\omega' = \omega + \delta\omega$, and it decays with a damping constant κ. These parameters are given by the expression

$$\delta\omega - i\kappa = \lim_{\epsilon \to 0^+} \sum_k \frac{|\lambda_k|^2}{\omega - \omega_k + i\epsilon}, \qquad (38)$$

which is indeed what we might find from second-order perturbation theory, but they occur to all orders in Eq. (37). The damping coefficient which emerges from this expression is given by:

$$\kappa = \pi \sum_k |\lambda_k|^2 \delta(\omega - \omega_k) = \pi |\lambda_\omega|^2 g(\omega), \qquad (39)$$

where $g(\omega)$ is the heat-bath spectral density.

To find the dispersion, as a function of time, let us recall the conservation law, Eq. (2), which fixes the total number of quanta. In the present case it implies in fact that

$$|u(t)|^2 + \sum_k |\vartheta_k(t)|^2 = 1. \qquad (40)$$

The dispersion is a weighted sum over the amplitudes $\vartheta_k(t)$

$$D(t) = \sum_k \langle n_k \rangle |\vartheta_k(t)|^2. \qquad (41)$$

However the functions $\vartheta_k(t)$ have a sharp resonant peak at $\omega_k = \omega + \delta\omega = \omega'$. So the strongest coupling by far of the central oscillator is to those oscillators in the heat bath which have nearly the same frequency. If n_k is a smoothly varying function of k, then the sum in Eq. (41) can be approximated by:

$$D(t) \simeq \langle n_{\omega'} \rangle \sum_k |\vartheta_k|^2 \simeq \langle n_{\omega'} \rangle (1 - e^{-2\kappa t}), \qquad (42)$$

where the last equality follows from Eqs. (40) and (37). This completes the description in the complex plane of the stochastic process which describes attenuation. The quasi-probability is thus:

$$P(\alpha\, 0 | \gamma\, t) = \frac{1}{\pi D(t)} \exp\left[\frac{-|\gamma - \alpha e^{-i(\omega' - i\kappa)t}|^2}{D(t)}\right]. \qquad (43)$$

This result can be easily understood by referring to the figures. Figure 1 shows the function P plotted above the complex α-plane. The path $\alpha = e^{-i\omega t}$ is represented by a circle. If there is no coupling whatever, the point representing the oscillator amplitude just moves around such a circle. The function P, which is initially a delta function remains a delta function and simply circles about with that point. Let us suppose now that there is some coupling present, but there are no quanta in the heat bath. The initial delta function will still remain a delta function, but it will go round in an exponential spiral shown in Fig. 1. In other words the pure state of the attenuator always remains a pure state but there is damping present. Due to the fact that we used the Weisskopf-Wigner approximation (i.e. we have thrown away Poincaré recurrences) the delta function will never come back again to its initial state.

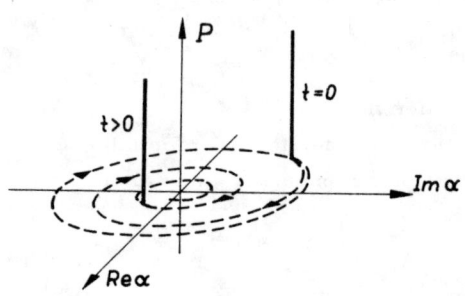

Fig. 1

Let us suppose now that we have a heat bath with a temperature $T > 0$. What happens of course is a certain amount of spreading; the delta function does not remain a delta function because the dispersion of the Gaussian function P becomes different from zero. The amplitude of the central oscillator goes around on the same exponential spiral, but its distribution spreads with time, and eventually it assumes an equilibrium distribution (Fig. 2), as did the zero temperature attenuator, but now the equilibrium is represented by a Gaussian distribution of finite width which sits ultimately at the origin.

There are, of course, other ways of deriving these results.

We could, for example, use the Fokker-Planck equation for the P-distribution or the master equation for the density matrix more directly. Such approaches have been described by many authors and all of that material is ancient in a sense.

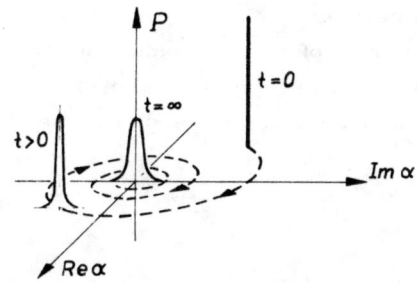

Fig. 2

Let me describe now something more recent. This is work done in collaboration with Prof. V. I. Mańko of the Lebedev Institute. We are concerned with the question: what happens if oscillators are coupled to more than one heat bath? Let us suppose for example, that we have two different heat baths coupled to a single central oscillator; let us suppose also that these two heat baths have different temperatures, so that we have different expectation values for the number of quanta in each mode Fig. 3. This situation is not really different from coupling the oscillator to one, single compound heat bath. The mean number of quanta in the central oscillator just eventually comes to equilibrium at the mean number of quanta in those modes of the compound heat bath which degenerate with the central oscillator. That is not very surprising. Now let us make the problem a little more interesting.

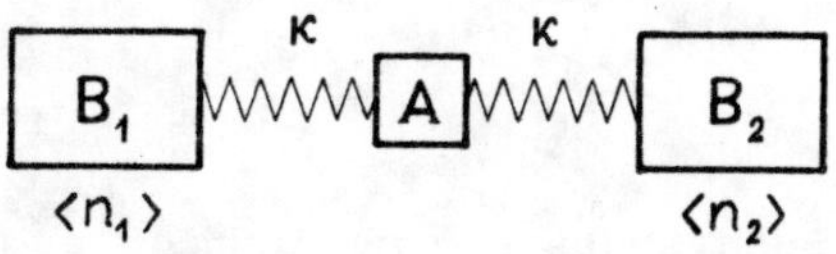

Fig. 3

Let us suppose we have two similar A oscillators and two different heat baths and we couple them by the scheme presented in Fig. 4. The two oscillators labelled A_1 and A_2 are coupled with a strength λ and each is also coupled to its own heat bath with a damping constant κ. Let us ask: how many quanta are present at equilibrium in the first of these, the A_1 oscillators? The general answer is following:

$$\langle a_1^+ a_1 \rangle \xrightarrow[t \to \infty]{} \tfrac{1}{2}(\langle n_1 \rangle + \langle n_2 \rangle) +$$
$$+ \tfrac{1}{2}(\langle n_1 \rangle - \langle n_2 \rangle) \frac{\kappa^2}{\lambda^2 + \kappa^2}, \qquad (44)$$

where $\langle n_1 \rangle$ and $\langle n_2 \rangle$ are the average occupation numbers characteristic of the two baths at the appropriate frequencies. There are three very simple limits of the formula (44). For no coupling between the A oscillators we have of course

$$\langle a_1^+ a_1 \rangle \longrightarrow \langle n_1 \rangle \qquad \text{for } \lambda = 0. \qquad (45)$$

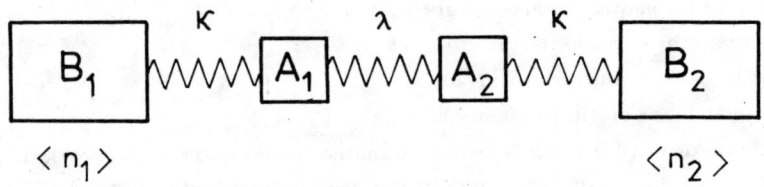

Fig. 4

If the A_1 and A_2 oscillators are coupled very strongly, we have

$$\langle a_1^+ a_1 \rangle \longrightarrow \tfrac{1}{2}(\langle n_1 \rangle + \langle n_2 \rangle) \qquad \text{for } \lambda \gg \kappa \qquad (46)$$

and both oscillators then go to the same equilibrium. For other values of (e.g. $\lambda = k$) we obtain an interesting intermediate equilibrium

$$\langle a_1^+ a_1 \rangle \longrightarrow \tfrac{3}{4} \langle n_1 \rangle + \tfrac{1}{4} \langle n_2 \rangle \qquad \text{for } \lambda = \kappa. \quad (47)$$

Such states are not ultimate equilibrium states because in some distant future equipartition must be reached between the two heat baths. However, such a process is always much slower, because there is only one degree of freedom connecting the two enormous heat baths. The kind of equilibrium indicated in Eq. (47) should be rather called a provisional equilibrium; it cannot in reality be infinitely long lived.

III. Model Amplifier

How do we get the harmonic oscillator to amplify instead of damping? We can accomplish that quite simply by turning the harmonic oscillator upside down in an energetic sense [6]. The Hamiltonian of such a system has the form

$$H = -\tfrac{1}{2}(p^2 + \omega^2 q^2) = -\hbar\omega(a^+ a + \tfrac{1}{2}). \qquad (48)$$

Note that both the kinetic and the potential energy have been inverted in sign. The oscillator still oscillates but it is an unstable system.

The differential equation for the operator a now takes the form

$$\dot{a} = i\omega a$$

which has the solution

$$a(t) = a(0)e^{i\omega t}.$$

It is still, in a sense, an annihilation operator; it decreases the amplitude of excitation. But it does that by putting operator energy into the oscillator. It creates a quantum of the energy. The role of the creation and annihilation operators in that sense is reversed. The n-quantum states are still defined by

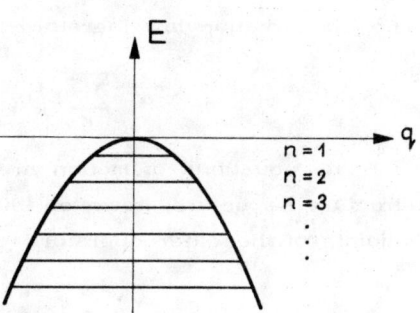

Fig. 5

$$|n\rangle = \frac{1}{\sqrt{n!}}(a^+)^n|0\rangle, \qquad (49)$$

where the 0-th state $|0\rangle$, which is defined by $a|0\rangle = 0$, is now the state of highest energy of the oscillator. The discussion of this system is then not different from the discussion of the usual oscillator except for the two interchanges:

$$\hbar\omega \longrightarrow -\hbar\omega, \qquad a \rightleftarrows a^+.$$

Let us now write down the Hamiltonian for the amplifier:

$$H = -\hbar\omega a^+ a + \sum_k \hbar\omega_k b_k^+ b_k + \\ + \hbar \sum_k (\lambda_k a b_k + \lambda_k^* a^+ b_k^+). \qquad (50)$$

The first term again describes the central oscillator - that is going to be the one that amplifies. The second term describes the heat-bath, while the third one describes the coupling between the two. The coupling is still the rotating wave or resonant coupling, i.e. it tends to conserve energy term by term. When the operator b annihilates a quantum in the heat bath, the operator a puts a quantum of energy into the central oscillator. This coupling is thus rather analogous to the one discussed in the previous section. Its invariance law however is slightly different

$$a \longrightarrow ae^{i\theta}, \qquad b_k \longrightarrow b_k e^{-i\theta}. \qquad (51)$$

The conservation law therefore takes also a slightly different form:

$$a^+ a - \sum_k b_k^+ b_k = \text{const.} \qquad (52)$$

The equations of motion are again linear and have much the same structure as before, however this time the operators are coupled to the adjoints of the other operators

$$\dot{a} = i\omega a - i \sum_k \lambda_k^* b_k^+,$$
$$\dot{b}_k = -i\omega_k b_k - i\lambda_k^* a^+. \qquad (53)$$

The mixing in these equations of the $\{b_k^+\}$ with a, and a^+ with the $\{b_k\}$ is going to change the behavior we found before, namely that a coherent state always remains coherent. If a coherent state for the inverted oscillator remained coherent, it would mean that minimum uncertainity is preserved in time. Here we find radically different behavior: the uncertainty amplifies, in fact it increases explosively as a function of time. That is the sad effect of quantum noise. Indeed all quantum amplifiers that approach the ideal level of sensitivity produce a great deal of quantum noise, which just consists in fact of amplified zero point oscillations.

The solution to the equations (53) can be written in the same form as before

$$a(t) = a(0)U(t) + \sum_k b_k^+(0) V_k(t), \qquad (54)$$

where $U(0) = 1$, $V_k(0) = 0$ etc. We again use the Weisskopf-Wigner approximation to find the functions $U(t)$, $V_k(t)$. The technique for finding the density operator must however be a more general one.

We define the normally ordered characteristic function as

$$\chi_N(\mu) = \text{Tr}_{\text{all}} \left\{ \rho(t) e^{\mu a^+} e^{-\mu a} \right\} =$$
$$= \text{Tr}_{\text{all}} \left\{ \rho(0) e^{\mu a^+(t)} e^{-\mu a(t)} \right\}. \tag{55}$$

By taking the Fourier transform of Eq. (55) we can solve for the P-distribution in the complex plane [7]. We again write the density operator in the form

$$\rho(t) = \int P(\gamma, t) |\gamma\rangle\langle\gamma| d^2\gamma. \tag{56}$$

Then for an initial coherent state

$$\rho(0) = |\alpha\rangle\langle\alpha|$$

we find the conditioned quasi-probability distribution

$$P(\alpha 0 | \gamma t) = \frac{1}{\pi n(t)} \exp\left[\frac{-|\gamma - \alpha U(t)|^2}{n(t)} \right] \tag{57}$$

with

$$n(t) = |U(t)|^2 - 1 + \sum_k \langle n_k \rangle |V_k(t)|^2 =$$
$$= \sum_k (1 + \langle n_k \rangle) |V_k(t)|^2. \tag{58}$$

By using the Weisskopf-Wigner approximation we find a solution for $U(t)$ which blows up exponentially

$$U(t) = e^{i\omega''t + Kt}. \tag{59}$$

It oscillates again at a shifted frequency $\omega'' = \omega - \delta\omega$ with $\delta\omega$, K given by a formula analogous to Eq. (38).

In order to find the dispersion as a function of time, we may once more make use of the fact that $\{V_k\}$ have a resonant character by writing

$$\sum_k \langle n_k \rangle |V_k|^2 = \langle n_{\omega''} \rangle \sum_k |V_k|^2 = \langle n_{\omega''} \rangle (|U(t)|^2 - 1). \tag{60}$$

The dispersion is therefore

$$n(t) = (1 + \langle n_{\omega''} \rangle)(e^{2\kappa t} - 1). \tag{61}$$

As we see, it increases rapidly with time. At temperature $T = 0$ we have $\langle n_{\omega''} \rangle = 0$ but we still have noise which increases in strength exponentially. There is inevitably zero point noise present initially and that is what blows up.

For an initial coherent state we again find Gaussian behavior for the function P but with noise that increases exponentially along with the amplification of the coherent state amplitude;

$$P(\alpha 0 | \gamma t) = \frac{1}{\pi n(t)} \exp\left[\frac{-|\gamma - \alpha e^{i\omega''t + \kappa t}|^2}{n(t)}\right]. \tag{62}$$

The picture analaogous to Fig. 2 is shown in Fig. 6.
The mean coherent state amplitude moves around an exponentially increasing spiral and at the same time the delta function spreads out and becomes a Gaussian function. There is no equilibrium - this Gaussian function keeps expanding in its width and in its distance from the origin. That is how the amplifier develops both its signal and its noise.

Let us talk about the gain and noise properties of this amplifier. If we start with the initial state

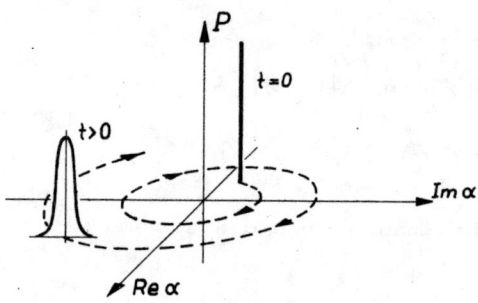

Fig. 6

$$P(\alpha 0 | \gamma 0) = \delta^{(2)}(\gamma - \alpha) \tag{63}$$

we see that the gain of the mean amplitude is given by

$$U(t) = e^{\kappa t}. \tag{64}$$

The power gain is thus

$$G = |U(t)|^2 = e^{2\kappa t}. \tag{65}$$

Now how should we define the noisiness of our amplifier? We have to do it in practical terms by asking in effect what kinds of input signals the amplifier is capable of distinguishing. A standard way of doing that is to imagine that we have a perfectly noiseless amplifier and then to ask what is the degree of randomness of the signal that we would have to put into that device in order to obtain the random output we actually find with the noisy amplifier. We would have to use as the input for our hypothetical noiseless amplifier an initial probability distribution, which contains a certain amount of noise. How much noise has it? The initial distribution, it is easy to see, would have to be a Gaussian function centered on α corresponding to a number of noise quanta equal to the ratio of the function $n(t)$ to the power gain $|U(t)|^2$:

$$N_{noise} = \frac{n(t)}{|U(t)|^2} = \frac{1}{|U|^2}\left\{|U|^2 - 1 + \sum_k \langle n_k \rangle |V_k|^2\right\} \quad (66)$$

$$= \left\{1 + \langle n_{\omega''} \rangle\right\}(1 - 1/G).$$

There is a rather general theorem for linear amplifiers due to C. Caves [8], which says that

$$N_{noise} \geq (1 + 1/G). \quad (67)$$

From Eq. (66) we can see exactly how much N_{noise} exceeds that limiting value.

An interesting game may be played with our two devices - our amplifier and our attenuator. A game very much like this one is in fact being played by all of the people who build optical transmission lines these days. In such a line we have first of all a certain amount of attenuation. An input signal is sent into the line and over some number of kilometers, say 100 or so, the signal loses most of its strength, but then it is fed into an optical amplifier, revitalised and sent over another 100 kilometers. One has then repeated cycles of amplification and attenuation. We can easily model such cycles with our two devices.

Let us begin with an amplitude distribution $P_0(\alpha)$ and amplify for a certain time t. We come out with an altered quasi-probability distribution

$$P_1(\gamma, t) = \frac{1}{\pi n(t)} \int e^{-\frac{|\gamma - \alpha U(t)|^2}{n(t)}} P_0(\alpha) d^2\alpha. \quad (68)$$

After we have amplified for time t let us reinvert the A oscillator and allow it to attenuate for the same time t. To make things simple let us assume zero temperature, i.e. $\langle n_k \rangle = 0$. The attenuation at $T = 0$ is an absolutely noiseless process, so in fact all it does is to scale down the function P. The resulting quasi-probability distribution is then

$$P_2(\beta,t) = \int \delta^{(2)}(\beta - \gamma u(t)) P_1(\gamma,t) d^2\gamma =$$

$$= \frac{1}{|u(t)|^2} P_1(\beta/u(t)) = \frac{1}{\pi |u(t)|^2 n(t)} \int e^{-\frac{|\beta - \alpha U(t)u(t)|^2}{|u(t)|^2 n(t)}} P_0(\alpha) d^2\alpha. \quad (69)$$

Now suppose we put both of these processes into balance; i.e. we assume the second one to produce just as much attenuation as we had for amplification in the first one. We have a cycle then which we could hope to carry out many times without too greatly distorting the input signal and that is more or less what is supposed to happen in optical transmission lines.

The product of the amplification and the attenuation factors in that case should be unity:

$$U(t) u(t) = 1, \quad (70)$$

so that

$$|u(t)|^2 n(t) = 1 - e^{-2\kappa t}. \quad (71)$$

According to Eq. (69) then the quasi-probability distribution after completing the cycle is

$$P_2(\beta,t) = \frac{1}{\pi(1 - e^{-2\kappa t})} \int e^{-\frac{|\beta - \alpha|^2}{1 - e^{-2\kappa t}}} P_0(\alpha) d^2\alpha. \quad (72)$$

For the case $\kappa t \gg 1$, in which the amplification is large this function is simply

$$P_2(\beta,t) = \frac{1}{\pi} \int e^{-|\beta - \alpha|^2} P_0(\alpha) d^2\alpha. \quad (73)$$

Such a function $P_2(\beta,t)$ corresponds to adding exactly one quantum of noise to the initial state. These two processes in succession, the irreversible amplification and reversible attenuation lead to the addition to the signal of precisely one quantum of Gaussian noise. For temperatures $T \neq 0$ and arbitrary values of κt the number of quanta of noise added by the cycle is found to be given more generally by

$$(1 + 2 \langle n_\omega \rangle)(1 - e^{-2\kappa t}) . \qquad (74)$$

IV. Is the Amplifier Model Realistic?

I do not think I really have to emphasize that the model I have been discussing is in one sense not too realistic. To be in actual possession of an inverted oscillator would be a fantastic thing: it would solve the world's energy problem. One such oscillator with no lower boud to its energy is all you would need. Over limited energy ranges, on the other hand, there do exist very good approximations to such systems. Whenever you have systems with equally spaced energy levels, which have an upper bound, their algebraic descriptions may have just the properties I am talking about. A famous example is the set of Zeeman levels of a magnetic moment in a magnetic field. If the magnetic moment is attached to a fixed spin $\vec{J}$ of magnitude j then there are $2j + 1$ equally spaced orientation states. If j is fairly large then the operator $\frac{1}{\sqrt{2j}} J_+ = \frac{1}{\sqrt{2j}} (J_x + i J_y)$ is algebraically quite analogous in its action upon the states with magnetic quantum numbers near j to the action of the annihilation operator a on the states of the inverted oscillator. Hence, as long as we deal only with the states which are not too far from maximum excitation, our precessing magnetic moment when coupled to the electromagnetic field behaves as a linear amplifying system.

Of course we must ask what happens when the system begins to lose a large fraction of its excitation. The answer is that the system then begins to behave nonlinearly, and the deviations of the operators $\frac{1}{\sqrt{2j}} J_\pm$ from the simple behavior of a and a^+ becomes significant. Eventually, when the angular momentum tilts too far from the vertical, this nonlinear behavior becomes dominant. When the magnetic moment has lost nearly all the energy it can and is pointing close to vertically downward it once again shows linear behavior. But this time the analogy connects $\frac{1}{\sqrt{2j}} J_-$ with the amplitude a for an ordinary oscillator. It describes in other words the behavior of an attenuating rather than an amplifying system. A system with fixed angular momentum coupled to the electromagnetic field thus interpolates in a sense between these two models.

The same sort of behavior is also observed, if you have a large number of two-level atoms which have identical coupling to the electromagnetic field. Such a system obviously has equally spaced energy levels. The symmetrical states of a system of N such atoms are equivalent to those of a magnetic moment attached to a spin N/2 in a magnetic field. For N large enough and for excitation not too far from the maximum the atoms constitute just the type of amplifier we have discussed. This is, in fact, just the process that F. Haake and I wrote about several years ago [9]. It is the early stage of superfluorescent light emission. Spontaneously emitted photons are rapidly amplifield into an intense superfluorescent pulse. All but the most intense part of this emission process can be described by means of our linear models.

References

1. W. H. Louisell, "Quantum Statistical Properties of Radiation" (John Wiley, N. Y., 1973).
2. I. R. Senitzki, Phys. Rev. 119, 670 (1960).
3. R. J. Glauber in Proceedings of the International School of Physics "Enrico Fermi", Course 42, Quantum Optics (Academic Press, N. Y. 1969).
4. R. J. Glauber, Phys. Rev. 131, 2766 (1963).
5. W. Weisskopf and E. Wigner, Z. Phys. 63, 54 (1930).
6. F. Schwabl and W. Thirring, Ergeb. d. Exakt. Naturwiss. 36, 219 (1964); R. J. Glauber in "Group Theoretical Methods in Physics", Proceeding of the International Seminar, Zvenigorod, 1982 (Nauka, Moscow, 1983), Vol. 2, p. 165.
7. R. J. Glauber in "Fundamental Problems of Statistical Mechanics II" ed. E. G. D. Cohen, (North Holland Publ. Co., Amsterdam 1968) p. 172.
8. C. M. Caves, Phys. Rev. D 26, p. 1157 (1982).
9. R. J. Glauber and F. Haake, Phys. Rev. A 13, 357 (1976).

R. GRAHAM
Fachbereich Physik
Universität Essen GHS
Essen, F. R. Germany

CHAOS IN QUANTUM OPTICS

Abstract

In this lecture a review is presented over the quantum theory of dynamical systems exhibiting chaos in the classical limit. Autonomous and periodically forced conservative and dissipative systems are considered and applications to quantum optics are stressed. The material presented in this lecture has been published in

R. Graham, Quantized Chaotic Systems, in Proceeding of the Topical Meeting on Instabilities and Dynamics of Lasers and Nonlinear Optical Systems, Rochester, New York, June 1985, to appear.

More detailed presentations of the particular topics of this lecture can be found in the following references. An introduction to chaos in dynamical systems is given in

A. J. Lichtenberg, M. A. Liebermann: Regular and Stochastic Motion, Springer, New York 1983

B. V. Chirikov, Phys. Reports 52, 263 (1979)

An introduction to the field of quantum chaos is provided by

G. M. Zaslavsky, Phys. Reports 80, 157 (1981); Chaotic Behavior in Quantum Systems, ed. G. Casati, Plenum, New York 1985

Important work on the quantized kicked rotator has appeared in

F. M. Izrailev, G. Casati, J. Ford, B. V. Chirikov, in Lecture Notes in Physics, vol. 93, ed. G. Casati, J. Ford, Springer, New York 1979

B. V. Chirikov, F. M. Izrailev, D. L. Sphelyansky, Sov. Sci. Rev. C2, 209 (1981)

D. L. Shepelyansky, Teor. i. Mat. Fizica 49, 117 (1981)

S. Fishman, D. R. Grempel, R. E. Prange, Phys. Rev. Lett. 49, 509 (1982)

D. L. Shepelyansky, Physica 8D, 208 (1983)

E. Ott, T. M. Antonsen Jr., J. D. Hanson, Phys. Rev. Lett. 53, 2187 (1984)

Work on chaos in the classical Jaynes Cummings model appeared in

P. I. Beloborov, G. M. Zaslavsky, G. Kh. Tartakovsky, Sov. Phys. JETP 44, 945 (1976)

P. W. Milonni, J. R. Ackerhalt, H. W. Galbraith, Phys. Rev. Lett. 50, 966 (1983)

The quantum case was studied in

R. Graham, M. Höhnerbach, Phys. Lett. 101A, 61 (1984)

R. Graham, M. Höhnerbach, Acta Phys. Austr. 56, 45 (1984)

R. Graham, M. Höhnerbach, Z. Phys. B57, 233 (1984)

M. Höhnerbach, thesis, University of Essen (unpublished)

An exactly solvable kicked quantum rotator with dissipation was analyzed in

    R. Graham, Phys. Lett. $\underline{99A}$, 131 (1983)

    R. Graham, Phys. Rep. $\underline{103}$, 143 (1984)

    R. Graham, Z. Physik $\underline{B59}$, 75 (1985)

A nonlinearly kicked quantized harmonic oscillator with dissipation is considered in

    R. Graham, T. Tél, Z. Physik $\underline{B}$, to appear.

The kicked quantum rotator with dissipation is considered in

    T. Dittrich, R. Graham, Z. Physik $\underline{B}$, to appear

The Lorenz model as a model for chaos in lasers was proposed in

    H. Haken, Phys. Lett. $\underline{53A}$, 77 (1975)

    R. Graham, Phys. Lett. $\underline{58A}$, 440 (1976)

Work on the steady state ensemble of the Lorenz model appeared in

    R. Graham, H. J. Scholz, Phys. Rev. $\underline{A22}$, 1198 (1980)

    M. Dörfle, R. Graham, Phys. Rev. $\underline{A27}$, 1096 (1983)

For the Wigner distribution of the quantized Lorenz model see

    R. Graham, Phys. Rev. Lett. $\underline{53}$, 2020 (1984)

M. V. FEDOROV

RESONANCES AND SATURATION IN THE CONTINUUM

1. Introduction

Talking about resonances in the continuum in this lecture we will discuss only the resonance field-induced bound-free atomic transitions. Resonances in free-free transitions of electrons here will not be considered (some of them have been described e.g. in the review [1]). The physical processes under consideration are the multiphoton resonance ionization of atoms and multiphoton resonance excitation of autoionizing states. Talking about saturation in the continuum we will mean in a general case the conditions under which the dependence of the probability of ionization w on the electromagnetic field intensity I deviates from the dependence $w(I)$ in the weak-field limit $I \to 0$ when the lowest order perturbation theory calculations are applicable. The effect of saturation in bound-free transitions can be most pronounced under the near-threshold conditions when the energy of photoelectrons is very small (see below sec. 4).

2. Multiphoton Resonance Ionization

The diagram of the process is described in fig. 1. A theoretical description is supposed to describe the dependence $w(\omega)$ where ω is the frequency of the electromagnetic wave for various ranges of the intensity I and for various numbers of photon n_1 and n_2. The main physical approaches which can be used in this problem are: 1) the restriction $I < I_{at} \approx 10^{16} W/cm^2$ (I_{at} is the intra-atomic field intensity) and 2) the so called resonance approximation which is correct if both the detuning $\Delta = E_0 + n_1 \hbar\omega - E_1$ (E_0 and E_1 are the ground and resonance level energies) and the width Γ of the curve $w(\omega)$ are small in comparison with the distance from E_1 to any other discrete level. The process of resonance ionization has been widely investigated both theoretically and experimentally (see the reviews [2, 3]). The results of a theoretical investigation are briefly summarized below.

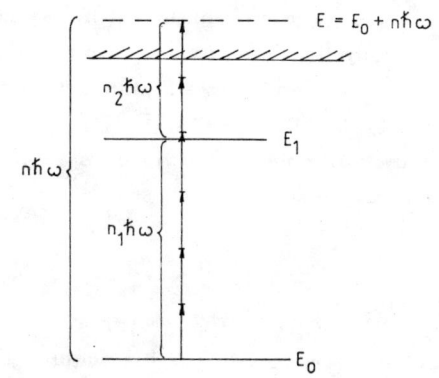

Fig. 1. Scheme of multiphoton resonance ionization

1) The weak-field limit. There is no saturation and $w \propto I^{n_1 + n_2}$. The width Γ of the curve $w(\omega)$ can be determined either by the field-free radiative width Γ_r of the level E_1 or by the inverse pulse duration $1/\tau$ ($\hbar = 1$). A specific form of the dispersion curve $w(\omega)$ depends on a character of the switching-on and -off processes (suddenly or adiabatically). In the simplest case of a suddenly switching-on interaction when $\Gamma_r \tau \ll 1$

$$w = \frac{\Gamma_i \Gamma_f^2 \tau}{8 \Delta^2} \left(1 - \frac{\sin \tau \Delta}{\tau \Delta}\right), \tag{1}$$

where $\Gamma_i = 2\pi |V_{1E}^{(n_2)}|^2$, $\Gamma_f = 4|V_{01}^{(n_1)}|$ are the so called ionization- and field-widths, V_{ij} are the matrix elements of the interaction operator $V = -\frac{1}{2}\vec{d}\vec{\mathcal{E}}_0$, $\vec{d}$ is the atomic dipole momentum, $\vec{\mathcal{E}}_0$ is the field strength amplitude.

2) $n = 1$, $\Gamma_f \tau \gg 1$, $\Gamma_f \gg \Gamma_r$, $\Gamma_i \tau \ll 1$. In this case the dominating process is the "field-broadening" which originates from the well known Rabi-splitting in the two-level system. The width of the curve $w(\omega)$ is equal to Γ_f now. It has the Lorentzian shape in the case of a suddenly switching-on interaction. If vice versa the time-dependent amplitude envelope of the pulse $f(t)$ is smooth the curve $w(\omega)$ is given by

$$w = \frac{\Gamma_i}{2} \int_{-\infty}^{+\infty} dt\, f^{2n}(t) \left(1 - \frac{1}{\Delta^2 + \frac{1}{4}\Gamma_f^2 f^2(t)}\right) \tag{2}$$

3) $n_1 > 2$, $n_2 = 1$, $\Gamma_i \gg \Gamma_f$, $\Gamma_i \tau \gg 1$, $\Gamma_i \gg \Gamma_r$, $\Gamma_i \Gamma_f^{-2} \ll \tau$.
This case corresponds to the ionization broadening of the curve $w(\omega)$. Its width is equal to Γ_i and the shape is the Lorentzian one both for smooth and rectangular envelope $f(t)$, $w_{max} \propto I^{n_1 - n_2}$.

4) If $\Gamma_i \tau \gg 1$, $\Gamma_i^{-1} \Gamma_f^2 \tau \gg 1$ the pulse duration is very long so that at small Δ there is an almost complete ionization ($w_{max} \approx 1$). The dispersion curve $w(\omega)$ is given by

$$w = 1 - \exp\left(-\frac{\Gamma_i \Gamma_f^2 \tau}{16 \Delta^2}\right) \qquad (3)$$

and it corresponds to the width $\Gamma_{tot} = \Gamma_j \sqrt{\Gamma_i \tau}$.

5) If the pulse envelope $f(t)$ is smooth there are some conditions under which the mechanism of ionization is connected with adiabatic inversion of the levels E_0 and E_1 due to their time-dependent Stark shifts $\delta E_{0,1} = -\frac{1}{4} \alpha_{0,1} f^2(t) \mathcal{E}_0^2$ where $\alpha_{0,1}$ are the polarizabilities of the levels $E_{0,1}$.

This shift and the population of the level $E_{0,1}$ are described in fig. 2.

The conditions of adiabatic inversion are given by

$$\Gamma_i \ll \tau^{-1} \ll \Gamma_f \ll \alpha \mathcal{E}_0^2 \ll \Gamma_f^2 \tau \qquad (4)$$

where $\alpha = \frac{1}{4} |\alpha_0 - \alpha_1|$.

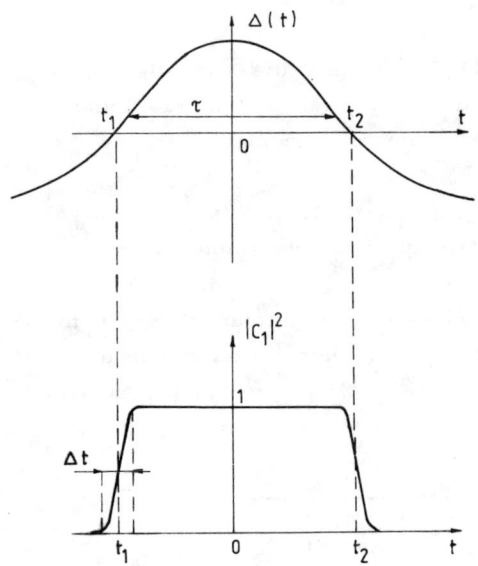

Fig. 2. Time-dependent detuning (including the time-dependent a. c. Stark shift) and population of resonant level under the conditions of adiabatic inversion

The dispersion curve $w(\omega)$ is qualitatively described in fig. 3, it has the width $\Gamma_{St} = \alpha \mathcal{E}_0^2$ and $w_{max} \propto I^{n_2}$. The same width Γ_{St} can appear also when one takes into account an inhomogeneity of the field-strength amplitude in the focal region.

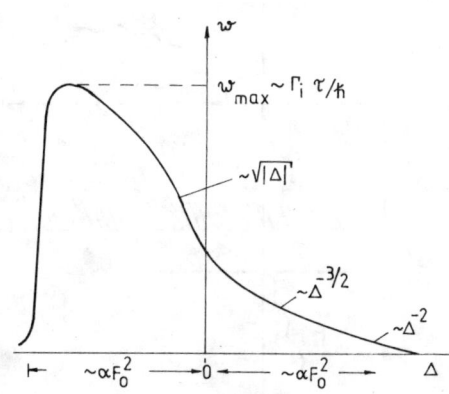

Fig. 3. Dispertion curve in the adiabatic inversion regime

As a resume we can say that the width of the dispersion curve $w(\omega)$ is given by

$$\Gamma = \max \left\{ \Gamma_r, \tau^{-1}, \Gamma_f, \Gamma_i, \Gamma_f \sqrt{\Gamma_i \tau}, \alpha \mathcal{E}_0^2 \right\}. \qquad (5)$$

The conditions under which one of these terms is larger than others determine the conditions of applicability of the corresponding mechanism of the field-induced broadening of the resonance dispersion curve $w(\omega)$.

3. Multiphoton Resonance Excitation of Autoionizing States

Two schemes of these processes are described in figs. 4a and 4b and they correspond to a single- and two-frequency electromagnetic field, respectively. The main physical peculiarity of these processes consists in a possibility of intereference of two channels of ionization: $E_1 \rightarrow E$ and $E_1 \rightarrow E_a \rightarrow E$ (E_1 and E_a are the energies of the resonant descrete and autoionizing levels). This fact has been described theoretically in many papers (see e.g. refs [4-17]). Very often this interference is considered as a reason of the field-induced narrowing of autoionizing resonances.

On the other hand in the processes described in fig. 4 there are also some noninterfering channels of ionization. One of them is connected with the photoionization of the autoionizing state $E_a \rightarrow \tilde{E} \approx E_a + \hbar\omega$. Some other noninterfering channels may appear due to a degen-

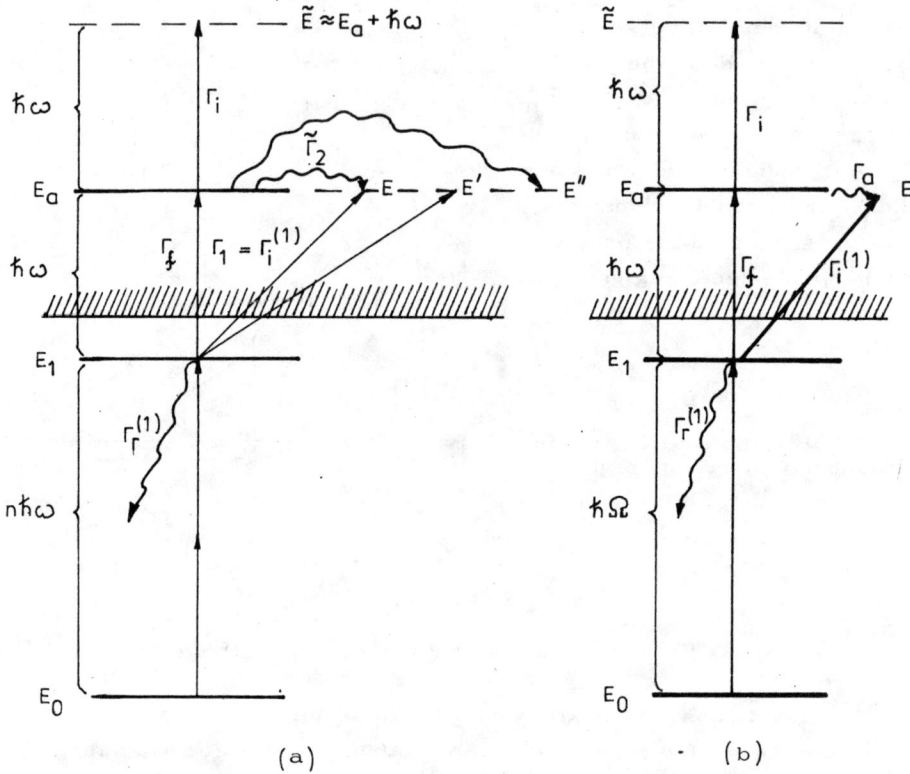

Fig. 4. Schemes of multiphoton excitation of autoionizing atomic states by a single- (a) and two- (b) frequency electromagnetic field

eration of the continuum (the noninterfering transitions $E_a \rightarrow E'' \approx E_a$ and $E_1 \rightarrow E' \approx E_a$ are also shown in fig. 4). For this reason the main problem which must be solved in a theoretical description is a proper description of a competition between interference and noninterfering channels of ionization. Now we will present the results of this analysis in application to some specific formulations of the problem.

1) Multiphoton resonance excitation of autoionizing state by a single-frequency electromagnetic field.

The dispersion curve $w(\omega)$ has two maxima. Their widths in the so called strong-field limit

$$\Gamma_f \gg \max(\Gamma_1, \Gamma_a)$$

are given by [14, 16]

$$\Gamma_{\pm} = \frac{1}{2}\frac{n\Gamma_2 + (n+1)\Gamma_1}{n(n+1)} \pm \frac{\Gamma_{12}}{\sqrt{n(n+1)}}, \quad (6)$$

where $\Gamma_f = 4|V_{1a}|$, Γ_1 and Γ_2 are the total widths of the levels E_1 and E_a, $\Gamma_{12} = \sqrt{\tilde{\Gamma}_a \tilde{\Gamma}_1}$, $\tilde{\Gamma}_{1,a}$ are the interfering parts of $\Gamma_{1,2}$. The dependence of $\Gamma_{\pm}$ on the field strength $\mathcal{E}_0$ is qualitatively described in fig. 5. The minimum of $\Gamma_-(\mathcal{E}_0)$ is achieved when

$$\Gamma_i^{(1)} = \Gamma_a \frac{n(n+1)}{[(\xi+1)n+1]^2} \quad (7)$$

and under this condition

$$\Gamma_- = (\Gamma_-)_{min} = \Gamma_a \frac{n^2\xi + \eta[n^2(1+\xi) + n(2+\xi) + 1]}{2n(n+1)[n(1+\xi)+1]}. \quad (8)$$

The dimensionless parameters ξ and η introduced here are given by $\eta = \Gamma_r^{(1)}/\Gamma_a \ll 1$ and $\xi = \Gamma_i/\Gamma_i^{(1)}$. If $\xi \sim 1$, $(\Gamma_-)_{min} \sim \Gamma_a$ and there is no considerable field-induced narrowing. Only if $\xi \ll 1$ (or $\Gamma_i \ll \Gamma_r^{(1)}$) $(\Gamma_-)_{min} \ll \Gamma_a$. The effect of narrowing occur only when all the noninterfering channels give a small contribution to the process of ionization. (The noninterfering transitions to the continua E' and E" are ignored in eqs. (7), (8)).

2) Heller-Popov scheme [5, 6]. This scheme corresponds to the case of a two-frequency electromagnetic field (fig. 4b) when the weak field frequency Ω = const. and the ionization probability is considered as a function of the strong resonance field $\omega = \omega(\omega)$. This function has only one maximum with the width [14, 16]

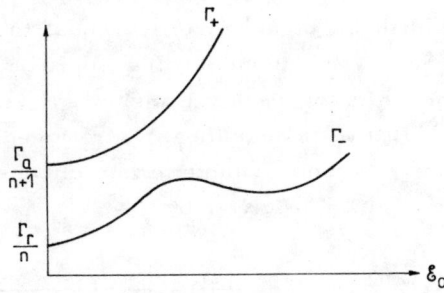

Fig. 5. The widths Γ_+ and Γ_- of the maxima of the dispersion curve $w(\omega)$ versus the field strength amplitude $\mathcal{E}_0$.

$$\Gamma = \Gamma_2 + \frac{\Gamma_1(\Gamma_f^2 - 4\Gamma_{12}^2) - 8\mathcal{E}_0 \Gamma_f \Gamma_{12}}{16(\mathcal{E}_0^2 + \frac{1}{4}\Gamma_1^2)}, \quad (9)$$

where $\mathcal{E}_0 = E_1 - E_0 - \Omega$.

In the weak-field limit $\Gamma = \Gamma_a$. Minimizing the width Γ (9) with respect to $\mathcal{E}_0$ and then with respect to the field strength $\mathcal{E}_0$ we find that the minimum occurs when

$$\mathcal{E}_0 = \frac{\Gamma_f \Gamma_1}{4 \Gamma_{12}} \gg \Gamma_1 \quad \text{and} \quad \Gamma_i \Gamma_i^{(1)} = \Gamma_a \Gamma_r^{(1)} \qquad (10)$$

and under these conditions

$$\Gamma = \Gamma_{min} = 2 \Gamma_a \sqrt{\xi \eta} \ll \Gamma_a. \qquad (11)$$

In this case there is a very pronounced narrowing of the autoionizing resonance for any ξ because $\eta \ll 1 \, (\Gamma_\eta^{(1)} \ll \Gamma_a)$. This narrowing occurs when the noninterfering channel does not broaden the resonanse because the corresponding field strength amplitude is small enough ($\Gamma_i \ll \Gamma_a$, eq. (10)).

3) Confluence of coherences (photoelectron spectrum). The simplest scheme for studying the photoelectron spectrum near an autoionizing state is described in fig. 6a. This spectrum shown in fig. 6b contains two maxima one of which (with the width Γ_a) is much broader than the other one. When the noninterfering channels are ignored [8] near the narrower maximum occurs the effect of confluence of coherences: under some conditions the width of this narrow maximum tends to zero and its position can coincide with the point where the probability of ionization is equal to zero itself. The resulting probability is not well defined in this approximation and the result depends on the order of limiting transitions (over energy, frequency, field strength etc.). This indefinitness is removed when one takes into account at any rate one noninterfering channel [15, 16]. The density of electron distribution is given by

$$\frac{dw}{dE} = \frac{\Gamma_a^{-1}}{2\pi} \frac{x}{(y-q)^2 + \frac{1}{4}} F(y; x, \delta), \qquad (12)$$

where $x = \Gamma_i^{(1)} / \Gamma_a \ll 1$ is the dimensionless field strength (squared), $y = \frac{E - E_a}{\Gamma_a} + q$ is the dimensionless photoelectron energy, $q = \Gamma_f / 4 \sqrt{\Gamma_a \Gamma_i^{(1)}}$ is the Fano-parameter, $q \gg 1$, $\delta = \frac{E_a - E_1 - \omega}{q \Gamma_a} + 1$

is the dimensionless detuning ($|\delta| \ll 1$) and $F(y; x, \delta)$ is the narrow maximum form-factor:

$$F(y; x, \delta) = \frac{y^2 + \frac{1}{4}\xi^2 x^2}{\left[y - q(\delta - x)\right]^2 + \frac{1}{4}\gamma^2(x, \delta)} \quad . \tag{13}$$

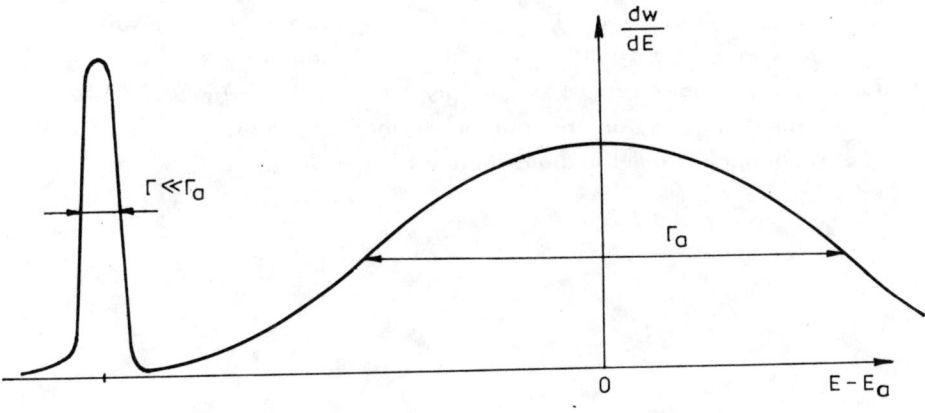

Fig. 6. The scheme of the process for studying the confluence of coherences (a); the photoelectron spectrum (b)

Here γ is the dimensionless width of the narrow maximum

$$\gamma(x, \delta) = \frac{\Gamma}{\Gamma_a} = \xi x^2 + x\left[\delta - (1-\xi)x\right]^2. \qquad (14)$$

Eqs. (12)-(14) take into account the noninterfering channel corresponding to photoionization of the autoionizing state E_a. The structure of the form-factor F is qualitatively described in fig. 7 for $\delta < x$,

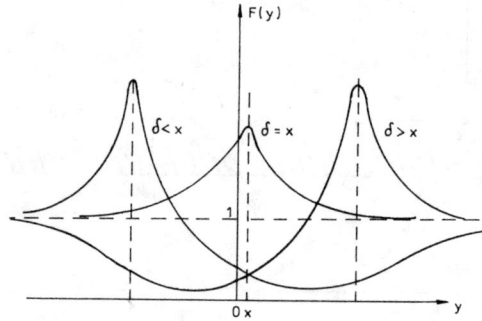

Fig. 7. The form-factor of the narrower maximum F(y) for various detunings δ

$\delta = x$ and $\delta > x$. Both F_{max} and F_{min} are finite and well defined for any δ. The width γ (14) grows monotonously with the increase of the field strength x if only $\xi > |\delta|$ where $|\delta| \ll 1$. Hence in this sense again the noninterfering channels prevent autoionizing resonance from the field-induced narrowing.

4. Near-Threshold Photodetachment of Electrons from Negative Ions

There is at any rate one feature of negative ions differing them essentially from neutral atoms. If a neutral atom contains an infinite number of subthreshold descrete levels a negative ion has only one (or maybe two) discrete level at all. For this reason a threshold is much better defined in negative ions than in neutral atoms. This is the reason why the photodetachment of an electron from a negative ion is much more sensitive to the a.c. Stark shift of the ground level and to the shape of the light pulse envelope than the near-threshold photoionization of neutral atoms.

The near-threshold photodetachment of an electron from a negative ion has many nontrivial features connected with the time-evolution of the probability (the nonexponential decay) [18-23]. We will not dwell upon these results here. Another and probably more interesting effect will be briefly described below instead. This effect is the predicted in refs. [19, 21, 22] saturation of the photodetachment probability. This effect can exist only if the pulse envelope $f(t)$ is smooth and if the a.c. Stark shift is large enough $\frac{1}{4}\alpha(\omega)\mathcal{E}_0^2\tau \gg 1$. Under these conditions the growth of the field strength amplitude $\mathcal{E}_0 f(t)$ can invert the situation from above- to below-threshold one (see fig. 8a). It means that not all the parts of the pulse $f(t)$ give equal contribution to the photodetachment.

Explicit expressions for the photodetachment probability depend on the quantum number of the ground state orbital member l (l = 0 or 1) and on the explicit form of the envelope $f(t)$. For example let us take $l = \frac{1}{2}$ (photodetachment from the S-state) and $f(t) = e^{-|t|/\tau}$. The process of photodetachment is characterized by the constant β determining the near-threshold behaviour of the matrix element $|V_{0E}|^2 = \beta E^{3/2}$, $\beta \infty \mathcal{E}_0^2$. Under the formulated conditions in the region of a small detuning from the threshold $\tau|\Delta| \ll 1$ (τ is the pulse duration, $\Delta = E_0 + \omega$) the photodetachment probability has the form [21]

$$w = \frac{3\beta}{\pi \mathcal{E}_0^2 \alpha(\omega) \tau^{3/2}} (2 - \frac{1}{\sqrt{2}}) \zeta(\frac{5}{2}), \qquad (15)$$

where ζ is the Riman ζ-function.

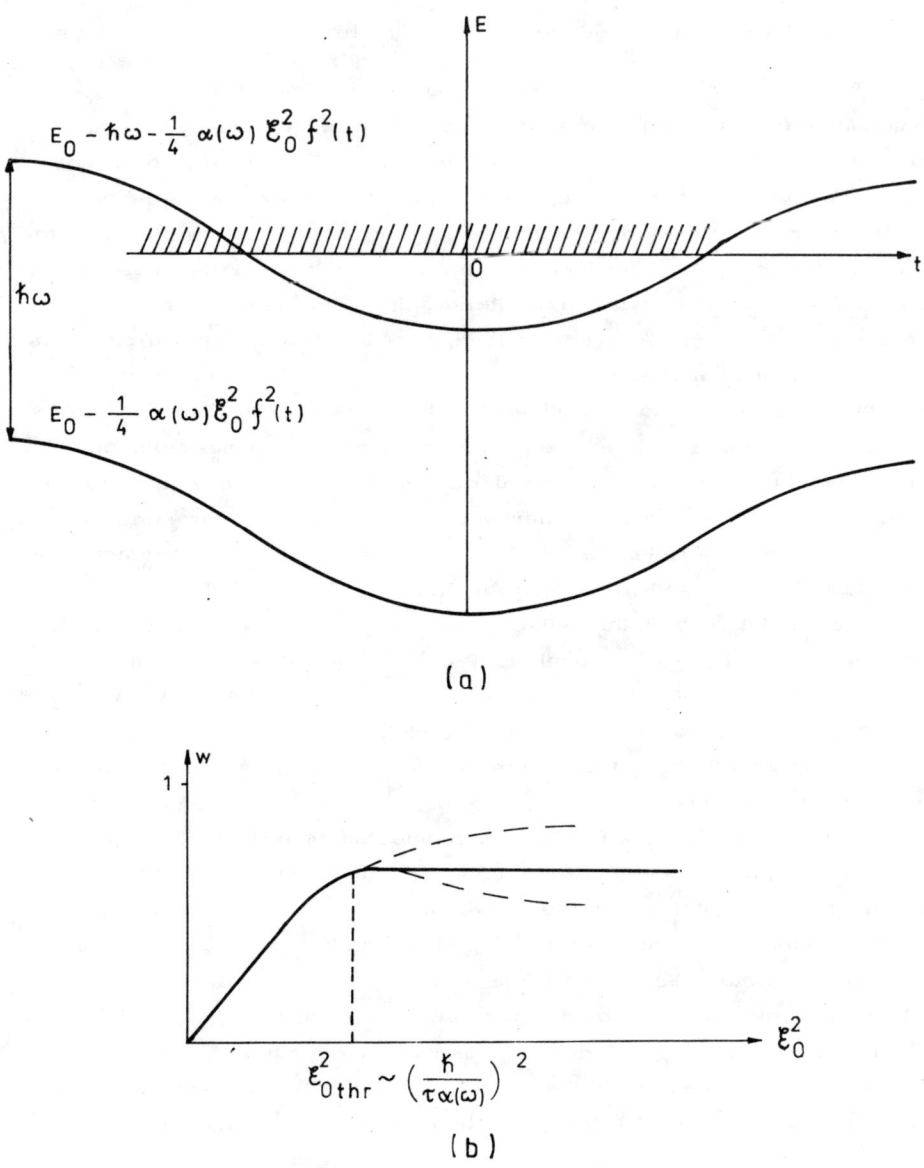

Fig. 8. The time-dependent a.c. Stark shift of the ground level of a negative ion and the scheme of near-threshold photodetachment process (a); saturation of the photodetachment probability $w(\mathcal{E}_0)$ (b)

If $\tau|\Delta| \gg 1$

$$w = \frac{16\,\pi\beta\tau}{5\,\mathcal{E}_0^2\,\alpha(\omega)}\,\Delta^{5/2} \qquad \text{for } \Delta > 0$$

$$w = \frac{6\beta}{\pi\,\mathcal{E}_0^2\,\alpha(\omega)\,\tau^{3/2}}\,e^{-\pi|\Delta|\tau} \qquad \text{for } \Delta > 0 \tag{16}$$

These equations describe the saturation of the photodetachment probability: $w(\mathcal{E}_0)$ = const. (the solid line in fig. 8b).

If the pulse envelope $f(t)$ is different from the simple exponent $e^{-|t|/\tau}$ there appears some dependence of w on $\mathcal{E}_0$ which however can be weak enough. E.g. if $f(t) = e^{-t^2/\tau^2}$ (the Gaussian envelope) the results of calculations [22] are given by

$$w \sim \tau^{-3/2}\left(\ln\frac{\mathcal{E}_0\,\tau\,\alpha(\omega)}{4}\right)^{3/4}, \quad \tau|\Delta| \ll \left(\ln\frac{\mathcal{E}_0\,\tau\,\alpha(\omega)}{4}\right)^{1/2}$$

$$w = \frac{8\pi\tau\sqrt{2}\,\beta\,\Delta^{5/2}}{5\,\alpha(\omega)\,\mathcal{E}_0^2}\left(\ln\frac{\mathcal{E}_0\,\alpha(\omega)}{4\Delta}\right)^{-1/2},\ \tau\Delta \gg \left(\ln\frac{\alpha(\omega)\,\mathcal{E}_0\,\tau}{4}\right)^{1/2} \tag{17}$$

$$w = \frac{12(2\ln\frac{1}{4}\mathcal{E}_0\,\alpha(\omega)\,\tau)^{3/4}\,\beta}{\mathcal{E}_0^2\,\alpha(\omega)\,(\pi\tau)^{3/2}}\,\Gamma(\tfrac{3}{2})\,e^{-\frac{\pi\tau|\Delta|}{2(\ln\frac{1}{4}\mathcal{E}_0\,\tau\,\alpha(\omega))^{1/2}}},\ \Delta < 0.$$

In all these cases the photodetachment probability w depends on $\mathcal{E}_0$ only logarithmically, i.e. there is almost a saturation (the dotted lines in fig. 8b). This effect appears when the field strength $\mathcal{E}_0$ becomes large enough $\alpha(\omega)\,\mathcal{E}_0\,\tau > 1$ (fig. 8b). Measuring the threshold of saturation experimentally (i.e. $(\mathcal{E}_0)_{th\Gamma} \sim (\alpha(\omega)\,\tau)^{-1}$). One can measure the a.c. polarizability of the ground states of negative ions $\alpha(\omega)$. This possibility can become an interesting and useful application of the described effect of saturation in the near-threshold region.

References

1. M. V. Fedorov, Progr. in Quant. Electr., 1981, v. 7, p. 73.
2. N. B. Delone, M. V. Fedorov, in "Multiphoton ionization of atoms", Trudi FIAN, 1979, v. 115, p. 42.
3. M. Crance, in "Multiphoton ionization of atom", 1982, p. 1153 ed. by S. Chin and P. Lambropulos, Ac. Press, N. Y.

4. A. M. F. Lau, C. K. Rhodes, Phys. Rev. Lett., 1977, v. A16, p. 2392.
5. Yu. I. Heller, Preprint IFSO-140 F, 1982, Krasnoyarsk, USSR.
6. Yu. I. Heller, A. K. Popov, Opt. Commun, 1981, v. 38, p. 345.
7. P. Lambropoulos, P. Zoller, Phys. Rev., 1981, v. A24, p. 379.
8. K. Rzazewski, J. H. Eberly, Phys. Rev. Lett., 1981, v. 47, p. 408.
9. G. S. Agarwal, S. L. Haan, K. Burnett, J. Cooper, Phys. Rev. Lett., 1982, v. 48, p. 1164.
10. Y. S. Kim, P. Lambropoulos, Phys. Rev. Lett., 1982, v. 49, p. 1698.
11. A. M. F. Lau, Phys. Rev. Lett., 1982, v. A35, p. 363; Adv. in Chem. Phys., 1982, v. 50, p. 191.
12. A. I. Andrjushin, A. E. Kazakov, M. V. Fedorov, Zh. Eksp. Teor. Fiz., 1982, v. 82, p. 91.
13. A. I. Andrjushin, M. V. Fedorov, A. E. Kazakov, J. Phys. B, 1982, v. 15, p. 2815.
14. A. I. Andrjushin, M. V. Fedorov, A. E. Kazakov, FIAN Preprint N 128, 1983, Moscow, USSR.
15. A. I. Andrjushin, M. V. Fedorov, A. E. Kazakov, Opt. Comm., 1984, v. 49, p. 120.
16. A. I. Andrjushin, A. E. Kazakov, M. V. Fedorov, Zh. Eksp. Teor. Fiz., 1985, v. 88, p. 1153.
17. A. I. Andrjushin, A. E. Kazakov, J. Phys. B, 1985, v. 18, p. 1501.
18. S. E. Kumekov, V. I. Perel, Zh. Eksp. Teor. Fiz., 1981, v. 81, p. 1693.
19. A. E. Kazakov, M. V. Fedorov, Zh. Eksp. Teor. Fiz., 1982, v. 83, p. 2035.
20. K. Rzazewski, M. Lewenstein, J. H. Eberly, J. Phys. B, 1982, v. 15, p. L661.
21. M. V. Fedorov, A. E. Kazakov, J. Phys. B, 1983, v. 16, p. 3641, 3653.
22. A. I. Andrjushin, M. V. Fedorov, A. E. Kazakov, J. Phys. B, 1984, v. 17, p. 3469.
23. S. L. Haan, J. Cooper, J. Phys. B, 1984, v. 17, p. 3481.

P. FILIPOWICZ, J. JAVANAINEN, and P. MEYSTRE

Max-Planck Institut für Quantenoptik,
D-8046 Garching,
West-Germany

THE ULTIMATE MASER

Recent experimental advances make possible for the first time the operation of a maser obeying purely microscopic dynamics. We present the theory of such a device, and show that it exhibits novel features not associated with masers and lasers generally. We find that the coherent output of conventional masers and lasers is due to the presence of a sufficient amount of incoherence in the pump and loss mechanisms.

Recent experimental advances in Rydberg-atom spectroscopy, together with the availability of superconducting high-Q microwave resonators [1-3], open up for the first time the study of isolated atoms in tailored electromagnetic environments. These microscopic systems are expected to exhibit the genuine quantum-mechanical dynamics generally masked by unavoidable fluctuations in their macroscopic counterparts.

We present the theory of a truly microscopic maser consisting of a single-mode high-Q resonator in which a monoenergetic beam of excited two-level atoms is injected at such a low flux that at most one atom at a time is present inside the resonator. We show that it exhibits a number of novel features that are averaged out in usual masers and lasers [4]. First, the field is in general sub-Poissonian [5] which reflects the quantization of both the field and its sources. Second, the onset of maser oscillations may be followed by a succession of abrupt transitions in the state of the field. Finally, we find that it is the incoherence of pump and loss mechanisms that is responsible for the coherent output of conventional laser and maser systems [6].

We consider the situation where at most one atom at a time is present inside the resonator, but the atomic flux is large enough for many atoms to traverse the cavity during the lifetime of the field [1]. The interaction time of the atoms with the cavity t_{int}, the mean interval between the arrivals of the atoms $\bar{t}_p$ and the cavity damping rate γ thus satisfy $t_{int} \ll \bar{t}_p \ll 1/\gamma$. Under these inequalities, the strategy to describe this "ultimate maser" is straightforward: multiple atom events are ignored, and cavity damping is neglected during the interaction of a single atom with the field. The calculation can then be divided into two stages: if an atom is present inside the cavity, the dynamics of the system is described by the Jaynes-Cummings model (JCM) [7-9], and if no atom is present, the field dynamics is given by the standard master equation describing the coupling of a single harmonic oscillator to a thermal bath [10].

The JCM hamiltonian is

$$H = (\hbar \omega_0/2) S_3 + \hbar \omega a^\dagger a + (\hbar \kappa/2)(S_+ a + S_- a^\dagger), \qquad (1)$$

where ω_0 is the energy difference between the two atomic levels, κ the electric dipole coupling constant, a and $a^\dagger$ the field annihilation and creation operators and ω its frequency, and S_3, $S_\pm$ the standard Pauli spin operators.

The JCM eigenstates can be separated into two-dimensional manifolds involving n photons plus an excited atom and n + 1 photons plus the lower-state atom (n = 0, 1, 2,...). In this way an analytic expression can be found for the time evolution operator $U(t)$ for the density operator of the combined system atom plus field. At time t_i an atom in the state ρ_a enters the cavity containing the field ρ_f^i, so the density operator of the combined system is $\rho = \rho_a \rho_f^i$. After the interaction time the state of the system is $\rho = U(t_{int}) \rho_a \rho_f^i$, and when the atom leaves the cavity the field is left in the state obtained by tracing over the atom's internal degrees of freedom,

$$\rho_f(t_i + t_{int}) = \text{Tr}_a \left[U(t_{int}) \rho_f^i \rho_a \right] \equiv A(t_{int}) \rho_f^i. \qquad (2)$$

The linear operator A gives the evolution of the density matrix of the field over the time the atom spends in the cavity.

Between $t_i + t_{int}$ and the time t_{i+1} at which the next atom is injected, the field is governed by the master equation [10]

$$\dot{\rho} = L\rho = (\gamma/2)(n_b + 1)\left[2a\rho a^\dagger - a^\dagger a\rho - \rho a^\dagger a\right] +$$
$$+ (\gamma/2)n_b\left[2a^\dagger \rho a - aa^\dagger \rho - \rho aa^\dagger\right]. \tag{3}$$

Here n_b is the number of blackbody photons in the resonator mode at finite temperature. At time t_{i+1}, the field is therefore described by the density matrix

$$\rho_f^{i+1} = \exp(Lt_p) A(t_{int}) \rho_f^i, \tag{4}$$

where $t_p = t_{i+1} - t_i - t_{int} \simeq t_{i+1} - t_i$ is the time interval between atom i leaving the resonator and atom i + 1 entering it.

If the field is initially described by a diagonal density matrix

$$\rho_f = \sum_n \rho_{fn} |n\rangle\langle n|, \tag{5}$$

it is easily shown that when atoms in the excited state are injected (no initial atomic coherence), the field always remains diagonal. Iteration of Eq. (4) eventually leads to a steady-state photon density matrix ρ_{st} which is the solution of Eq. (4) with $\rho_f^{i+1} = \rho_f^i$.

Our aim is to solve the steady state of the field from the version of Eq. (4) restricted to diagonal photon states. Since we only treat the case when many atoms traverse the cavity during the lifetime of the field, $N_{ex} = 1/\bar{t}_p \gamma \gg 1$, the results are nearly independent of the statistics of the possibly random arrivals of the atoms, as long as the atoms do not come in bursts separated by an interval $\geq \gamma^{-1}$. Note that this is a <u>classical</u> source of stochasticity in the maser, and that it could be removed in principle. But to comply with the experimental necessities, we keep it here and invoke the technical assumption that the arrivals obey a Poisson process with mean spacing $\bar{t}_p$ between events. The stochastic average of ρ_{st} can be carried out exactly by replacing the operator $\exp(Lt_p)$ by its average $(1 - L\bar{t}_p)^{-1}$ over the exponential distribution of the intervals between arrivals, and the steady-state field occupation probabilities can be solved <u>analytically</u> up to a normalization constant:

$$\rho_{fn} = C \cdot \left(\frac{n_b}{1+n_b}\right)^n \prod_{k=1}^n \left(1 + \frac{N_{ex}}{n_b} \frac{\kappa^2 \sin^2(\sqrt{\kappa^2 k + \Delta^2}\, t_{int}/2)}{\kappa^2 k + \Delta^2}\right). \tag{6}$$

In the following we concentrate on exact resonance, $\Delta = \omega - \omega_0 = 0$.

Fig. 1 shows the mean photon number $\langle n \rangle$ inside the resonator as a function of the scaled interaction time $\tau_{int} \equiv \sqrt{N_{ex}} \kappa t_{int}/2$, with $N_{ex} = 200$ and the number of thermal photons being $n_b = 0.1$. For small τ_{int} $\langle n \rangle$ is nearly zero. But at $\tau_{int} = 1$, a finite $\langle n \rangle$ emerges, $\langle n \rangle$ growing first rapidly for increasing τ_{int} to later decrease and reach a minimum at about $\tau_{int} \simeq 2\pi$. At this point, the field abruptly jumps to a higher intensity. This general behaviour recurs roughly at integer multiples of 2π, but becomes less pronounced for increasing τ_{int}. Finally, a stationary regime with $\langle n \rangle$ independent of τ_{int} is reached. Outside the time scale of Fig. 1 there is still additional structure reminiscent of the JCM revivals [9].

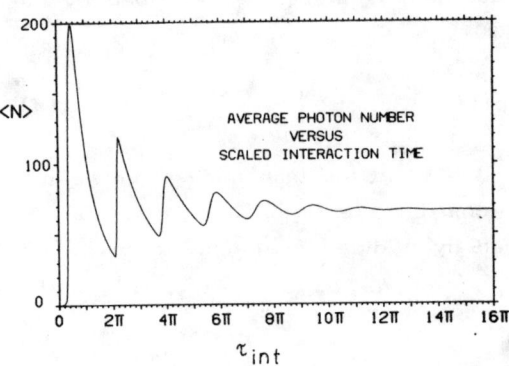

Fig. 1. Intracavity mean photon number $\langle n \rangle$ as a function of the dimensionless interaction time τ_{int} for $N_{ex} = 200$, $n_b = 0.1$ and $\Delta = 0$, in the case of a monoenergetic atomic beam.

We found that for N_{ex} not too small, the very first transition of the function $\langle n(\tau_{int}) \rangle / N_{ex}$ essentially did not depend on N_{ex}. In contrast, the subsequent transitions become sharper for increasing N_{ex}. We conclude that in the limit $N_{ex} \to \infty$ the first transition can be interpreted as a continuous phase transition, while the others are analogous to first-order phase transitions. This conjecture has been corroborated by a Fokker-Planck analysis which will be published elsewhere [11].

The origin of the phase transitions can be understood intuitively via the phenomenological stochastic evolution equation for the photon number n

$$\dot{n} = \gamma N_{ex} \sin^2(\kappa \sqrt{n+1}\, t_{int}/2) - \gamma n + \xi(n, N_{ex}, t). \tag{7}$$

The first term in Eq. (7) is the gain due to the change in atomic inversion as deduced from JCM and the second term describes cavity losses (here $n_b = 0$). The third term is a stochastic force reflecting both the <u>classical</u> random character of the arrival times of the atoms and the <u>quantum</u> fluctuations of the state. An involved analysis is needed [11] to obtain $\xi(n, N_{ex}, t)$, but for $N_{ex} \gg 1$ it can be regarded as small since the number of photons in the cavity is on the order of (but less than) N_{ex}, with standard deviation roughly $\sqrt{N_{ex}} \ll N_{ex}$. Furthermore, we stress that although the weak classical fluctuations associated with the intervals t_p influence the details of the results, they do not change any qualitative features and are therefore irrelevant to the interpretation. We will return to this very important point later on to show that stronger fluctuations in the system lead to a smoothing of the quantum mechanical phases in Eq. (6), and paradoxically to a concomitent increase in the coherence of the maser.

The possible photon numbers n in the maser are approximately given by the stable stationary solutions of Eq. (7) with $\xi = 0$. In the following discussion we assume $\xi(n, N_{ex}, t)$ to be n-independent white noise. Although not strictly correct, this Ansatz is sufficient for our purposes.

For $\tau_{int} < 1$ the only solution for the field is $n \propto \tau_{int}^2 \ll 1$. The maser threshold $\tau_{int} = 1$ occurs when the linearized gain for $n \simeq 0$ compensates the cavity losses:

$$\frac{1}{t_p} \frac{d}{dn} \sin^2(\kappa \sqrt{n+1}\, t_{int}/2)\bigg|_{n=0} \simeq (\kappa t_{int})^2/2 t_p = \gamma . \qquad (8)$$

The threshold condition is thus recovered from a simple gain-loss point of view.

When τ_{int} is further increased, Eq. (7) exhibits an increasing number of metastable steady states, and the fluctuating force ξ causes transitions between them. The white-noise model suggests that the system predominantly tends to reside in the <u>global</u> minimum of the potential

$$V(n) = -\int dn\, \gamma \left\{ N_{ex} \sin^2(\kappa \sqrt{n+1}\, \frac{t_{int}}{2}) - n \right\}. \qquad (9)$$

An inspection of Eq. (9) shows that the value of n giving the global minimum is a discontinuous function of τ_{int}, which explains the first-order phase transitions of Fig. 1.

Fig. 2 shows the standard deviation of the photon distribution as a function of τ_{int} for the parameters of Fig. 1. Above the threshold $\tau_{int} = 1$ the photon statistics is first strongly super-Poissonian, with $\sigma = \sqrt{<n^2> - <n>^2}/\sqrt{<n>} \simeq 4$. Further super-Poissonian peaks occur at the positions of the first-order phase transitions. In the remaining intervals of τ_{int}, σ is typically of the order of 0,5, a signature

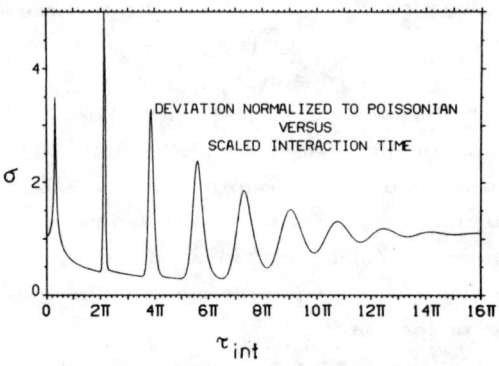

Fig. 2. Normalized second moment σ of the photon distribution for the conditions of Fig. 1. For a sub-Poissonian distribution, $\sigma < 1$

of the sub-Poissonian nature of the field. Because in the neighborhood of the phase transitions the photon distribution ρ_{fn} shows more than one peak, the first two moments are not sufficient to give an adequate description of the field. Nonetheless, they clearly illustrate that above threshold the ultimate maser cannot in general produce coherent radiation.

In view of these results obtained in the absence of significant sources of decay and/or noise, one should ask why standard lasers and masers produce coherent radiation, or more specifically, radiation with Poisson statistics. The answer to this question is paradoxically found in the presence of incoherent processes in such systems. These incoherent processes have the effect of smoothing the phases on the RHS of Eq. (6). For instance, for large enough fluctuations of the interaction times, $\sin^2(\kappa \sqrt{k}\, t_{int}/2)$ averages to 1/2, and far above threshold the photon statistics becomes Poissonian.

In the original experimental realization of a truly microscopic maser [1], atoms entered the cavity with a Maxwellian velocity distribution. In the theory the velocity distribution is taken into account by averaging $A(t_{int})$ in Eq. (4) over the distribution of the interaction times. We found numerically that the first few transitions survive a velocity spread as large as $\Delta v/v \simeq 10\%$, but all features except the maser threshold are washed out when the full Maxwellian width is kept. In this case, we recover essentially the results of conventional laser theory [4].

In a new version [1-2] of the experiment of Ref. 1, a Fizeau-type velocity selector has been installed to considerably narrow the atomic velocity distribution, and Jaynes-Cummings revivals [9] have been observed. In future work, the resonator will be cooled to a temperature of 0.3 K, which is expected to give a damping rate $\gamma \simeq 10\ s^{-1}$. An atomic flux of $1/\bar{t}_p = 2000\ s^{-1}$ would then realize the parameters of Fig. 1. For $\kappa = 2\pi \cdot 2.6$ kHz appropriate for the transition used in Ref. 1, the interaction time corresponding to the maser threshold $\tau_{int} = 1$

becomes $t_{int} = 9$ μs and the first discontinuous transition occurs at $t_{int} = 60$ μs, encouragingly close to the average interaction time 80 μs for the thermal beam in Ref. 1. It is very difficult to access the field inside the cavity, but evidence of the effects described here could be obtained by monitoring the atomic inversion at the exit of the resonator; it would also display sharp jumps as a function of the velocity selected.

In conclusion, we have shown that the ultimate maser exhibits sub-Poissonian statistics and a sequence of first-order-like phase transitions attributed to the quantized nature of the interaction between a single atom and a single field mode. Paradoxically, it is incoherent processes that produce the smoothing necessary to achieve a coherent output of the system. Ultimate masers such as considered here are within the range of present-day experimental techniques and dwell in the border zone between ordinary masers or lasers and the Jaynes--Cummings model. Their further study, both experimental and theoretical, will shed further light on the onset of macroscopic behaviour in microscopic radiative systems.

We acknowledge numerous discussions and suggestions from J. H. Eberly, G. Rempe, and H. Walther.

References

1. D. Meschede, H. Walther, and G. Müller, Phys. Rev. Lett. 54, 551 (1985).
2. A. Vaidyanathan, W. Spencer, and D. Kleppner, Phys. Rev. Lett. 47, 1592 (1981).
3. P. Goy, J. Raimond, M. Gross, and S. Haroche, Phys. Rev. Lett. 50, 1903 (1983).
4. M. Sargent III, M. O. Scully, W. E. Lamb, Jr. Laser Physics (Addison--Wesley, Reading, 1974). See in particular Chapter 17.
5. P. Filipowicz, J. Javanainen, and P. Meystre, submitted for publication.
6. P. Filipowicz, J. Javanainen, and P. Meystre, in the R. J. Glauber 60th Birthday Festschrift, Ed. by F. Haake, L. M. Narducci, and D. F. Walls, Cambridge University Press, to be published.
7. E. T. Jaynes and F. W. Cummings, Proc. IEEE 51, 89 (1963).
8. P. Meystre, E. Geneux, A. Faist, and A. Quattropani, Nuovo Cimento 25 B, 521 (1975); T. von Foerster, J. Phys. A8, 95 (1975); S. Stenholm, Phys. Rep. 6, 1 (1973).

9. J. H. Eberly, N. B. Narozhny, and J. J. Sanchez-Mondragon, Phys. Rev. Lett. <u>44</u>, 1323 (1980); H. I. Yoo and J. H. Eberly, Phys. Rep. <u>118</u>, 239 (1985).
10. W. H. Louisell, <u>Quantum Statistical Properties of Radiation</u> (John Wiley and Sons, New York 1973).
11. P. Filipowicz, J. Javanainen, and P. Meystre, to be published.
12. H. Walther, these Proceedings.

A. P. KAZANTSEV

L. D. Landau Institute of Theoretical Physics,
Moscow,
USSR

KINETIC PHENOMENA OF ATOMIC MOTION IN A LIGHT FIELD

Max SCHUBERT

Friedrich-Schiller-University
Jena, Dept. of Physics, DDR
6900 Jena, Max-Wien-Platz 1, G.D.R.

NEW PHOTON STATES AND QUANTUM-STATISTICAL ANALYSIS OF MULTIPHOTON PROCESSES

SUMMARY: This paper concentrates on new photon states that are connected with reduction of field fluctuations called "squeezing". In a squeezed state the field-strength fluctuations are, in a certain space-time interval, less than those in a coherent state whose fluctuations equal the photon vacuum fluctuations. Because of this surprising fundamental property and because of the possibility of a noise reduction in important applications (e.g. in optical communication systems and in the detection of gravitational waves), squeezed states received a great deal of attention. Squeezed states have been introduced in the literature also as generalized coherent states and two-photon coherent states. We shall treat definitions and basic properties (Section 1.) of squeezed states, potential physical processes for their production (Section 2.), and detection as well as application aspects (Section 3.).

1. Definitions and basic properties of squeezed states

First we will deal with the formal construction of squeezed states by the generalization of coherent photon states. The operator of the electric field strength of a (nearly) monochromatic wave can be expressed by

$$\hat{E}(r, t) = i g \hat{a} \exp\left[i(kr - \omega t)\right] + \text{h.c.}, \qquad (1)$$

where $k, \omega, \hat{a}, \hat{a}^\dagger$ are the wave number, the frequency, the destruction and creation operators of a single radiation mode with given directions of polarization and wave number vector; $g = (\hbar\omega/2\varepsilon_0 V)^{1/2}$.

As well-known a coherent state $|\alpha\rangle$ of this mode can be equivalently defined by

$$|\alpha\rangle = \hat{D}(\alpha)|0\rangle \quad \text{or} \quad \hat{a}|\alpha\rangle = \alpha|\alpha\rangle, \qquad (2)$$

with $\hat{D}(\alpha) = \exp\left[\alpha \hat{a}^+ - \alpha^* \hat{a}\right]$ being the displacement operator. Following Stoler [1] the state

$$|\gamma,\eta\rangle_s = \hat{D}(\gamma)\hat{S}(\eta)|0\rangle \qquad (3)$$

can be obtained by generalizing the first relation in (2), where now a "squeeze operator" $\hat{S}(\eta) = \exp\left[(1/2)\eta^* \hat{a}^2 - (1/2)\eta \hat{a}^{+2}\right]$ with quadratic terms in $\hat{a}$, $\hat{a}^+$ is used; η is a complex number. Following Yuen [2] the eigenstate $|\beta,\mu,\nu\rangle_y$ can be constructed by using a generalized form of the second relation in (2), namely

$$\hat{b}|\beta,\mu,\nu\rangle_y = \beta|\beta,\mu,\nu\rangle_y, \quad \text{where} \quad \hat{b} = \mu\hat{a} + \nu\hat{a}^+. \qquad (4)$$

The complex numbers μ, ν are assumed to obey the condition $|\mu|^2 = 1 + |\nu|^2$; there holds $\left[\hat{b},\hat{b}^+\right] = \hat{1}$. Both the states $|\gamma,\eta\rangle_s$ and $|\beta,\mu,\nu\rangle_y$ are called squeezed states. This naming as well as the meaning of the numbers γ, η and β, μ, ν will become clear in the following discussion on physically relevant expectation values.

The decomposition of $\hat{E}$ into <u>quadrature components</u> with the phase difference $\pi/2$ yields

$$\hat{E}(r,t) = -2g\hat{a}_1 \sin(kr - \omega t) - 2g\hat{a}_2 \sin(kr - \omega t + \frac{\pi}{2}), \qquad (5)$$

where the hermitian <u>quadrature operators</u> $\hat{a}_1$ (the "in-phase" operator) and $\hat{a}_2$ (the "out-of-phase" operator) are given by

$$\hat{a}_1 = (1/2)(\hat{a} + \hat{a}^+), \qquad \hat{a}_2 = (i/2)(\hat{a}^+ - \hat{a}). \qquad (6)$$

The variances for a coherent state $|\alpha\rangle$ are

$$\langle(\Delta\hat{a}_1)^2\rangle_{coh} = \langle(\Delta\hat{a}_2)^2\rangle_{coh} = 1/4, \qquad (7)$$

whereas the corresponding values for the squeezed state $|\beta, \mu, \nu>_y$ are given by

$$\langle(\Delta \hat{a}_1)^2\rangle_y = (1/4)|\mu - \nu|^2, \quad \langle(\Delta \hat{a}_2)^2\rangle_y = (1/4)|\mu + \nu|^2. \quad (8)$$

These variances obviously depend on the numbers μ, ν, but not on the eigenvalue β. A typical feature of squeezed states becomes clear by considering simple relations between the phases of μ and ν. In the case of $\varphi_\mu = \varphi_\nu$ there holds

$$\langle(\Delta \hat{a}_1)^2\rangle_y = (1/4)\left[1 - 2|\nu|\left[|\mu| - |\nu|\right]\right],$$
$$\langle(\Delta \hat{a}_2)^2\rangle_y = (1/4)\left[1 + 2|\nu|(|\mu| + |\nu|)\right]; \quad (9)$$

there are less in-phase fluctuations than in the coherent state at the expense of higher out-of-phase fluctuations. The reduction of fluctuations (with respect to one quadrature component) led to the naming squeezed states. For $\varphi_\eta = 0$ the state yields the variances

$$\langle(\Delta \hat{a}_1)^2\rangle_s = (1/4)\exp[-2\eta],$$
$$\langle(\Delta \hat{a}_2)^2\rangle_s = (1/4)\exp[+2\eta]. \quad (10)$$

While the expectation values of the photon number and of the coherent field amplitude for a coherent state $|\alpha\rangle$ are given by $\langle \hat{N}\rangle_{coh} = |\alpha|^2$ and $\langle \hat{a}\rangle_{coh} = \alpha$, the corresponding values for the squeezed states $|...\rangle_y$ and $|...\rangle_s$ are expressed by

$$\langle \hat{N}\rangle_y = |\nu|^2 + \langle \hat{a}\rangle_y \langle \hat{a}^+\rangle_y, \quad \langle \hat{N}\rangle_s = \sinh^2|\eta| + \langle \hat{a}\rangle_s \langle \hat{a}^+\rangle_s, \quad (11)$$
$$\langle \hat{a}\rangle_y = \mu^*\beta - \nu\beta^*, \quad \langle \hat{a}\rangle_s = \gamma.$$

There follows the relation $|\nu|^2 = \sinh^2|\eta|$ between the parameters of the two kind of states.

Let us now discuss the connection between <u>squeezing and bunching behaviour</u>. The calculation of the bunching excess $\langle(\Delta \hat{N})^2\rangle - \langle \hat{N}\rangle^2$ (which attains positive values for bunching and negative values for antibunching) exhibits that there exists no general relation between antibunching and squeezing behaviour. Examples are shown in

Fig. 1 for squeezed states with special mean values and variances of $\hat{a}_1$, $\hat{a}_2$. Whereas in the case of large coherent field amplitudes both bunching and antibunching occurs, in the case of the so-called squeezed vacuum states ($|\langle\beta|\hat{a}|\beta\rangle_y|^2 \ll |\nu|^2$) only bunching appears.

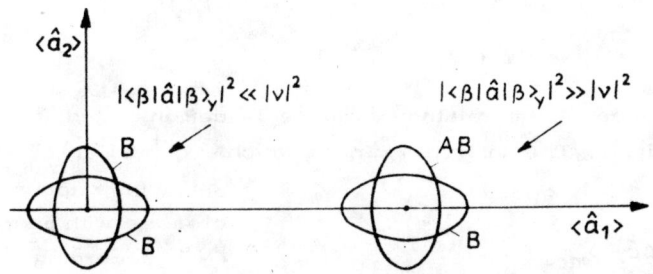

Fig. 1. Special squeezed states that exhibit bunching (B) or antibunching (AB): $|\langle\beta|\hat{a}|\beta\rangle_y|^2 \gg |\nu|^2$ (squeezed coherent states); $|\langle\beta|\hat{a}|\beta\rangle_y|^2 \ll |\nu|^2$ (squeezed vacuum states)

The utilization of the specific properties of squeezed states requires the knowledge of their <u>spatial and temporal as well as their probability behaviour</u>. This <u>phase-dependent behaviour</u> was studied by Schubert and Vogel [3] (however, in 1978 not under the name of squeezed states). Using the phase $\varphi \equiv kr - \omega t - \varphi_0$ (φ_0 being the initial phase) the field strength expresses as

$$\hat{E}(\varphi) = ig\,\hat{a}\,\exp[i\varphi] + \text{h.c.} \qquad (12)$$

There results a Gaussian probability distribution for the probability of finding the measurable field-strength value [4] in a squeezed state. The Gaussian distribution is characterized by the mean value and the variance of the field strength

$$\langle \hat{E}(\varphi)\rangle_y = 2g|\langle\hat{a}\rangle_y|\cos\left[\varphi + (\pi/2) - \arg\langle\hat{a}\rangle_y\right],$$
$$\langle[\Delta\hat{E}(\varphi)]^2\rangle_y = g^2\left\{1 + 2|\nu|^2 + 2|\nu||\mu|\cos\left[2\varphi + \varphi_\nu - \varphi_\mu\right]\right\}; \qquad (13)$$

$g^2 = \langle(\Delta\hat{E})^2\rangle_{vac}$ characterizes the fluctuations of the photon vacuum. Salient features can well be illustrated by consideration of the spatial behaviour at fixed time (say $t = 0$) and vanishing initial phase φ_0. There results

$$\langle [\Delta \hat{E}(r)]^2 \rangle_y - \langle (\Delta \hat{E})^2 \rangle_{vac} < 0, \text{ where } |\nu| + |\mu|\cos[2kr + \varphi_\nu - \varphi_\mu] < 0 \quad (14)$$

$$\text{Min}\{\langle [\Delta \hat{E}(r)]^2 \rangle_y\} =$$

$$= \langle (\Delta \hat{E})^2 \rangle_{vac}(|\mu| - |\nu|)^2 \text{ at } 2kr + \varphi_\nu - \varphi_\mu = \quad (15)$$

$$= n(1 + 2\pi).$$

The r-dependence of the relative variance is demonstrated for certain r-values in Fig. 2. The interval Δr, in which $\langle [\Delta \hat{E}(r)]^2 \rangle_y$ falls below the vacuum value decreases with increasing $|\nu|$ according to $\Delta r \simeq \lambda/2\pi|\nu|$, where λ is the wave length. The minimum of the relative variance decreases asymptotically according to $1/4|\nu|^2$. If for instance a $|\nu|$-value with a maximum squeezing gain of five is chosen, the relative squeezing interval $\Delta r/\lambda$ is 0,14.

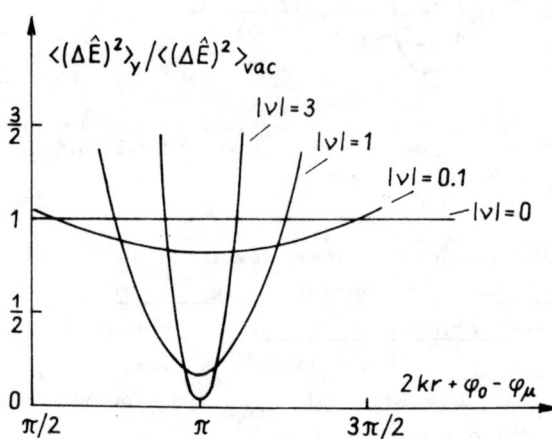

Fig. 2. Relative mean-square field fluctuations as a function of the spatial coordinate r. After Schubert and Vogel [3]

Note that the space average value of the variance

$$\overline{\langle [\Delta \hat{E}(r)]^2 \rangle_y^r} = \langle (\Delta \hat{E})^2 \rangle_{vac} \cdot (1 + |\nu|^2) \quad (16)$$

exceeds the vacuum value for proper squeezed states ($|\nu| > 0$).

In the framework of the field-theoretical formalism the vector potential $\hat{A}$ (as the field variable) and its canonically conjugated field- $\varepsilon_0 \hat{E}$ characterize the dynamical state of the radiation field at every instant. The general uncertainty relation between $\hat{A}$ and $-\varepsilon_0 \hat{E}$ reads in the special case of the state $|\beta, \mu, \nu\rangle_y$:

$$\langle(\Delta\hat{E}(r))^2\rangle_y \cdot \langle(\Delta\hat{A}(r))^2\rangle_y =$$
$$= g^4\omega^{-2}\left\{1 + 4(|\nu|^2 + |\nu|^4)\sin^2\left[2kr + \varphi_\nu - \varphi_\mu\right]\right\}. \quad (17)$$

The right-hand side is a function of r (or more general of the phase ϕ and may exceed the value $g^4\omega^{-2}$ of a coherent state. In this sense squeezed states cannot quite generally be regarded as <u>minimum uncertainty states</u>. They have the minimum value of the uncertainty product $g^4\omega^{-2}$ only at $2kr + \varphi_\nu - \varphi_\mu = n\pi$. Denoting the positive and negative frequency part of $\hat{E}$ by $\hat{E}^{(+)}$ and $\hat{E}^{(-)}$ we have

$$\hat{E} = \hat{E}^{(+)} + \hat{E}^{(-)},$$
$$\omega\hat{A} = \hat{E}^{(+)}\exp[-i\pi/2] + \hat{E}^{(-)}\exp[+i\pi/2]. \quad (18)$$

Obviously $\hat{A}$ describes a wave with the constant phase difference $\pi/2$, compared with the $\hat{E}$-wave. The eigenvalues of the operator of the vector potential (!) are not directly measurable, therefore we shall formulate the situation in terms of two $\hat{E}$-waves instead of $\hat{E}$ and $\hat{A}$.

Let us consider both the phase-shifted waves

$$\hat{E} = \hat{E}^{(+)}(\phi) + \hat{E}^{(-)}(\phi), \quad \hat{E}_{\frac{\pi}{2}} = \hat{E}^{(+)}(\phi - \frac{\pi}{2}) + \hat{E}^{(-)}(\phi - \frac{\pi}{2}), \quad (19)$$

where the following commutation and uncertainty relations hold

$$\left[\hat{E}^{(+)}, \hat{E}^{(-)}\right] = G\hat{I}, \quad \left[\hat{E}, \hat{E}_{\frac{\pi}{2}}\right] = i2G\hat{I}, \quad \langle(\Delta\hat{E})^2\rangle \cdot \langle(\Delta\hat{E}_{\frac{\pi}{2}})^2\rangle \geq G^2. \quad (20)$$

G coincides with the variance in the coherent state equal to the variance in the vacuum state. The discussion of (14) suggests the following general definition of a squeezed state (with the density operator $\hat{\rho}_{squ}$). A state is squeezed, if there exist phase values ϕ, for which

$$\langle[\Delta\hat{E}(\phi)]^2\rangle_{squ} - G < 0,$$
$$\text{where } \langle[\Delta\hat{E}(\phi)]^2\rangle_{squ} = \text{Tr}\left\{\hat{\rho}_{squ}[\Delta\hat{E}(\phi)]^2\right\}. \quad (21)$$

One may say that (21) is a general, representation-free definition being independent of a special construction of squeezed states (as e.g. $|\gamma, \eta\rangle_s$ or $|\beta, \mu, \nu\rangle_y$). A survey of possible squeezed states is given by the ensemble of operators $\hat{\rho}_{squ}$ that satisfy the squeezing condition (21). Using the Glauber-Sudarshan P-representation $P_{squ}(\alpha)$ for the characterization of $\hat{\rho}_{squ}$, the condition (21) leads to

$$W\left[P_{squ}(\alpha), \phi\right] \equiv \int d^2\alpha\, P_{squ}(\alpha) \left[(e^{i\phi}\alpha + c.c.) - (e^{i\phi}\int d^2\alpha'\, P_{squ}(\alpha')\alpha' + c.c.)\right]^2 < 0. \tag{22}$$

The minimum value $W_{min}(\phi)$ is assigned to the maximum squeezing gain $\left[1 + W_{min}(\phi)\right]^{-1}$. Obviously the requirement $W < 0$ leads to nonpositive definite functions $P_{squ}(\alpha)$, which cannot be interpreted as a proper probability distribution and are in general not well-behaved.

So far we have discussed squeezing with the help of second-order moments. Hong and Mandel [5] have given a natural generalization from second-order squeezing in (21) to <u>higher-order squeezing</u> (2n--order with n > 1). States are squeezed to the 2n-th order if there exist phase values ϕ, for which holds

$$\left\langle \left[\Delta\hat{E}(\phi)\right]^{2n}\right\rangle_{2n,squ} - (2n-1)!!\, G^n < 0. \tag{23}$$

Note that an equivalent definition for higher-order (and second-order) squeezing can be derived from (21) by using normally-ordered higher order moments; e.g. second-order and fourth-order squeezing exists if

$$\left\langle :\left[\Delta\hat{E}(\phi)\right]^2:\right\rangle_{2,squ} < 0,$$
$$\left\langle :\left[\Delta\hat{E}(\phi)\right]^4:\right\rangle_{4,squ} + 6G\left\langle :\left[\Delta\hat{E}(\phi)\right]^2:\right\rangle_{4,squ} < 0. \tag{24}$$

Certain physical processes, which yield squeezing to higher than second order, may be particularly important from the point of noise reduction. On the basis of the above general definition we can examine whether the state $|\beta, \mu, \nu\rangle_y$ is squeezed to higher order. The fourth--order calculation yields

$$\left\langle \left[\Delta\hat{E}(\phi)\right]^4\right\rangle_y = 3G^2\left\{1 + 2|\nu|^2 + 2|\nu||\mu|\cos\left[2\phi + \varphi_\nu - \varphi_\mu\right]\right\}^2. \tag{25}$$

The states $|\beta, \mu, \nu\rangle_y$ are higher-order squeezed states with the same phase dependence as in the case of second-order squeezing [see (13)].

At the end of this section we will sketch the connection of squeezing with other attributes. <u>Aspects of symmetry</u> (Lie groups and commutation relations) have been treated by Milburn [6] and Wodkiewicz and Eberly [7]. On this basis in [6] a concept of generalized coherent states was presented that includes multimode squeezed states, while in [7] the reduction of fluctuations of both atomic and photon states has been studied.

So far we have characterized light fields by the attributes: nonsqueezed/squeezed and bunched/antibunched. Now we add a characterization concerning photon-counting statistics, namely super-Poisson/sub-Poisson, which was recently studied by Teich, Saleh, and Peřina [8]. Thermal (chaotic) light is nonsqueezed, bunched, and super--Poisson. Its Glauber-Sudarshan P-function can be interpreted as a proper (positive definite) probability. These photon states have a classical counterpart and are called classical light. Coherent light (laser light well above the threshold) is nonsqueezed, neither bunched nor antibunched, and has a Poissonian distribution; it can also be regarded as classical light. If the light exhibits one or more of the attributes squeezed, antibunched, sub-Poisson, it has no positive definite, well--behaved P-function and is called nonclassical light. While antibunched radiation (in resonance fluorescence measurements) and sub-Poisson radiation (in the investigation of Franck-Hertz light) have already been found experimentally, up to now (September 1985) a publication on a successful detection of squeezed light is not known.

2. Potential ways for producing squeezed states

First we will deal with some squeezing processes on the basis of nonlinear optical interactions, which can be described with the coupling Hamiltonian

$$\hat{H}_c = \hbar \left[\chi \hat{a}^2 + \text{h.c.} \right] \tag{26}$$

χ is a c-number containing (classical) nonlinear optical susceptibilities and the (classical) pump amplitude (s).

<u>Degenerate four-wave mixing</u> was studied by Yuen and Shapiro [9]. Here $\chi \propto \varkappa^{(3)} A_1 A_2$ holds, where A_1, A_2 are the pump amplitudes and $\varkappa^{(3)}$ is the appropriate third-order susceptibility. In a standard

degenerate four-wave mixer with the interaction length L an object wave (associated with the destruction operator $\hat{a}_0$) and a phase-conjugated image wave ($\hat{a}_p$) are amplified. Both the input waves are assumed to be in coherent states. After having traversed an optical delay line the amplified object wave is combined with the amplified image wave at a beam splitter. By properly setting up the optical delay lines and the beam splitter one can obtain two waves associated with the modal destruction operators

$$\hat{b}_\pm = \sec(\tilde{\chi} L)\, \hat{c}_\pm - \tan(\tilde{\chi} L)\, \hat{c}_\pm^+, \text{ where } \hat{c}_\pm = (1/\sqrt{2})\left[\hat{a}_p \pm i\hat{a}_0\right], \quad (27)$$

Here $\tilde{\chi}$ is a reciprocal length proportional to $A_1 A_2 \varkappa^{(3)}$. The modes associated with the operators $\hat{c}_\pm$ are in a coherent state. Comparison with (4) reveals that the combination modes characterize squeezed states $|\beta, \mu, \nu\rangle_y$, where

$$\mu = \sec(\tilde{\chi} L), \quad \nu = -\tan(\tilde{\chi} L). \quad (28)$$

Both the combination modes have unequal uncertainties for $\tilde{\chi} L > 0$. If (as it can be expected at least in pulse experiments) $\tilde{\chi} L \approx 1$ can be achieved, a squeezing gain of 10 would be reached.

In the case of <u>degenerate parametric down-conversion</u> [1] the strong incident wave of frequency 2ω may be described classically with the complex amplitude A, while the generated subharmonic wave of frequency ω has to be treated as quantized (destruction operator $\hat{a}$). Now the term χ in (26) is proportional to $A\varkappa^{(2)}$, where $\varkappa^{(2)}$ is the appropriate second-order susceptibility. Assuming an initial vacuum state, there results (via the equation of motion derivable from (26)), the following expression for the subharmonic wave

$$\langle [\hat{E}(t,\varphi_0)]^{2n}\rangle = (2n-1)!! \exp\left[-2n(\varkappa^{(2)}/\pi)|A|t\right], \quad (29)$$

if the phase relation $2\omega t + 2\varphi_0 - \arg A - (\pi/2) = 0$ holds. Comparison with (23) shows that the exponential factor implies second- and higher-order squeezing for the parametrically down converted field.

Yuen [2] treated <u>two-photon-lasing</u> (TPL) on the basis of the coupling Hamiltonian (26), where the destruction operator $\hat{a}$ is associated with the generated field and χ is proportional to the nonlinear susceptibility $\varkappa^{(2)}$ characterizing two-photon absorption (TPA) and emission (TPE), respectively. Yuen pointed out that the use of c-numbers $\varkappa^{(2)}$ and $\varkappa^{(2)*}$ instead of atomic operators should be justified, since TPL might have sufficiently small atomic fluctuations far above threshold. The equation of motion from $\hat{H}_C$ yields the result, that an initially coherent state $|\alpha(0)\rangle$ passes to the squeezed state $|\beta(t), \mu(t), \nu(t)\rangle_y$ which is a right-hand eigenstate of the operator

$$\hat{b}(t) = \mu(t)\hat{a}(0) + \nu(t)\hat{a}^+(0), \text{ where } |\mu(t)|^2 = 1 + |\nu(t)|^2 \quad (30)$$

for all positive t (the functions $\mu(t)$, $\nu(t)$ depend on the parameters contained in the coupling Hamiltonian). This state was introduced by Yuen under the name two-photon coherent state (TPCS); as shown in Section 1, the TPCS's reveal second- and higher-order squeezing.

So far we have discussed the production of squeezed states on the basis of multiphoton processes under rather idealized conditions. Several objections to the used procedure can be raised: the neglect of the quantization of the pump field, the neglect of relaxation processes, the neglect of the fluctuations of the nonlinear medium. The neglected effects might strongly counter the squeezing effect in reality.

Therefore we will now deal with the problem of production of squeezing under <u>more realistic physical conditions</u>.

To treat TPA and TPE processes Schubert and Vogel [10] have generalized the Hamiltonian (26) used for the construction of TPCS's by taking into account the <u>quantization of the atomic systems</u>. Assuming two-level atomic systems the coupling Hamiltonian

$$\hat{H}_C = \hbar \sum_{k=1}^{N} (\varkappa \hat{B}_{21}^{(k)} \hat{a}^2 + \varkappa^* \hat{B}_{12}^{(k)} \hat{a}^{+2}) \quad (31)$$

now contains, besides the field operators $\hat{a}$, $\hat{a}^+$, the flip operators $\hat{B}_{21}^{(k)}$, $\hat{B}_{12}^{(k)}$ of the k-th atomic system; $\varkappa$ is the two-photon coupling constant. The free Hamiltonian of the atomic system contains the energies $\hbar\omega_1$ and $\hbar\omega_2$ of the ground and excited states as well as the corresponding occupation operators $\hat{B}_{11}^{(k)}$ and $\hat{B}_{22}^{(k)}$. The equations of motion for the operators $\hat{a}$, $\hat{B}_{12}^{(k)}$ and the inversion $\hat{\Delta}^{(k)} = \hat{B}_{22}^{(k)} - \hat{B}_{11}^{(k)}$ can be derived; this set of equations had been exactly solved up to the second order. Using this short-time approximation and starting from a fixed but arbitrarily chosen preparation of the atomic system we can in any case calculate the tendency of the transient behaviour of coherence and squeezing properties. There result qualitative discrepancies from the TPCS solution mentioned above: nonlinear terms enter the formula for $\hat{a}(t)$; an initial coherent state does not automatically pass to a TPCS with increasing interaction time; the mean photon number does not depend on the phase of the field as in the case of the TPCS; the bunching excess (calculated for an initial coherent state) attains other values than those in the TPCS solution; if the medium is partially inverted, that means $\sum_k \langle \hat{\Delta}^{(k)} \rangle > 0$ (so that TPE is realized) the initial coherent state tends with increasing time to a bunched state; under the condition $\langle \hat{N}(0) \rangle \gg 1$ antibunching can only occur in the case of a dominant TPA process, namely if the ground-state occupation $\tilde{N}_1$ exceeds the excited-state occupation $\tilde{N}_2$ by the factor 5. To discuss squeezing we will study the variance of the electric field strength under the condition of an initial coherent state and a small coupling constant $|\varkappa|$ compared with the radiation frequency ω:

$$\langle [\Delta \hat{E}(r,t)]^2 \rangle =$$

$$= g^2 + 8g^2(1 + |\alpha|^2) |\varkappa|^2 t^2 \sum_{k=1}^{N} \langle \hat{B}_{22}^{(k)}(0) \rangle \qquad (32)$$

$$-g^2 |\alpha|^2 |\varkappa|^2 t^2 \cos[2kr - 2\omega_0 t + \varphi_\alpha] \cdot \sum_{k=1}^{N} \langle \hat{\Delta}^{(k)}(0) \rangle + o(|\varkappa|^3 t^3).$$

The first (time-independent) summand represents the field fluctuations of the initial coherent field. The second term grows with increasing time, if some atomic systems are initially excited. The third summand exhibits an oscillatory behaviour depending on the space and time coordinate. The amplitude of these oscillations increases with the interaction time. In the case of a dominant TPE process the fluctuations cannot fall below the vacuum fluctuations, which is in disagreement with the TPCS solution. However, for a dominant TPA process squeezing occurs, namely if the condition $\tilde{N}_1 > 5 N_2$ (that agrees with the above antibunching condition) holds. In contrast to the behaviour of TPCSs the uncertainty product does not oscillate. Furthermore, the minimum uncertainty state character of the initial coherent state is only preserved in the case of ideal TPA. In summary the more realistic description with atomic operators leads to results that differ strongly from the TPCS solution. Somewhat later some other authors reported on similar results: Lugiato and Strini [11] and Reid and Walls [12] concerning nonsqueezing in the two-photon laser (TPE) case; Loudon [13] concerning squeezing of coherent light subjected to TPA.

Although in the case of a dominant TPE process squeezing cannot be expected, it may occur on the basis of <u>two-photon absorptive optical bistability</u> as shown by Lugiato and Strini [11].

Generalizing the above given description of degenerate <u>four-wave mixing</u>, Reid and Walls [14] presented a treatment of this process, where the medium is modeled as an ensemble of quantized two-level atomic systems in which the effects of loss and spontaneous emission on squeezing are accounted for. The pump modes are treated classically. Only in a certain limit (sufficiently large detuning and relatively small pump intensities) the derived quantum noise agrees with that of the ideal Yuen-Shapiro model [9] sketched above.

Hong and Mandel [5] have studied squeezing in the process of the <u>generation of a harmonic wave</u> under the condition of a quantized pump field. The initial state was assumed to be the direct product of the coherent pump state and the vacuum state of the SH-wave. In the short-time approximation (keeping exactly all contributions up to the second order) they found that the fundamental wave of harmonic generation exhibited squeezing in the second and fourth order.

An analysis of the <u>degenerate parametric oscillator (amplifier)</u> including the quantization of the pump field has been carried out by Milburn and Walls [15]. The squeezing in the idler mode characterized by $\langle (\Delta \hat{a}_2)^2 \rangle$ is shown as function of the reduced pump field amplitude $\langle \hat{E}_p \rangle$ in Fig. 3; a nonzero pump amplitude obviously leads to squeezing. The squeezing effect has a maximum (corresponding to $\langle (\Delta \hat{a}_2)^2 \rangle = 1/8$) close to the threshold value of the pump field; increasing power above the threshold decreases the squeezing effect. The dashed line exhibits the monotonously increasing squeezing effect of an ideal parametric amplifier.

As a rule <u>nonlinear intracavity devices</u> give a maximum squeezing factor near two [16], since the coupling of the cavity modes to the vacuum

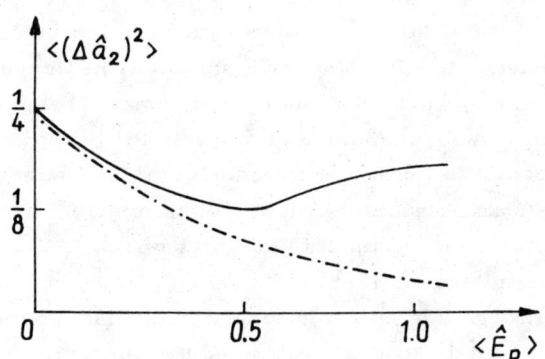

Fig. 3. Squeezing in a parametric oscillator (———) compared with an ideal parametric amplifier (—·—). After Milburn and Walls [15]

modes of the extracavity modes counters the squeezing produced by the nonlinear interaction.

Let us now discuss processes on the basis of <u>resonance fluorescence</u>. As shown by Walls and Zoller [17] fluorescent radiation from a coherently excited two-level atom is capable of yielding second-order squeezing. The electric field at a distant point is

$$\hat{E}^{(+)}(\vec{r}, t) = K(\vec{r})\, \hat{b}(t - r/c) + \hat{E}^{(+)}_{free}(\vec{r}, t), \qquad (33)$$

where $\hat{b}$ is the atomic lowering operator and $K(\vec{r})$ a geometrical factor. If one is only interested in the normally ordered moments of the field at point $\vec{r}$ and the driving field is assumed to vanish at $\vec{r}$, one may proceed as if $\hat{E}^{(+)}_{free}$ is absent in (33). The stationary solution of $\langle \hat{b}(t) \rangle$ can be explicitly calculated; its time dependence agrees with that of the incident coherent laser light and its amplitude depends on the Rabi frequency Ω, the natural atomic line width 2β, and the relative laser-atom detuning θ. More recently Ficek, Tanas, and Kielich [18] treated squeezed states in the transient regime of resonance fluorescence. Under the conditions given in [17], Hong and Mandel [5] calculated the time-dependent normally ordered moment of the field strength at point r. These quantities depend on the phase $\phi = \omega_{las} t + \varphi_{las} + \arg K - \varphi_0 + \tan^{-1}\theta + \pi$. With the help of the normally ordered moments the squeezing condition can be formulated [see (24)]. The choice $\phi = n\pi$ ensures second- and higher-order squeezing, if $\Omega^2/\beta^2 < 2 + 2\theta^2$ holds. However, the squeezing factor of a single atom is rather small.

If a large number of atoms contributes to the fluorescent light, the effect of squeezing is lost, when the atoms are located at random positions, since the appropriate phase relations between the light parts are absent. Vogel and Welsch [19] showed that a squeezing pattern could be produced in resonance fluorescence from an ensemble of N atomic systems regularly distributed in a linear chain. In this case the first term on the right-hand side of (33) must be replaced by a sum of all atomic terms. The effect of the mutual interaction of the atoms through the electromagnetic field can be neglected, if there holds $\lambda/|\vec{d}| < 0{,}1$, where $\vec{d}$ is the distance vector between neighbouring atoms. The remaining interference effect yields a squeezing pattern expressed by formula

$$(\vec{k}' - \vec{k}_n)\,\vec{d}/\pi = \begin{cases} n=0, \pm 2,\ldots \text{ (squeezed coherent states)} \\ n=\pm 1, \pm 3,\ldots \text{ (squeezed vacuum states)}, \end{cases} \qquad (34)$$

where $\vec{k}$ is the wave vector of the incident laser light and $\vec{k}_n$ the wave vector of the observed light. The intensity of the even-order coherent radiation is proportional to N^2, while that of the odd-order incoherent radiation is proportional to N. The above calculations can be extended to the somewhat more complicated three-dimensional case.

3. Detection and applications of squeezed states

Let us first sketch <u>detection aspects</u>. Squeezing can be detected by <u>homodyning the signal field</u> with a local oscillator (see, e.g. Loudon [20]). The signal radiation to be investigated, say a laser beam traversing a two-photon absorber, excites a single mode (operator $\hat{s}$, frequency ω). By a beam splitter this wave is combined with the wave of the local oscillator (operator $\hat{a}$, state $|\alpha\rangle$ of ideal laser light, frequency ω, $|\alpha|^2 \gg \langle \hat{s}^+ \hat{s}\rangle$). One of the combination waves enters a photon-counting detector. Its modal operator is

$$\hat{d} = (1/\sqrt{2})\,(\hat{s} + i\hat{a})\,\exp\left[i\varphi_{t-r}\right], \qquad (35)$$

where φ_{t-r} accounts for the phase changes on transmission and reflection. Mandel [21] has shown that the measurable photon fluctuations at the homodyne detector, characterized by $Q = \langle :(\Delta \hat{N}_d)^2:\rangle / \sqrt{\langle \hat{N}_d\rangle}$, can be expressed by the quadrature operators $\hat{s}_1, \hat{s}_2$ of the signal mode and by the phase φ_α of the laser beam:

$$Q = 2\langle :(\Delta \hat{F})^2:\rangle, \qquad (36)$$

where $\hat{F} = \hat{s}_1 \cos\left[\varphi_\alpha + (\pi/2)\right] + \hat{s}_2 \sin\left[\varphi_\alpha + \pi/2\right]$.

If there are phases φ_α, for which Q is negative, the signal mode is in a squeezed state.

An interesting application field is the reduction of <u>noise in optical communication systems</u> (see, e.g., Yuen [22]). In the classical case, a linear amplifier can be represented by the relation $\langle \hat{d}\rangle = G^{1/2}\langle \hat{c}\rangle$ between the mean values $\langle \hat{c}\rangle$ and $\langle \hat{d}\rangle$ assigned to the input and output field; here (under ideal conditions) noise is neglected. The gain

G is positive; $\hat{c}$ and $\hat{d}$ are the modal operators of the input and output field. The <u>quantized relation</u> is

$$\hat{d} = G^{1/2}\hat{c} + (G - 1)^{1/2}\hat{r}^+, \qquad (37)$$

where noise is taken into account (modal operator $\hat{r}$, $\langle \hat{r} \rangle = V$, $[\hat{r}, \hat{r}^+] = \hat{1}$, $\hat{r}$ and $\hat{c}$ modes uncorrelated) and there holds $[\hat{c}, \hat{c}^+] = \hat{1}$. The form (37) must be chosen to ensure linear amplification and the boson commutator relation $[\hat{d}, \hat{d}^+] = \hat{1}$ for the output. Let $\hat{d}_1$, $\hat{d}_2$, $\hat{c}_1$, $\hat{c}_2$ be the quadrature operators of the output and input mode. Then the quantized linear amplification scheme under ideal conditions (without noise)

$$\hat{d}_1 = G^{1/2}\hat{c}_1, \qquad \hat{d}_2 = G^{-1/2}\hat{c}_2 \qquad (38)$$

can be introduced; the quadrature $\hat{c}_1$ is amplified, while the quadrature $\hat{c}_2$ is attenuated. This amplification of one single quadrature preserves the signal-quantum noise-ratio:

$$S - QN - R = \langle \hat{d}_1 \rangle^2 / \langle (\Delta \hat{d}_1)^2 \rangle = \langle \hat{c}_1 \rangle^2 / \langle (\Delta \hat{c}_1)^2 \rangle. \qquad (39)$$

The amplification scheme (38) is linear with respect to $\hat{c}_1$, but nonlinear with respect to $\hat{c}$. Nonlinear processes involving phase-dependent amplification are candidates for the realization of this amplification--attenuation-mechanism; relation (39) reveals the advantage, in case squeezed states are used.

Another possible application for squeezed states is connected with the <u>detection of gravitational waves</u>. One promising technique uses an interferometer (see Fig. 4) to measure relatively small position changes $L = L_x - L_y$ of widely separated masses m of the end mirrors. Changes in L exhibit the passing of a gravitational wave. A two--arm, multireflection Michelson system powered by a laser (modal operator $\hat{a}$, coherent state $|\alpha\rangle$) is considered. The mass M of the beam splitter (and the inner mirrors) is assumed to be much greater than m, so that this inner part of the interferometer can be regarded as at rest. There are two basic error sources limiting the accuracy of L: the radiation-pressure error (rp) due to fluctuating radiation forces on the end mirrors and the photon-counting error (pc) due to output

number fluctuations. The "squeezed-
-state technique" proposed by Caves [23] requires the modification
of the electromagnetic excitation
entering the usually unused port
(UUP), where the modal operator
of the radiation is denoted by $\hat{d}$.
The radiation with normally entering
vacuum state is replaced by a
squeezed state whose fluctuations
in one quadrature component are
less than the fluctuations in the
vacuum state. The operator char-
acterizing the difference between
the photon momenta transferred to
both the end mirrors

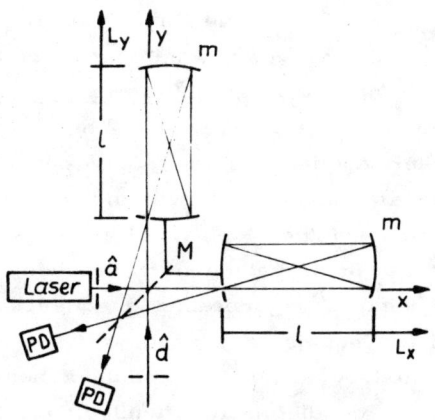

Fig. 4. Schematic diagram of Mi-
chelson interferometer. After Caves [23]

$$\hat{P} = i 2n \hbar \omega c^{-1}(\hat{a}^+\hat{d} - \hat{d}^+\hat{a}), \tag{40}$$

depends on $\hat{a}$ and $\hat{d}$. $\hat{P}$ leads to the radiation-pressure error
$(\Delta L)^2_{rp} = \tau^2 m^{-2} \langle (\Delta \hat{P})^2 \rangle$; τ is the measuring duration. The
photon-counting error and the radiation-pressure error for various states
of the light entering the port UUP are given by

$$(\Delta L)^2_{rp} = C_{rp} \langle \hat{N}_a \rangle, \quad (\Delta L)^2_{pc} = C_{pc} \langle \hat{N}_a \rangle^{-1} \quad \text{for} \mid ... \rangle_{vac} \tag{41}$$

$$(\Delta L)^2_{rp} = C_{rp} \langle \hat{N}_a \rangle \exp[2|\eta|],$$
$$(\Delta L)^2_{pc} = C_{pc} \langle \hat{N}_a \rangle^{-1} \exp[-2|\eta|] \quad \text{for} \mid ... \rangle_{squ}. \tag{42}$$

While $(\Delta L)^2_{rp}$ increases, $(\Delta L)^2_{pc}$ decreases with growing mean number
of laser photons. The formulas in (41) hold for a vacuum state at UUP.
The formulas in (42) show that by squeezing the vacuum (squeezing
factor $|\eta|$) the photon-counting error can be reduced at the cost of
the radiation pressure. The minimum total error is reached at the opti-
mum laser input power $\tilde{P}_{opt} = \tilde{P} \exp[-2|\eta|]$. $\tilde{P}$ depends on m,
ω, τ, and the number of bounces; a value of 10^4 W(cw) can be assumed. Obviously the required highly-stabilized input laser power can
be reduced by squeezing.

Schubert, Vogel, and Welsch [24] proposed to generate the squeezed vacuum states, which are useful for interferometric detection of gravitational waves, by <u>interference of squeezed coherent states</u>. Let two waves being incident in x- and y-direction with equal frequencies and polarizations. A semi-transparent mirror mixes both the modes. Under the assumption that both the incident waves are in squeezed coherent states (of the Yuen-type) physically relevant values of the combination mode travelling in x-direction have been calculated. The optimum condition for generating a squeezed vacuum state is the superposition of two equally squeezed waves of equal moduli, where the phases are shifted by π. If the moduli of both ν-values are equal and appropriate phase conditions are fulfilled, one has $\langle \hat{E}(x,t) \rangle = 0$ and the corresponding variance is

$$\langle [\Delta\hat{E}(x,t)]^2 \rangle = $$
$$= g^2 + g^2 \{ 2|\nu|^2 + 2|\nu||\mu| \cos[2kr - 2\omega t + \varphi_\nu - \varphi_\mu] \} . \quad (43)$$

Comparison with (13) exhibits that the squeezed vacuum state generated by superimposing two squeezed coherent states has the same degree of squeezing as the original squeezed coherent states. The realization of such a squeezed vacuum requires special experimental efforts, in particular, with respect to a sufficiently high stabilization between the original squeezed coherent states. The simplest way to achieve this stabilization seems to be splitting of one squeezed coherent state and the subsequent combination of both the parts (this procedure reduces the squeezing gain by 1/2).

I acknowledge valuable discussions with Dr W. Vogel and Dr D. G. Welsch.

References

1. D. Stoler, Phys. Rev. <u>D1</u> 3217 (1970).
2. H. P. Yuen, Phys. Rev. <u>A13</u>, 2226 (1976).
3. M. Schubert, W. Vogel, Phys. Lett. <u>A68</u>, 321 (1978).
4. M. Schubert, W. Vogel, Wiss. Zeitschr. d. Univ. Jena, Math.-Nat. <u>27</u>, 179 (1978).
5. C. K. Hong, L. Mandel, Phys. Rev. Lett. <u>54</u>, 323 (1985).

6. G. J. Milburn, J. Phys. A: Math. Gen. <u>17</u>, 737 (1984).
7. K. Wódkiewicz, J. H. Eberly, J.O.S.A. <u>B2</u>, 458 (1985).
8. M. C. Teich, B. E. A. Saleh, J. Peřina, J.O.S.A. B1, 366 (1984).
9. H. P. Yuen, J. H. Shapiro, Opt. Lett. <u>4</u>, 334 (1979).
10. M. Schubert, W. Vogel, Opt. Comm. <u>36</u>, 164 (1981).
11. L. A. Lugiato, G. Strini, Opt. Comm. <u>41</u>, 374 (1982).
12. M. D. Reid, D. F. Walls, Phys. Rev. <u>A28</u>, 332 (1983).
13. R. Loudon, Opt. Comm. <u>49</u>, 67 (1984).
14. M. D. Reid, D. F. Walls, Phys. Rev. <u>A31</u>, 1622 (1985).
15. G. Milburn, D. F. Walls, Opt. Comm. <u>39</u>, 401 (1981).
16. D. F. Walls, Nature <u>306</u>, 141 (1983).
17. D. F. Walls, P. Zoller, Phys. Rev. Lett. <u>54</u>, 1802 (1985).
18. Z. Ficek, R. Tanas, S. Kielich, J.O.S.A. <u>B1</u>, 882 (1984).
19. W. Vogel, D. G. Welsch, Phys. Rev. Lett. <u>54</u>, 1802 (1985).
20. R. Loudon, Opt. Comm. <u>49</u>, 67 (1984).
21. L. Mandel, Opt. Comm. <u>42</u>, 437 (1982).
22. H. P. Yuen, p. 249 in "Quantum Optics, Exp. Grav., and Measurement Theory", 1983, Plenum Press.
23. C. M. Caves, Phys. Rev. <u>A23</u>, 1693 (1981).
24. M. Schubert, W. Vogel and D. G. Welsch, Opt. Comm. <u>52</u>, 247 (1984).

Stig STENHOLM

Research Institute for Theoretical Physics,
University of Helsinki,
Siltavuorenpenger 20 C, SF-00170 Helsinki, Finland

THEORY OF LASER COOLING

Lecture I: Cooling of free particles

When an atomic particle absorbs or emits radiative energy, its state of motion must be changed to take care of the momentum carried by the field. In ordinary optical spectroscopy these effects are too small to be observable, but the recent progress in high resolution measurements has made them important. They are also utilized by laser researchers to achieve special effects, of which the most popular one now is cooling. In this lecture I will review the basic theoretical ideas of the cooling of free particles and mention the main experiments that have been carried out recently. I will not attempt to give a complete history of the field; for the early references I refer the readers to my recent review article [1].

The earliest traces of atomic recoil were detected in the high resolution work needed for laser frequency standards. Both in Novosibirsk (Chebotaev et alii) and in Boulder (Hall et alii) there were spectra observed which showed a splitting due to the photon recoil.

The suggestion to use lasers to cool an atomic sample was presented by Hänsch and Schawlow [2]. The static part of the light pressure force had been derived earlier by Ashkin [3].

If we take a dipole μ in the electric field E its energy is μE. Its gradient in the laser field of wave vector q gives the force $\sim q \mu E$ as an estimate for the static light pressure. On the other hand, near saturation conditions and optimal detuning both the Rabi flipping rate

$\mu E/\hbar$ and the natural line width Γ have to nearly equal the Doppler detuning and hence we have

$$\frac{\mu E}{\hbar} \sim \Gamma \sim qv. \tag{1}$$

From these estimates we obtain an estimate for the rate of change of the kinetic energy K in the form

$$\frac{dK}{dt} = Mv\dot{v} = -\varepsilon K. \tag{2}$$

Thus the time scale needed to change the energy of the particle appreciably is given by the recoil energy ε in frequency units

$$\varepsilon = \frac{\hbar q^2}{2M}. \tag{3}$$

During this time scale many spontaneous emission events take place and hence the fraction

$$\xi = \frac{\varepsilon}{\Gamma} \tag{4}$$

is a small parameter of the order of 10^{-5}. It is often convenient to use this as the expansion parameter in the theoretical treatments.

When we attempt to cool a gas of freely moving atoms, we must tune the laser to resonance with some velocity group. These atoms are then cooled and decrease their velocity. To keep on cooling them we must take some special precautions; either the laser is following the atomic resonance frequency [4] or the atomic energy levels are made to follow in the change of the Doppler tuning [5].

The cooling cycle does however utilize the spontaneous decay as an important step. Because this is a stochastic process, the outgoing photons give a random recoil to the atom, and there appears a Brownian-motion like random drift in momentum space, which can be estimated simply. The step of each spontaneous recoil kick is of the order of $\hbar q$ and their number is determined by the rate of spontaneous emission Γt. For a reasonably strong laser field the atom is rapidly excited after each decay, and the bottleneck of the process is given by the decay rate. The spreading in the period t determines the diffusion coefficient D from

$$\overline{p^2} = Dt \sim (\hbar q)^2 \Gamma t. \tag{5}$$

The diffusion process defines a time scale T_D which is determined as the time in which the diffusive spreading equals the Doppler range over which atoms interact in accordance with (1). This gives

$$DT_D \sim (Mv)^2 \sim \left(\frac{M\Gamma}{q}\right)^2 \tag{6}$$

$$T_D \sim \left(\frac{\Gamma}{\varepsilon}\right) \varepsilon^{-1}. \tag{7}$$

Because the parameter ξ in (4) is small the diffusive time scale is long compared with the time scale set by the static force. Hence one can often forget the influence of diffusion and calculate with only the cooling force. The diffusive heating is, however, unavoidable, and for instance the transverse heating due to diffusion may seriously deteriorate the beam quality during cooling.

The diffusive heating also sets the ultimate limit to the cooling procedure. In a standing light wave the two counter-propagating waves cancel their static forces and only a net cooling remains. When the atomic velocity distribution reaches a state where the remaining cooling is exactly counteracted by the diffusive spreading there is no more evolution; the steady state is reached. This can be used to obtain an estimate of the ultimate cooling temperature reachable with laser cooling. There exists only one single velocity scale in this situation, and we assume it to be given by the diffusion as

$$v_\infty^2 \sim \frac{Dt}{M^2}. \tag{8}$$

At the same time the cooling force acts and we suppose it to decrease the kinetic energy by the corresponding amount. This in its turn determines the time scale from (2) and we have $t \sim \varepsilon^{-1}$. Inserted into Eq. (8) it gives

$$v_\infty^2 \sim \frac{\hbar\Gamma}{M}. \tag{9}$$

This implies that the ultimate cooling limit is given by the energy

$$E_f \sim \hbar\Gamma. \tag{10}$$

This can be understood if we remember that the time of spontaneous emission is uncertain within the time Γ^{-1}. Hence the energy of the corresponding state cannot be determined with a higher accuracy than $\hbar\Gamma$, which gives just the limit (10). It is also reasonable that the cooling due to the vacuum fluctuations of the fields should cool only to an energy equal to the energy density of the fluctuations at the frequency of the interaction. As these can be regarded as the cause of the spontaneous emission, their intensity is directly given by Γ and hence the result (10) follows again. For atomic systems the limit (10) gives an ultimate temperature of about 10^{-3} K, and it consenquently does not impose too strict limitations on the possibilities of cooling

Because the basic interaction utilized in the cooling process combines the absorption of a laser photon with the subsequent spontaneous emission of another, there appears a strict correlation between the statistical distribution of the cooled momentum and the outgoing photons. This was suggested first by Mandel [6] and developed further by Cook [7]. The fact, that consecutive photon emission processes are coupled through an intervening laser excitation process, makes them acquire a correlation which causes their statistical properties to deviate from the totally random Poisson distribution. They emerge with a more regular spacing than a random decay, and hence they are sub-Poissonian. In the momentum diffusion the same is seen as an anomalous contribution to the spreading, which makes it less than in the totally random case. This was first noted by Gordon and Ashkin [8], Minogin [9] and Cook [10]. Experimentally only the single observation by Mandel and Short [11] supports the sub-Poissonian distribution of the outgoing photons.

The recent experimental progress in the field of laser cooling has been directed towards the cooling of the longitudinal velocity distribution of an atomic beam. The earliest experiments were reported by Balykin et alii [12], who used frequency sweeping to achieve cooling. Phillips et alii [13] have continued this work, partly using the Zeeman effect to keep the atoms in resonance. Results have also been reported by Hall et alii [14].

In addition to the induced light pressure force, which can be envisaged as deriving from the photon recoil momentum, there are gradient forces owing to the variation of the field intensity. These are usually smaller than the resonanant light pressure forces, but they provide a better chanse to trap the interacting particle by pure light forces. In the resonant case the trapping potential is upset by the large fluctuations due to resonant transitions between the levels. It now seems to be the general opinion that no simple resonant trap exists. For special schemes it may be possible to overcome the difficulties. Combining laser cooling with an external magnetic field Migdall et alii [15] have confined neutral particles in space.

The picture of laser cooling given above is oversimplified in many aspects. To obtain a more satisfactory theory of the cooling some detailed quantum mechanical calculations are needed. These have to identify the small parameter of the problem and use a systematic approximation scheme to separate the various effects under discussion. In the atomic case the suitable small parameter is the one in Eq. (4). It is proportional to the photon wave vector q squared and to the first power of Planck's constant. When photon recoil is neglected there is no modification of an initially imposed atomic momentum distribution. This corresponds to the fact that classically the photon momentum is zero and the atomic momentum is conserved in the radiative interaction. The solution is then highly degenerate and only the photon recoil provides the light pressure force given in Eq. (2). This is indeed proportional to Planck's constant $\hbar$. The drift of the atomic momentum distribution in phase space can now be described by a local conservation law, with a continuity equation determined by the light pressure, which assumes that the photon impacts are so small and frequent that a continuum limit can be taken.

To second order in the small parameter we have to recognize that the single events are discrete and constitute a Brownian motion; the result (5) was derived from such considerations. The ensuing quantum correction to the light pressure force is represented by the diffusion coefficient D, which is indeed proportional to the second power of Planck's constant. Thus we see that the expansion includes successive quantum corrections to the classically conserved quantities, and it may hence be called a semi-classical theory [16]. The theoretical approach has been shown to be related to the Chapman-Enskog singular perturbation theory used in statistical physics.

In calculating the consecutive approximations to the theory, one must remember that the diffusion has to be considered in a frame moving under the influence of the light pressure force, which was obtained in the previous order of the approximation. This gives rise to correlations between successive photon emission events, which is manifest in the anomalous diffusion and also in the photon statistics as discussed above. The actual calculations are rather involved, and they have to be looked up in the references. The anomalous contribution to the diffusion is found to be a higher order coherent process, which disappears for large intensities too. In this case the atom jumps so fast between its different levels that no specific restrictions are imposed on the time of spontaneous emission; the atom finds itself at the upper level often enough to be able to emit radiation at almost any time. Hence the photon correlations disappear.

By considering the small parameter to be the photon wave vector, Dalibard and Cohen-Tannoudji [17] find both the diffusion (5) and the damping rate (2) in the same order. They also show that they are related by a fluctuation-dissipation theorem just like their equilibrium counterparts.

Lecture II: Cooling of trapped particles

One of the main aims of laser spectroscopy is to overcome the restrictions imposed by the Doppler shift. There are several methods to achieve this with respect to the linear one, but the quadratic Doppler shift can be eliminated only by bringing the investigated particles to rest in the laboratory. This is the main motivation behind the attempts to carry out spectroscopy on trapped particles. This also allows the use of long interaction times, which in its turn improves the resolution. Because the particles' mutual interaction is to be eliminated the considerations lead to the requirement that only one single trapped particle is to be investigated. As already mentioned no stable trap exists for neutral particles yet, but for charged particles it has been possible to devise satisfactory traps. These cannot be achieved by static electric fields (the Earnshaw theorem), but it is possible with radiofrequency fields [18] or by adding magnetic forces (the Penning trap) [19].

As a theoretical model for a trapped particle we use a harmonically bound two-level system. This does not correspond exactly to either of the two traps used in experiments, but it is close to the conditions in the radio-frequency trap towards the end of a cooling process.

If the particle is oscillating only slowly, its position and velocity are not much changed during the absorption and subsequent emission of a photon. The cooling processes proceed much as in the free particle case, and the instantaneous velocity serves only as a parameter. This is the heavy particle limit, where we can take over most results directly from the theory of cooling the free particle. When the motion becomes faster, the trap confines the particles more tightly and the oscillational quanta increase in size. Towards the end of a successful cooling cycle, we then find the particle confined within one wave length of the radiation and the discrete nature of the trapping energy levels cannot be neglected. This quantum limit of the theory is called the Lamb-Dicke limit, and it is the one which has received the most experimental and theoretical interest.

If we take as our small parameter η^2 the inverse of the number of recoil energies needed to make up a single oscillational quantum, we find the parameter

$$\eta^2 = \frac{\hbar \varepsilon}{\hbar \nu} = \left(\frac{\pi a_0}{\lambda}\right), \qquad (11)$$

where a_0 is the size of the ground state of the harmonic potential. Towards the end of the cooling a small value of (11) guarantees that the Lamb-Dicke condition is satisfied.

If we define the cooling time scale T_{os} by the time it takes to cool an energy equal to the oscillator quantum $\hbar \nu$ using the rate (2) and the energy estimate (10) we obtain

$$\frac{\hbar \nu}{T_{os}} \sim \varepsilon \hbar \Gamma, \qquad (12)$$

and consequently

$$T_{os} = \eta^{-2} \Gamma^{-1}. \qquad (13)$$

This shows that the factor η^2 tells exactly how much slower the cooling process is than the spontaneous decay process proceeding at the rate Γ. The time scale T_{os} defines a rate of change which is determined by a master equation for the cooling in the Lamb-Dicke limit. This has been derived in Refs. [20], and it shows that in contrast to the free particle case this limit contains only one slow rate given by (13). Thus both the quantum coherences and the nonequilibrium distributions disappear at the same rate [21].

The derivation of the master equation is a modification of the adiabatic elimination method used for the free particle. The eigenvalue spectrum of the system is found to be highly degenerate when no recoil momenta are considered. A degenerate perturbation theory is used to split the energy levels, and the ensuing effective Hamiltonian is used as the time evolution operator for the master equation. All degeneracies are found to disappear, and the states are found to decay, except a single state that remains at its eigenvalue zero. This is the unique steady state which is approached towards the end of the cooling process.

A detailed solution [21] of the steady state proves it to be of the same type as the Planck black-body radiation, and hence the final energy E_f corresponds to a temperature T_f. These are however connected by the quantum formula relevant for Bose statistics

$$E_f = \hbar \nu \left[\frac{1}{\exp(\hbar \nu / k_B T_f) - 1} + \frac{1}{2} \right]. \quad (14)$$

The cooling itself progresses by the aid of quanta obtained from the sidebands imposed on the atomic resonance by the periodic motion in the trap. The final excitation energy can be made to be close to the ground state, when the energy spacing in the harmonic oscillator ladder $\hbar \nu$ greatly exceeds the final energy $\hbar \Gamma$ imposed by the quantum fluctuations. This "freezing out of degrees of freedom" is a phenomenon well known from other branches of quantum statistics. In this case the final excitation energy is given by

$$E_f - \frac{\hbar \nu}{2} \sim \left(\frac{\Gamma}{\nu}\right) \hbar \Gamma \ll \hbar \Gamma, \quad (15)$$

which is better than the estimate (10), but from (14) we find easily that the temperature satisfies the relation

$$k_B T_f > \hbar \Gamma. \quad (16)$$

Also here the final cooling energy is found to have anomalous contributions, but their isolation is harder because we only have one slow time scale to deal with. The evaluation of the final cooling limit is a complicated problem, the solution of which is found in the references. No information about the photon statistics is available for this case.

Both the group at Boulder [19] and the one in Heidelberg-Hamburg [18] have reported successful laser cooling of trapped ions. These experiments give the first example of an atomic particle trapped for hours in a specific place, where it can be observed in the light it itself scatters. The arrangements provide exciting possibilities to carry out fundamental experiments on single atomic particles.

References

1. S. Stenholm, Submitted to Rev. Mod. Phys.
2. T. Hänsch and A. Schawlow, Optics Comm., 13, 68 (1975).
3. A. Ashkin, Phys. Rev. Letters, 25, 1321 (1970).
4. V. S. Letokhov, V. G. Minogin and B. D. Pavlik, Optics Comm., 19, 72 (1976) and Sov. Phys. JETP, 45, 698 (1977).
5. W. D. Phillips and H. Metcalf, Phys. Rev. Letters, 48, 596 (1982).
6. L. Mandel, J. Optics (Paris), 10, 51 (1979).
7. R. J. Cook, Optics Comm., 35, 347 (1980) and Phys. Rev., A23, 1243 (1981).
8. J. P. Gordon and A. Ashkin, Phys. Rev., A21, 606 (1980).
9. V. G. Minogin, Sov. Phys. JETP, 52, 1032 (1980).
10. R. J. Cook, Phys. Rev., A22, 1078 (1980).
11. R. Short and L. Mandel, Phys. Rev. Letters, 51, 384 (1983).
12. V. I. Balykin, V. S. Letokhov and V. I. Mushin, JETP Letters, 29, 560 (1979) and Sov. Phys. JETP, 51, 692 (1980); V. I. Balykin, V. S. Letokhov and A. I. Sidorov, Soviet Physics JETP, 59, 1174 (1984) and Optics Comm., 49, 248 (1984).
13. J. V. Prodan, W. D. Phillips and H. Metcalf, Phys. Rev. Letters, 49, 1148 (1982) and J. Opt. Soc. Am. B (to be published 1985); J. V. Prodan, A. Migdall, W. D. Phillips, I. So, H. Metcalf and J. Dalibard, Phys. Rev. Letters, 54, 992 (1985).
14. W. Ertmer, R. Blatt, J. L. Hall and M. Zhu, Phys. Rev. Letters, 54, 996 (1985).
15. A. L. Migdall, J. V. Prodan, W. D. Phillips, T. H. Bergman and H. J. Metcalf, Phys. Rev. Letters, 54, 2596 (1985).
16. S. Stenholm, in <u>Quantum Electrodynamics and Quantum Optics</u>, Ed. A. O. Barut (Plenum Publ. Co. 1984).
17. J. Dalibard and C. Cohen-Tannoudji, J. Opt. Soc. Am. B (to be published 1985).
18. W. Neuhauser, M. Hohenstatt, P. E. Toschek and H. G. Dehmelt, Phys. Rev. Letters, 41, 233 (1978).
19. D. J. Wineland, R. E. Drullinger and F. L. Walls, Phys. Rev. Letters, 40, 1639 (1978).
20. M. Lindberg and S. Stenholm, J. Phys. B: At. Mol. Phys., 17, 3375 (1984).
21. S. Stenholm, J. Opt. Soc. Am. B (to be published 1985).

P. W. MILONNI

Department of Physics,
University of Arkansas,
Fayetteville, AR 72701

CHAOS IN QUANTUM OPTICS

The Greeks at the time of Ptolemy thought all motion could be decomposed into "perfect" circular motions. With the usual acuity of hindsight we might phrase their theory this way: All motion is quasi-periodic, such that the Fourier transform of any coordinate consists of sharp spikes. Poincaré, near the turn of the century, was perhaps the first person to understand that there are systems whose spectra do not have this form, that there are non-integrable systems. Such systems have a broadband, continuous component to their spectra. Non-quasi-periodic motion is associated with nonlinearity.

Deterministic chaos may be defined in terms of "very sensitive dependence on initial conditions". More specifically, a dynamical system is chaotic if it has a positive Lyapunov exponent. With this definition, chaos implies non-quasi-periodicity and a broadband spectrum.

The great current interest in chaotic behavior seems due to the following factors: (a) the realization that chaos can occur even in "simple", low-dimensional systems; (b) the discovery of a few prevalent routes from orderly motion to chaos; and (c) the increasing use of digital computers as laboratories for "experimental" studies. But the idea that there are classical systems which are practically indeterminable because of their sensitivity to initial conditions is an old one, going back at least as far as Maxwell, who wrote of a "prejudice" in favor of Laplacian determinism due to a lack of adequate attention to unstable motion. And Max Born wrote that "It is misleading to compare quantum mechanics with deterministically formulated classical mechanics; instead, one should first reformulate the classical theory, even for a single particle, in an indeterministic, statistical manner. Then some of the distinctions between the two theories disappear, others emerge with great clarity. Amongst the first is the feature of quantum mechanics, that each measurement interrupts the automatic flow of events and introduces new initial conditions (so-called "reduction of probability"); this is true just as well for a statistically formulated classical theory..." Similar views about the "indeterminism" of classical physics are expressed in the Feynman Lectures.

The recent interest in chaos should lead to a more widespread appreciation for the fact that classical physics, as a practical matter, is not really "deterministic" in the sense of being able to make detailed, long-term predictions about the future. The big conceptual difference between classical and quantum physics lies in the fact that quantum mechanics does not allow for an objective, Einsteinian reality. Experimental distinctions between locally objective and quantum-mechanical conceptions of reality have been tested in quantum-optical experiments of Clauser, Fry, and Aspect and their collaborators.

Quantum-optical experiments have more recently contributed to the growing awareness that chaos is not an unusual mode of behavior in physical systems. This is especially clear from the observations of chaos in lasers and optically bistable systems by Arecchi and Gibbs, respectively, and their collaborators. Ackerhalt, Shih, and I have written a very introductory review of chaos and some of its manifestations in quantum optics [1]. I have discussed elsewhere some implications of chaos in a somewhat more general context [2].

Is chaos a classical phenomenon, appearing only in systems well described by a semiclassical approximation? It appears, at least in some cases, that quantum effects act to suppress certain aspects of chaos in classical Hamiltonian systems. In the kicked pendulum example, for instance, the diffusive energy growth associated with the classical model tends to be suppressed after a certain time due to a sort of Anderson localization in phase space [3]. This feature is found in the numerical experiments of Casati, et al. [4]. A similar diffusive energy growth is found in a model of multiple-photon excitation of molecular vibrations in an infrared laser field, and has been invoked to explain the fluence dependence of the process in large polyatomic molecules [5]. This fluence dependence seems to be a general experimental trend [6]. How well will the classical model survive quantum corrections and, in particular, will the diffusive energy deposition survive or will it be severely suppressed? Similar questions arise in connection with classical models of the microwave ionization of highly excited hydrogen atoms [7].

These problems once again rise questions about the classical limit of quantum mechanics. In the multiple-photon excitation problem just mentioned it is reasonable that, once the molecule has been pumped high enough up a vibrational ladder, a classical model ought to be accurate. This argument is a familiar one and has been used by several workers [8]. On several occasions supporting reference has been made to computations of Preston and Walker [9], who studied Heisenberg versus classical equations of motion for a sinusoidally driven anharmonic oscillator. These are indeed quite interesting computations, showing a rather good classical-quantum correspondence. However, the computations were done only for short time intervals, in order to track the oscillations of the driving force (e.g., no rotating-wave approximation was made). It is precisely for such short times that the classical-quantum correspondence should be best, and in fact the authors themselves made no claims for long-term expectation values that would be of more interest for the problem at hand.

Part of the difficulty with the whole area of "quantum chaos" in driven systems is that there is no consensus as to what should be computed to determine whether a given system should be called "chaotic". (Classically, the problem is well defined, for we can compute the largest Lyapunov exponent and, if it is positive, the dynamics are chaotic.) Diffusive energy deposition in the kicked pendulum, or the hydrogen atom in an external field, or a model of multiple-photon vibrational excitation, seems at present to be one of the more interesting things in this field because it can be studied directly in real experiments [10].

References

1. J. R. Ackerhalt, P. W. Milonni, and M.-L. Shih, Physics Reports (in press).
2. P. W. Milonni, in <u>Proceedings of the Workshop on the Foundations of Physics</u> (University of Puerto Rico, in press).
3. S. Fishman, D. R. Grempel, and R. E. Prange, Phys. Rev. Lett., <u>49</u>, 509 (1982); Phys. Rev., <u>A29</u>, 1639 (1984).

4. G. Casati, B. V. Chirikov, F. M. Izrailev, and J. Ford, in *Stochastic Behavior in Classical and Quantum Hamiltonian Systems*, Springer Lecture Notes in Physics 93 (1979).
5. J. R. Ackerhalt, H. W. Galbraith, and P. W. Milonni, Phys. Rev. Lett., $\underline{51}$, 1259 (1983).
6. T. B. Simpson, J. G. Black, I. Burak, E. Yablonovitch, and N. Bloembergen, J. Chem. Phys., $\underline{83}$, 628 (1985).
7. J. G. Leopold and I. C. Percival, Phys. Rev. Lett., $\underline{41}$, 944 (1978).
8. See, for instance, W. E. Lamb, Jr, in *Chaotic Behavior in Quantum Systems*, ed. by G. Casati (Plenum, N.Y., 1985).
9. R. B. Walker and R. K. Preston, J. Chem. Phys., $\underline{67}$, 2017 (1977).
10. J. E. Bayfield and L. A. Pinnaduwage, American Physical Society DEAP meeting at Norman, Oklahoma, 29-31 May 1985.

Jan PEŘINA

Joint Laboratory of Optics,
Czech. Acad. Sci. and
Palacký University,
Olomouc, Czechoslovakia

SUB-POISSON LIGHT

1. Introduction

The use of coherent states makes it possible to treat quantum systems in a "classical" way although all quantum features of the systems are conserved. In this way it has been found that there exists a class of optical fields whose Glauber-Sudarshan diagonal representation of the density matrix in terms of the coherent states does not exist in a space of ordinary functions; it exists only in a space of generalized functions. The corresponding quasi-distribution may take on negative values and may be more singular than the Dirac function, which is just mathematical reflection of the quantum properties of systems. Such purely quantum fields having no classical analogues may exhibit antibunching of photons, sub-Poisson photon statistics and squeezing of fluctuations of the real or imaginary part of the complex field amplitude. These effects in light fields are purely quantal and need fully quantum approach in their description. They have fundamental meaning for validity of the quantum theory and may be expected to be used in optical interferometry with reduced quantum noise limits, in optical communication and processing, in research of biological systems, such as the human visual system, at the threshold of operation in terms of neural statistics, etc. Of course compared to laser light this quiet light cannot be too intensive otherwise the effects are smoothed out.

In principle two ways may be used to produce quiet light beams:

a) We can use resonance fluorescence light from one pumped atom, which is really a quantum system. Therefore emitted photons must be antibunched and sub-poissonian, as has been observed by Mandel and co-workers [1-3] and by Cresser et al. [4]. Such light is of course very weak, but these experimental results are fundamental for physics because they violate the so-called Bell inequalities and fully agree with quantum electrodynamics. Quite recently very nice experiment was realized by Teich and Saleh [5] to produce antibunched and sub-Poisson stationary light using the Franck-Hertz experiment. Such stationary quiet light may be applied. In this case a number of Hg atoms excited by a regular electron beam may emit sub--Poisson photons. Hence they use here the effect of primary excitation statistics on coherence of emitted light from the light source. A new solid-state version of the Franck-Hertz-Teich-Saleh experiment is in progress using a semiconductor [6].

b) The second way uses nonlinear optical phenomena as nonlinear transformations of ideal coherent beams in which some photons are "consumed" in the nonlinear interaction and the beams are successively regularized and quiet. For this many nonlinear optical phenomena have been analyzed and experiments have been suggested [7-9], such as optical parametric processes, Brillouin, Raman and hyper-Raman scattering, multiphoton absorption and emission, free-electron laser systems and four-wave mixing. Particularly using the four--wave mixing Bondurant et al. [10] and Slusher et al. [11] have obtained some preliminary experimental achievements in squeezed light generation finding the most convenient spectral regions for generation and limits given by spontaneous emission.

2. Definitions

Optical fields having no classical analogues are defined with the use of the Glauber-Sudarshan diagonal representation of the density matrix

$$\hat{\rho} = \int \Phi_N(\alpha) |\alpha\rangle\langle\alpha| d^2\alpha \qquad (1)$$

in terms of the projection operators $|\alpha\rangle\langle\alpha|$ onto the coherent state $|\alpha\rangle$. Here $\Phi_N(\alpha)$ is the quasi-distribution function related to normal ordering of field operators whose unregular behaviour reflects the quantum properties of optical fields, such as antibunching, sub-Poisson statistics and squeezing. For the coherent state it is a δ-function, light from natural sources is described by a gaussian distribution. This quasi-distribution is related to the photoelectron distribution $p(n)$ by the photodetection equation

$$p(n) = \int_0^\infty \Phi_N(\alpha) \frac{(\eta|\alpha|^2)^n}{n!} \exp(-\eta|\alpha|^2) \, d^2\alpha, \qquad (2)$$

where η is the photodetection effectivity and $|\alpha|^2 = I$ is intensity of the incident field. This is a consequence of the definitions of the photon number and normal correlations in the quantized electromagnetic field.

For a photon number distribution $\eta = 1$ and we obtain from (2) for the photon number fluctuations

$$\langle(\Delta n)^2\rangle = \langle n^2 \rangle - \langle n \rangle^2 = \langle n \rangle + \langle(\Delta I)^2\rangle, \qquad (3)$$

($\langle(\Delta I)^2\rangle = \langle I^2 \rangle - \langle I \rangle^2$ are intensity wave fluctuations) which expresses the particle-wave duality in the optical region where both the terms in (8) are equally important. The first term represents the variance of Poisson particles, the second one represents the wave fluctuations. The average is taken with Φ_N so that $\langle(\Delta I)^2\rangle$ is not necessarily non-negative.

For optical fields having classical analogues $\langle(\Delta I)^2\rangle \geq 0$ and we have

$\langle(\Delta n)^2\rangle = \langle n \rangle, \quad \langle(\Delta I)^2\rangle = 0 \quad$ for Poisson coherent light,

$\langle(\Delta n)^2\rangle > \langle n \rangle, \quad \langle(\Delta I)^2\rangle > 0 \quad$ for super-Poisson light.
(4)

For instance, for natural sources $\langle(\Delta I)^2\rangle = \langle n \rangle^2$ from the central limit theorem and $\langle(\Delta n)^2\rangle = \langle n \rangle(1 + \langle n \rangle)$, i.e. we have Bose--Einstein statistics in this case.

If Φ_N behaves non-classically, it may happen that $\langle(\Delta I)^2\rangle < 0$ and $\langle(\Delta n)^2\rangle < \langle n \rangle$; such fields are sub-poissonian. For instance, for the Fock state $|n\rangle$ with quite certain number of photons n, it holds that $\langle(\Delta n)^2\rangle = 0$ and $\langle(\Delta I)^2\rangle = -n < 0$; in this case Φ_N is proportional to the nth derivative of the δ-function.

Relating m to photons and n to photoelectrons, we obtain from (2) the quantum analogue of the Burgess theorem [12]

$$F_n - 1 = \eta (F_m - 1), \tag{5}$$

where $F_n = \langle (\Delta n)^2 \rangle / \langle n \rangle$ is the so-called Fano factor. Thus

$$\begin{aligned} F_m &= 1 \quad \text{for coherent poissonian light,} \\ F_m &> 1 \quad \text{for super-poissonian light,} \\ F_m &< 1 \quad \text{for sub-poissonian non-classical light.} \end{aligned} \tag{6}$$

If m-particles are sub-poissonian, $F_m - 1 < 0$ and also n-particles are sub-poissonian, $F_n - 1 < 0$ and vice versa. For instance, the Franck-Hertz-Teich-Saleh experiment can be understood as the inverse photodetection process and then m represents the exciting electrons and n emitted photons. However, one can see from (5) that a small value of the effectivity η can substantially diminish the effect. Moreover it was shown [12] that additional noise cannot change the sign of this relation.

As a consequence of the Bose-Einstein statistics and maximal entropy a system of natural chaotic photons exhibits bunching properties, i.e. photons have tendency to come to the photodetector in pairs. This effect is completely compensated by correlations in the source in laser light. These properties of optical beams are described by the intensity correlation function, which expresses the probability of separation of photons by τ seconds, appropriately normalized,

$$\gamma^{(2)}(\tau) = \langle I(t) I(t+\tau) \rangle / \langle I(t) \rangle^2, \tag{7}$$

fulfilling $\gamma^{(2)}(\tau) \geq 1$ for classical fields; for $\tau = 0$ it is maximal, which expresses the bunching properties. For ideal laser field, $\gamma^{(2)}(\tau) = 1$ for all τ. If Φ_N has unregular behaviour, it may happen that $\gamma^{(2)}(\tau) < 1$ and we speak of antibunching. In this case the classical inequality $\langle I^2 \rangle \geq \langle I \rangle^2$ is violated (in correspondence to violation of the Bell's inequality). In the experiments on antibunching in resonance fluorescence light Mandel and co-workers [1, 2] used the weaker definition of antibunching $\gamma^{(2)}(0) < \text{Max } \gamma^{(2)}(\tau)$, which means that it is sufficient that $\gamma^{(2)}(\tau)$ is increasing in the neighbourhood of $\tau = 0$ (while the classical correlation function has always its maximum at $\tau = 0$), which really reflects antibunching of photons.

Finally we define squeezing of fluctuations. Defining the hermitian operators proportional to the generalized coordinate and momentum (corresponding to real and imaginary parts of the complex field amplitude) as follows

$$\hat{a}_1 = \tfrac{1}{2}(\hat{a} + \hat{a}^+), \quad \hat{a}_2 = \tfrac{1}{2i}(\hat{a} - \hat{a}^+), \qquad (8)$$

fulfilling the uncertainty relation

$$\langle(\Delta \hat{a}_1)^2\rangle \langle(\Delta \hat{a}_2)^2\rangle \geq \tfrac{1}{16}, \qquad (9)$$

we speak of squeezing if

$$\langle(\Delta \hat{a}_1)^2\rangle < \tfrac{1}{4} \quad \text{or} \quad \langle(\Delta \hat{a}_2)^2\rangle < \tfrac{1}{4}; \qquad (10)$$

here $\hat{a}$ and $\hat{a}^+$ are the annihilation and creation operators of a photon respectively. This means that in the corresponding variable the fluctuations are reduced under the level of vacuum fluctuations; of course in the other variable they must increase to fulfil (9).

In general these three quantum effects are independent as being of different orders in the complex field amplitude. The squeezing is the second-order effect, antibunching is the fourth-order effect and the sub-poissonian statistics need all-order moments. Nevertheless, in many cases these effects occur together.

These non-classical effects are typical in the simplest nonlinear optical interaction with the interaction hamiltonian

$$\hat{H}_{int} = \hbar g\, \hat{a}^2 + \text{h.c.} \qquad (11)$$

g being the coupling constant. This may describe two-photon absorption and emission provided that the atomic system is described classically [13], or an optical parametric process, or a four-wave mixing process provided that the pump modes are strong and classical. A solution of the Heisenberg equations is represented by an operator in the form

$$\hat{b} = \mu \hat{a} + \nu \hat{a}^+, \quad \hat{a} = \mu^* \hat{b} - \nu \hat{b}^+,$$
$$[\hat{a}, \hat{a}^+] = [\hat{b}, \hat{b}^+] = \hat{1}, \quad |\mu|^2 - |\nu|^2 = 1, \qquad (12)$$

μ and ν being functions of time. Eigenstates of $\hat{b}$,

$$\hat{b}|\beta\rangle_g = \beta|\beta\rangle_g, \quad {}_g\langle\beta|\hat{b}^+ = {}_g\langle\beta|\beta^*, \qquad (13)$$

are called the two-photon coherent states [13] and it holds for them that

$$\langle(\Delta\hat{a}_1)^2\rangle = \tfrac{1}{4}|\mu-\nu|^2, \quad \langle(\Delta\hat{a}_2)^2\rangle = \tfrac{1}{4}|\mu+\nu|^2, \qquad (14)$$

i.e. really one of these variances may be less than $1/4$. For $\mu = 1$ and $\nu = 0$ we have the one-photon coherent state. It may be shown that an optical field generated in this interaction always exhibits antibunching and sub-poissonian behaviour if the initial phases are appropriate because Φ_N never exists [9].

3. Experiments

The first experiment to observe antibunching was performed by Kimble, Dagenais and Mandel [1, 2] with resonance fluorescence light. For single pumped atom the intensity corelation function is factorized to the form

$$\langle I(t) I(t+\tau)\rangle = \langle I(t)\rangle \langle I(\tau)\rangle_G, \qquad (15)$$

where $\langle I(t)\rangle$ is the steady-state mean intensity and $\langle I(\tau)\rangle_G$ is the mean intensity of light that is radiated by the atom driven by an external field at time τ if it starts in the ground state at time $t = 0$. As an atom cannot radiate in its ground state with respect to quantum electrodynamics, it follows that $\langle I(\tau)\rangle_G$ always starts from zero at $\tau = 0$ and then grows with τ and reaches its steady-state value $\langle I\rangle$ after a time larger than the natural life time, i.e. $\langle I(0)\rangle_G = 0$.

Using the Hanbury Brown-Twiss correlation arrangement, they measured the normalized correlation of intensity fluctuations on two photodetectors

$$\lambda(\tau) = \frac{\langle I_1(t) I_2(t+\tau)\rangle}{\langle I_1(t)\rangle \langle I_2(t+\tau)\rangle} - 1. \qquad (16)$$

With respect to (15) this correlation must have non-classical behaviour since it starts at $\lambda(0) = -1$, while any classical correlation function

$\langle I_1(t) I_2(t + \tau) \rangle$ is always maximal at $\tau = 0$. This just reflects the fact that the atom is a purely quantum system. In these experiments a beam of sodium atoms, optically prepumped in order to prepare a pure two-level system, is irradiated at right angles with light by a stabilized dye laser tuned on resonance with the transition $3^2S_{1/2}$ to $3^2P_{3/2}$ in sodium. The intensity of the atomic beam was reduced so that on the average no more than one atom was present in the observation volume in a time. The fluorescent light beam from the small observation volume was obtained in a direction orthogonal to both the atomic beam and the laser pump beam. Further the Hanbury Brown-Twiss coincidence arrangement was used, i.e. the fluorescent light was divided by a splitter and the arrival of photons in each beam was detected by two photomultipliers. The pulses from the two detectors were fed to the start and stop inputs of a time-to-digital converter, where the time intervals τ between the start and stop pulses were digitized in units of 0.5 ns and stored. The number of events $n(\tau)$ stored at address τ is a measure of the joint probability density of separation of photons by τ second. In this way antibunching was clearly observed.

These experiments demonstrate that fluorescent photons from one atom exhibit antibunching in time, which may be regarded as a reflection of the fact that the atom makes a quantum jump from the excited state to the ground state in the process of emitting a photon and is unable to radiate again immediately afterwards. No classical system can exhibit such a behaviour. Thus the photon coincidence technique makes it possible to test the bahaviour of the atom and to observe the antibunching effect which is a direct manifestation of the quantum nature of light in agreement with quantum electrodynamics.

Short and Mandel [3] improved this arrangement to control the observation volume. The measurement was performed if only one atom was present in the volume. In this way they were able to measure the sub-poissonian statistics with $F_n = 0.998$ in 24×10^6 realizations with very high statistical confidence. Also this experiment has fundamental physical meaning. The problem of very low level of intensity may be removed in the Franck-Hertz-Teich-Saleh experiment [5, 6].

Teich, Saleh and Stoler [14] suggested to use the Franck-Hertz experiment (performed 70 years ago!) to produce stationary antibunched and sub-Poisson light. For this a detailed quantum electrodynamical analysis has been performed [15] of the role of primary excitation statistics and individual emission statistics on coherence and photon statistics of light. Although this microscopic analysis is not simple, phenomenologically the quantum Burgess theorem (5) holds in this case. Thus, if a beam of pumping electrons interacting with Hg-atoms is sub-poissonian, then also the system of emitted photons is sub-poissonian.

Teich and Saleh have used a triode containing 0.75 g of Hg in their experiment [5]. The triode was heated in an oven to permit the vapor pressure of the Hg to be temperature controlled. By means of external voltage the cathode was heated to an appropriate temperature to produce thermionically emitted electrons and to form a space-charged cluod. The cathode is formed and the grid voltages are adjusted to provide a desired space-charge-limited electron current with controlled electron distance. In Hg atoms the transition $6^3P_1 \rightarrow 6^1S_0$ was used to obtain 253.7 nm Franck-Hertz light by inelastic collisions of Hg atoms with the space-charge-limited quiet electron beam. The most pronounced sub-poissonian behaviour was obtained at the temperature $26.6^\circ C$. The poissonian fluctuations of the electron beam were reduced to $F_m \approx 0.1$ (but also values 0.01 are achievable). The counting interval $T = 1 \mu s$ and $10 - 15 \times 10^6$ samples were used to construct the photocount distribution, 50 photocount distributions provided a value of F_n giving a global measure of the sub-poissonian behaviour of Franck-Hertz light. As the effectivity $\eta \approx 10^{-3}$, the sub-Poisson effect was small, $F_n \approx 0.998$, but the statistical confidence was high; the effect was by 2 to 3 standard deviations. For $\eta = 0.5$ and $F_m = 0.01$ one can obtain $F_n = 0.525$, i.e. the photon beam can be quieted of about a factor 2 relative to that of a coherent laser source. There are no obvious fundamental barriers to produce an arbitrarily quiet intense cw light which is also arbitrarily sub-poissonian. Thus the Franck-Hertz-Teich-Saleh experiment represents the first stationary source of sub-Poisson light. In principle, the method may be adopted to use atomic, ionic or molecular beams to produce sub-poissonian X-rays.

These authors suggested the solid-state version of the experiment to produce quiet light which is in progress [6]. One can use a region of semiconductor (intrinsic region) to provide space charge. Then the electrons flow from this region into a p-n junction where they create recombination radiation. The electron current entering the device is presumably poissonian, so by first passing through the intrinsic region the current becomes space-charge limited and quieted and then enters the junction region where each electron gives rise to a photon by recombination radiation (Franck-Hertz effect) and the light should be quiet. Of course the efficiency could be much higher and the sub-Poisson effect as well.

These experiments clearly demonstrate the validity of quantum electrodynamics. Stationary quiet Franck-Hertz light may be used in optical interferometry, optical communications and processing and in research of biological systems.

The above methods of generating quiet light exert an influence on the mechanism of the emission of light in the source. Another way is to produce such optical fields from an ideal coherent laser beam by non-linear transformations with the use of nonlinear optical processes [16--34]. Such experiments were suggested, but till now not performed. Some simulation experiment by Wagner et al. [35] simulated the second harmonic generation by an electro-optical filter and used an analogue device and a computer. They obtained the sub-Poisson behaviour of the fundamental and second-harmonic beams in accordance with the theory. Also the first attemps [10, 11] were reported to generate squeezed light using the four-wave mixing and it may be hoped that such experiments will be successful soon.

References

1. H. J. Kimble, M. Dagenais, L. Mandel, Phys. Rev. Lett., 39, 691 (1977); Phys. Rev., A18, 201 (1978).
2. M. Dagenais, L. Mandel, Phys. Rev., A18, 2217 (1978).
3. R. Short, L. Mandel, Phys. Rev. Lett., 51, 384 (1983).
4. J. D. Cresser, J. Häger, G. Leuchs, M. Rateike, H. Walther, Dissipative Systems in Quantum Optics, Ed. R. Bonifacio, Springer, Berlin 1982, p. 21.
5. M. C. Teich, B. E. A. Saleh, J. Opt. Soc. Am., B2, 275 (1985).
6. M. C. Teich, 1985, private communication.
7. H. Paul, Rev. Mod. Phys., 54, 1061 (1982).
8. D. F. Walls, Nature, 306, 141 (1983).
9. J. Peřina, Quantum Statistics of Linear and Nonlinear Optical Phenomena, D. Reidel, Dordrecht 1984; Coherence of Light, 2nd Ed., D. Reidel, Dordrecht 1985.
10. R. S. Bondurant, P. Kumar, J. H. Shapiro, M. Maeda, Phys. Rev., A30, 343 (1984).
11. R. E. Slusher, L. Hollberg, B. Yurke, J. C. Mertz, J. F. Valley, Phys. Rev., A31, 3512 (1985).

12. J. Peřina, B. E. A. Saleh, M. C. Teich, Opt. Comm., 48, 212 (1983).
13. H. P. Yuen, Phys. Rev., A13, 2226 (1976).
14. M. C. Teich, B. E. A. Saleh, D. Stoler, Opt. Comm., 46, 244 (1983).
15. M. C. Teich, B. E. A. Saleh, J. Peřina, J. Opt. Soc. Am., B1, 366 (1984).
16. D. Stoler, Phys. Rev. Lett., 33, 1397 (1974).
17. M. Kozierowski, R. Tanaś, Opt. Comm., 21, 229 (1977).
18. L. Mišta, J. Peřina, Czech. J. Phys., B28, 392 (1978).
19. H. Paul, W. Brunner, Opt. Acta, 27, 263 (1980).
20. P. Chmela, R. Horák, J. Peřina, Opt. Acta, 28, 1209 (1981).
21. M. Kozierowski, S. Kielich, Phys. Lett., 96A, 213 (1983).
22. V. Peřinová, J. Peřina, P. Szlachetka, S. Kielich, Acta Phys. Pol., A56, 267 and 275 (1979).
23. P. Szlachetka, S. Kielich, J. Peřina, V. Peřinová, Opt. Acta, 27, 1609 (1980).
24. J. Peřina, V. Peřinová, J. Koďousek, Opt. Comm., 49, 210 (1984).
25. J. Peřina, V. Peřinová, M. Bertolotti, C. Sibilia, Opt. Comm., 49, 285 (1984).
26. U. Mohr, H. Paul, Ann. Physic, 35, 461 (1978).
27. J. Bajer, J. Peřina, Czech. J. Phys. B, in print.
28. J. Peřina, V. Peřinová, J. Křepelka, A. Lukš, C. Sibilia, M. Bertolotti, Opt. Acta, 30, 959 (1983).
29. V. Peřinová, J. Peřina, A. Lukš, C. Sibilia, M. Bertolotti, Opt. Acta, 31, 735 (1984).
30. M. Bertolotti, C. Sibilia, J. Peřina, V. Peřinová, Phys. Rev., A30, 1353 (1984).
31. P. Kumar, J. H. Shapiro, Phys. Rev., A30, 1568 (1984).
32. S. Ya. Kilin, Opt. Comm., 53, 409 (1985).
33. M. D. Reid, D. F. Walls, Phys. Rev., A31, 1622 (1985).
34. B. Yurke, Phys. Rev., A29, 408 (1984).
35. J. Wagner, P. Kurowski, W. Martienssen, Z. Phys., B33, 391 (1979).

K. WÓDKIEWICZ

Institute of Theoretical Physics,
Warsaw University,
00-681 Warsaw, Poland

NOISE IN STRONG LASER-ATOM INTERACTIONS

I. Introduction

We present a general theoretical description of the influence of external noises in strong laser-atom interactions. A general theory of Gaussian external fluctuations with Ornstein-Uhlenbeck statistics is developed for a wide class of physical problems. We discuss the role and the impact of such external fluctuations in collisions, light fluctuations, N-photon processes, optical bistability and potential scattering. For Gaussian external fluctuations in strong laser-atom interactions it is possible to establish and to solve exactly the relevant Fokker-Planck equations. Such exact solutions can be obtained for linear and quadratic external Ornstein-Uhlenbeck fluctuations.

We can generalize our approach to a discussion of strong laser--atom interactions without the Gaussian assumption. We present a theory of the noise based on a special Markov-chain description of non--Gaussian external fluctuations. This chain, composed of n independent two-state jump processes, forms a non-Gaussian stochastic process from which a detailed and exact discussion of the atomic response can be given. We show that it is possible to establish and to solve exactly the atomic master equation, even with nonlinear external fluctuations. We discuss the role of such fluctuations in multiphoton processes and optical resonance equations.

II. Examples of External Fluctuations

We start our discussion by reviewing some examples of external fluctuations. When the impact of laser fluctuations is investigated one represents the laser light by a classical stochastic complex field:

$$E(t) = E_0(t) \exp(-i\omega_0 t - i\phi(t)), \qquad (2.1)$$

where depending on the source of the noise we can have amplitude $E_0(t)$, phase $\phi(t)$ or frequency $\mu(t) = d\phi(t)/dt$ fluctuations. Resonance interaction of laser light with atoms and molecules with such fluctuations has been discussed and studied in great details in many physically interesting cases [1].

For atomic systems with collisions the influence of the external perturbers can be included in the following stochastic way. The electric atomic dipole moment operator can be written in the following form:

$$d(t) = d_0 \exp(-i\omega_0 t - i\phi(t)), \qquad (2.2)$$

where depending now on the conditions and physical circumstances we can have collisional phase or frequency fluctuations. The dipole moment autocorrelation function reflects the velocity-velocity correlations induced by the fluctuations due to the external atomic perturbers. Both for collision and laser phase or frequency noises the relevant part of the external fluctuations is described by the following correlation function:

$$C(t) = \langle \exp(i\phi(t+s) - i\phi(s)) \rangle. \qquad (2.3)$$

It is quite obvious that in general, different models of phase $\phi(t)$ or frequency $\mu(t)$ fluctuations will lead to different forms of the correlation function given by Eq. (2.3). We can approach now the noise problem using two independent methods. The first method involves Gaussian stochastic processes (diffusion processes) as models of laser or collisional fluctuations. The second method deals with discrete non-Gaussian jump-like Markov noises.

For Ornstein-Uhlenbeck (diffusion) fluctuations of the frequency we obtain (we assume from now that $t > 0$):

$$C(t) = \exp\left\{-a^2 \tau_c(t + \tau_c(\exp(-t/\tau_c) - 1))\right\}. \qquad (2.4a)$$

For Ornstein-Uhlenbeck phase fluctuations $\phi(t)$ we have:

$$C(t) = \exp\left\{-a^2(1 - (\exp(-t/\tau_c)))\right\}. \qquad (2.4b)$$

For random telegraph jumps of the frequency we obtain:

$$C(t) = \frac{1}{2}\left(\frac{1}{2\tau_c \lambda} + 1\right)\exp\left(-\left(\frac{1}{2\tau_c} - \lambda\right)t\right) - \frac{1}{2}\left(\frac{1}{2\tau_c \lambda} - 1\right)\exp\left(-\left(\frac{1}{2\tau_c} + \lambda\right)\right) \quad (2.4c)$$

where $\lambda^2 = 1/4\tau_c^2 - a^2$.

For random telegraph jumps of the phase we have:

$$C(t) = \cos^2(a) + \sin^2(a)\exp(-t/\tau_c). \qquad (2.4d)$$

Eqs. (2.4) offer the basic property of the collisional and laser phase or frequency noise models based on the Gaussian or the jump-like assumption.

III. Noise in Interacting Systems

If we denote the possible source of noise by $x(t)$ we can write the relevant atomic equations of motion in the following form:

$$d\rho/dt = -iM(\rho, x(t)), \qquad (3.1)$$

where ρ is the density operator or the associated set of Heisenberg operators of the investigated system and M is accordingly the Liouville or the Hamilton operator. Because as usual only a limited number of degrees of freedom of the atomic system is investigated we allow for a partial elimination of other "irrelevant" degrees of freedom. This procedure can end up in some nonlinear equations of motion for the relevant components of ρ. This is reflected in Eq. (3.1) by the additional dependence of M on ρ (possibly nonlinearly). In optical bistability the nonlinearity in M results from the adiabatic elimination of atomic variables in the good cavity case. The external noise enters such dynamical equations in most of the cases in a multiplicative and possibly also in a nonlinear way. The methods of solving and averaging such equations with Gaussian or jump-like noises are described in this volume in the lecture of B.W. Shore. We limit here only our discussion to some results obtained after a stochastic average of Eq. (3.1) has been already performed.

Let us illustrate the different statistical models defined by Eq. (2.4) in the framework of multiphoton processes. Let us start first with a simple theory of N-photon absorption in the presence of a fluctuating light Lorentzian power spectrum but different statistics. We notice first that it is a rather simple exercise to establish for Gaussian and jump-like processes a common regime in which all the correlation functions given by Eqs. (2.4) will decay exponentially i.e., $C(t) = \exp(-\gamma t)$. This means that all these models predict a Lorentzian power-spectrum associated with the correlation function $C(t)$.

Despite this common property the N-photon absorption profile has different width depending on the source of fluctuations. For diffusive Gaussian noises and jump-like frequency fluctuations we have the absorption profile proportional to:

$$\text{Abs}(\Delta) \propto \frac{\beta + N^2\gamma}{\Delta^2 + (\beta + N^2\gamma)^2}, \qquad (3.2a)$$

where Δ is the atomic detuning and β and γ are the atomic and fluctuating linewidths, respectively. Eq. (3.2) shows that the N-photon absorption width increases as N^2.

For jump-like phase fluctuations we obtain the folowing result:

$$\text{Abs}(\Delta) \propto \left\{ \frac{\beta \cos\frac{2\pi N}{2}}{\Delta^2 + \beta^2} + \frac{(\beta + \gamma)\sin\frac{2\pi N}{2}}{\Delta^2 + (\beta + \gamma)^2} \right\} \qquad (3.2b)$$

i.e., for even number of photon transitions the accumulated phase is equivalent to an amplitude sign change and as a result no impact of fluctuations on the N-photon absorption is observed. For other transitions the linewidth is simply broadened by γ. In view of recent experiments (see the lecture of S. J. Smith in this volume) where correlation effects of phase-diffusing field on two-photon absorption have been measured, the PG models offer an interesting possibility of some other phase and frequency fluctuations with different impact on N-photon absorption profile.

A different example of nonequivalent impact of noise fluctuations in strong atom-laser interactions can be seen when one discusses the influence of frequency atomic dipole or laser fluctuations on spectral characteristics of a two-level atom resonance fluorescence. The common feature of all these cases is a nonperturbative interaction in the presence of strong laser field.

A comparison of the light scattering spectrum of the atom calculated in the two cases (laser frequency noise, collisional frequency noise) shows that the origin of the frequency noise makes a difference [2]. This difference can be traced back to the original form of the matrix M entering the basic equation of motion (3.1) involving the atomic variables needed in the calculations of the fluorescence spectrum. For collisional fluctuations the matrix M in equation (3.1) can be obtained from the coherent case by a simple shift of the constant detuning by an amount $\mu_c(t)$ i.e., collision frequency noise. For laser noise the situation is more complicated because the dipole-moment autocorrelation function $\langle d^*(t_1) d(t_2) \rangle$, crucial for the calculation of the fluorescence spectrum will be coupled to other atomic correlations which involve some extra phase factors built out of $\mu_L(t)$ i.e., laser frequency noise.

IV. Fluctuations and Nonlinear Dynamics

Multiphoton ionisation, cooperative effects like optical bistability are only examples of an entire variety of nonlinear effects where fluctuations are important and quite nontrivial to handle.

To illustrate the noise problem in a nonlinear dynamical case let us investigate the case of absorptive optical bistability (AOB) with laser-amplitude fluctuations. The steady state theory of AOB in a cavity filled with homogeneous broadened two-level atoms is particularly simple in the limit of low-transmission mirrors and weak enough absorption. In this limit the relation between the transmitted x and incident electric field y has the following form:

$$y = x + 2Cx/(1 + x^2), \qquad (4.1)$$

where C is the order parameter of AOB. In realistic experimental situation the injected electric-field amplitude should be replaced by y + $\delta y(t)$ where $\delta y(t)$ describes Gaussian fluctuations of the amplitude with correlation:

$$\langle \delta y(s) \delta y(s + t) \rangle = a^2 \exp(-t/b). \qquad (4.2)$$

For such a case the Fokker-Planck equation is very complicated due to the color and the nonlinearity of the operator M. For $b < 1$ one can obtain an equation with nonconstant diffusion. This equation leads to the following relation for the most probable value of the stationary probability distribution [3]:

$$y = x + 2Cx/(1 + x^2) + 4a^2b^2C(3 - x^2)/(1 + x^2)^3. \qquad (4.3)$$

This equation provides the generalization of the deterministic relation (4.1) for the case of laser amplitude fluctuations.

V. Pregaussian Noise

When the coherence time of the external noise is comparable with the intrinsic time scale of the atomic system, the solution of the stochastic equation (3.1) is a difficult task.

It is possible to consider a Markov chain consisting of a superposition of n independent simple two-state jump processes. Such a Markov chain provides a closed master equation for atomic variables. For more details see the lecture of B. W. Shore in this volume. For finite values of n such a noise is statistically non-Gaussian. Such an approach offers also a possibility of strong CW-laser-atom interations without a Gaussian assumption. For $n \to \infty$, our Markov chain tends to the Ornstein-Uhlenbeck stochastis process. This means that by increasing n, we can study in a systematic way the approach of our Markov chain to the Gaussian limit, with full control of the parameters that characterize the various interaction time scales. Because of this fundamental properties we call this n-component Markov chain a pre--Gaussian (PG) stochastic process [4]. Let us illustrate the useful feature of the PG noise for a chaotic light. In the simplest static case we have $\langle I^n \rangle = n! \langle I \rangle^n$. Such a model of fluctuations can be obtained assuming that the real and the imaginary part of the complex electric field amplitude each performs independent Ornstein-Uhlenbeck fluctuations. The chaotic model of laser fluctuations is widely used as a very good model of a multi-mode laser. In most of the cases the impact of such fluctuations in multiphoton transitions is a very difficult task to perform. Additional complications arise from Stark broadening which is proportional to the instantaneous intensity of the laser light I(t). Even in the rate equation regime the noise problem is quite complex. For example the simplest rate equation that incorporates the fluctuations of the N-photon process can be written in the following form:

$$d\rho/dt = -\alpha I^N(t)\rho, \qquad (5.1)$$

where ρ is the population of the ground state and α is a constant that characterizes the efficiency of the process. The useful character of the PG noise in such problem can be illustrated by calculating the normalized second-order intensity coherence function $g = \langle I^2 \rangle / \langle I \rangle^2$ of the driving field. The chaotic light can be now replaced by a PG light for which the real and the imaginary part of the electric field amplitude performs independent PG fluctuations composed of n telegraphs. For n = 1 we have g = 1 i.e., this light behaves like a coherent source. For PG noise with n = 2 we have g = 1.5 i.e., it has a fractional value. For three jump processes we obtain g = 5/3 and of course for n → ∞ we obtain g = 2! a quantity which is well known for chaotic fluctuations in two-photon processes. The nice property of the PG noise is the fact that one can solve in finite terms the stochastic rate equation (5.1) with a nonlinear multiplicative noise [5].

VI. Optical Resonanse Equations

It is well known that the conventional Optical Bloch Equations (OBE) have damping terms T_2 and T_1 that are independent of the strength of the incident field. This property is certainly true for weak fields when only a few photons are exchanged with the atomic system. For strong excitation involving a large number of exchanged photons, one might suspect that the damping terms would be affected by the incident field. Such a suspicion is not merely academic in light of recent experimental measurements of free induction decay rates as a function of the incident Rabi frequency (see the lecture of R. G. Brewer in this volume). These measurements indicate the failure of the conventional OBE, with T_1 and T_2 lifetimes independent of incident intensity [6]. In different language, these observations can be attributed to a breakdown of the usual substitution rules for the inhomogeneous decay rate at high incident intensities. These substitution rules consist of simple additive modifications of T_1 and T_2. For example for Gaussian frequency fluctuations with white noise type correlation:

$$\langle \mu(t+s)\mu(s) \rangle = 2\Gamma\delta(t) \qquad (6.1)$$

we have $1/T_1$ and $1/T_2 + \Gamma$.

The fundamental property of the frequency fluctuations model based on Eq. (6.1) is an infinite short time of such correlations. For correlations with finite coherence time in general no simple substitution rule can be implemented.

For the Free Induction Decay (FID) effect in solids the only source of fluctuations with finite coherence time comes from random changes of the atomic detuning. These fluctuations are equivalent in form to frequency fluctuations of the atomic dipoles in typical collisional effects. With such fluctuations one can show that the expectation value of the atomic density matrix forming the Bloch vector satisfies the following equation of motion:

$$d \langle \rho(t) \rangle /dt = (-iM - \Gamma_r + \sum) \langle \rho(t) \rangle, \qquad (6.2)$$

where the coherent evolution and purely radiative damping matrices M and Γ_r are joined by an additional "damping" matrix $\sum$ dependent explicitly on the reservoir parameters. The point is to show in what way $\sum$ also depends on M. For a general reservoir (even with Gaussian statistics) the expression for $\sum$ is very complicated and formally involves an infinite hierarchy of equations.

For FID in solids the magnetic field creates a random fluctuation $\delta\Delta(t)$ in the resonance detuning. We shall assume that $\delta\Delta(t)$ performs a random telegraph motion with the following correlation:

$$\langle \delta\Delta(t+s) \, \delta\Delta(s) \rangle = a^2 \exp(-t/\tau_c), \qquad (6.3)$$

where a^2 is the strength of the noise and τ_c is the characteristic time of these fluctuations. For such colored jump-like noise the damping matrix can be evaluated for all values of the driving field intensity. We notice that in this case the damping matrix is power dependent and has non-diagonal dampings. It is this equation (6.1) with such time-independent damping matrix $\sum$ that we call the optical resonance equation (ORE). This equation can be taken as a substitute for the OBE - if one makes a detailed quantitative discussion of the OBE. If one makes a detailed quantitative discussion of the recent experiments and our theory based on the ORE, we obtain a good agreement. For the FID experiment on $PR^{3+}:Laf_3$ at low temperatures we obtain agreement with the experiment for a $\tau_c = 8$ μsec [7].

VII. Relativistic Charged Particle Interactions in Chaotic Laser Field

Interactions of charged particles in the presence of chaotic laser field can be carried out for the case of the full relativistic Volkov solution, keeping track of the relativistic A^2 term. The laser light is described by a circularly polarized plane-wave that satisfies a stationary Gaussian stochastic process. For such a situation the cross section or the decay rate is proportional to objects that can be obtained from the following generating functional:

$$I(f, f^*) = \left\langle \exp\left\{i \int ds(f(s)A(s) + f^*(s)A^*(s) + QA(s)A^*(s))\right\}\right\rangle. \quad (7.1)$$

This generating functional leads to a stochastic equation with a quadratic external Ornstein-Uhlenbeck noise. For chaotic fluctuations of the electromagnetic vector potential $A(s)$ we have evaluated this generating function with the A^2 term exactly [8].

For a high-intensity Compton scattering in the presence of coherent light the cross section is a sum over partial sections for emission into (approximately) the stimulating frequency and harmonics. The partial cross section for the emission of the n-th harmonic is proportional to terms which are power dependent and Q-dependent. The so-called intensity-dependent shift can be traced back to the quadratic term Q in the Volkov solution. In chaotic light the higher harmonic emission will be enhanced by the familiar n!. Because the frequencies emitted at a fixed angle are intensity dependent, for chaotic light the line shapes of the individual harmonics will for scattering angles not equal to zero become broader and broader until they will finally overlap so that a separation becomes impossible.

Acknowledgement

I am indebted to my colleagues for their shared insights into noise problems. In connection with the topics mentioned in this review I thank W. Becker, L. V. Cao, J. H. Eberly, J. A. C. Gallas, M. Kuś, M. O. Scully, B. W. Shore and M. S. Zubairy. This work was supported by the Polish program MR-I-7.

References

1. See, for example J. H. Eberly in <u>Laser Spectroscopy IV</u> edited by H. Walther and K. Rothe (Springer, Heidelberg, 1979), p. 80.

2. K. Wódkiewicz, B. W. Shore and J. H. Eberly, Phys. Rev., $\underline{A30}$, 2390 (1984).
3. M. Kuś, K. Wódkiewicz and J. A. Gallas, Phys. Rev., $\underline{A28}$, 314 (1983).
4. K. Wódkiewicz, B. W. Shore and J. H. Eberly JOSA, $\underline{B1}$, 398 (1984).
5. Cao Long Van and K. Wódkiewicz J., Phys. B (in press).
6. R. G. DeVoe and R. G. Brewer, Phys. Rev. Lett., $\underline{50}$, 1269 (1983).
7. K. Wódkiewicz and J. H. Eberly, Phys. Rev., $\underline{A32}$, 992 (1985).
8. W. Becker, M. O. Scully, K. Wódkiewicz and M. S. Zubairy, Phys. Rev., $\underline{A30}$, 2245 (1984).

W. BRUNNER, R. FISCHER and H. PAUL

Zentralinstitut für Optik und Spektroskopie
der Akademie der Wissenschaften der DDR,
DDR - 1199 Berlin, Rudower Chaussee 6,
German Democratic Republic

REGULAR AND IRREGULAR BEHAVIOUR OF MULTIMODE LASER RADIATION

Abstract

The results of recent numerical studies of the output characteristics of standing-wave lasers, based on Lamb's gas laser theory, are presented. Both the cases of (i) large inhomogeneous line broadening, dominating homogeneous line broadening, and (ii) purely homogeneous line broadening are considered. In dependence on relevant physical parameters we found both regular and irregular behaviour in case (i), but only regular one in case (ii). Regularity means that the laser system approaches a phase-locked steady state. In an irregular regime, the evolution has been shown to sensitively depend on the initial conditions. This gives a strong hint that the irregularity is of the type termed "deterministic chaos".

1. Introduction

While the primary goal of laser theory was the description of steady-state regimes, it became obvious in the last decade, both experimentally and theoretically, that in suitable operating conditions lasers display an irregular behaviour. Specifically, it has been shown theoretically that there exists an instability threshold, called second laser threshold, for single-mode homogeneously broadened lasers. Unfortunately, this threshold is too high to be reached with available lasers. On the other hand, a lot of observations concerning multimode lasers has been reported. However, such systems involve so many coupled parameters that a complete theoretical investigation is difficult and an analysis can be done only numerically.

In the following, some results of numerical studies will be presented that indicate both regular and irregular multimode laser operation to occur, in dependence on physical parameters such as homogeneous and inhomogeneous line broadening, mode spacing, and excitation level.

2. Mode Competition

Different laser modes are normally coupled owing to the simple physical fact that one and the same group of atoms interacts with two, or even more, modes, supplying them with energy. Hence, strongest mode competition occurs in a ring laser with a purely homogeneously broadened emission line.

However, there exist several mechanisms that counteract mode competition, and hence favour mode coexistence. These mechanisms are mainly due to:
(a) spontaneous emission which prevents suppressed modes from being completely quenched,
(b) inhomogeneous line broadening,
(c) the spatially inhomogeneous intensity distribution of standing-wave type modes, as they are present in Fabry Perot type resonators (spatial hole burning).

While the effect of spontaneous emission is of importance only in the immediate neighbourhood of the laser threshold, the remaining two effects will be shown to significantly affect the spectral and temporal output characteristics of a multimode laser.

3. Basic Equations

Our study is based on Lamb's gas laser theory [1, 2]. We consider the case of (i) large inhomogeneous line broadening, dominating homogeneous line broadening (the so-called Doppler limit) as well as that of (ii) purely homogeneous line broadening in lasers with a Fabry Perot type resonator configuration. In both cases, the equations of motion can be cast in the following form

$$\dot{\tilde{E}}_n = \frac{1}{2}(\frac{g_n}{K} - 1)\tilde{E}_n - \sum_{\mu,\rho,\sigma} \tilde{E}_\mu \tilde{E}_\rho \tilde{E}_\sigma \; \delta_{\mu - \rho + \sigma, n} \\ \times \left[\cos \Psi_{\mu\rho\sigma n} \; G(\mu\rho\sigma) - \sin \Psi_{\mu\rho\sigma n} \; F(\mu\rho\sigma) \right]. \quad (1)$$

$$\dot{\phi}_n \tilde{E}_n = g_n \tilde{E}_n (\Omega - \omega_n) \tilde{E}_n - \sum_{\mu,\rho,\sigma} \tilde{E}_\mu \tilde{E}_\rho \tilde{E}_\sigma \, \delta_{\mu-\rho+\sigma,n}$$
$$\times \left[\cos \psi_{\mu\rho\sigma n} F(\mu\rho\sigma) + \sin \psi_{\mu\rho\sigma n} G(\mu\rho\sigma) \right]. \qquad (2)$$

Here, $\tilde{E}_n$ is the amplitude in the n-th mode. The normalization has been chosen as follows

$$\tilde{E}_n = \begin{cases} \left(\dfrac{g\chi}{k} \dfrac{\gamma_a + \gamma_b}{2\gamma_a \gamma_b \gamma} \right)^{\frac{1}{2}} E_n & \text{in case (i)} \\[1em] \left(\dfrac{g\chi}{2k\gamma^2} \right)^{\frac{1}{2}} E_n & \text{in case (ii),} \end{cases} \qquad (3)$$

where E_n is the slowly varying amplitude of the electric field strength in Lamb's notation [1], g is the gain at the line centre (see Eq. (5) below), and

$$\chi = \frac{p^2}{16\hbar^2} \qquad (4)$$

denotes the coupling constant, p (assumed real) being the nondiagonal matrix element for the electric dipole moment. The parameter γ has the meaning of the homogeneous linewidth Δ_{hom}, multiplied by π, while γ_a and γ_b are the reciprocal lifetimes for the upper and the lower atomic level, respectively. The quantity k represents the cavity losses (assumed equal in all modes), and g_n stands for the small-signal gain in the n-th mode

$$g_n = \begin{cases} g \exp\left\{ -\left(\dfrac{\Omega - \omega_n}{\varepsilon} \right)^2 \right\} & \text{in case (i)} \\[1em] g \left[1 + \left(\dfrac{\Omega - \omega_n}{\gamma} \right)^2 \right]^{-1} & \text{in case (ii),} \end{cases} \qquad (5)$$

where Ω is the frequency* at the line center, ω_n the oscillation frequency of the n-th mode, and ε is connected with the half width Γ of the Gaussian gain profile (Doppler line) through the relation $\Gamma = \pi^{-1}(\ln 2)^{1/2} \varepsilon$.

The quantity ϕ_n is defined as

$$\phi_n = (\omega_n - \Omega_n) t + \varphi_n(t), \qquad (6)$$

---

* By frequencies we generally mean circular frequencies. Linewidths, however, are given as ordinary frequencies.

where Ω_n is the cavity resonance frequency and $\varphi_n(t)$ the slowly varying phase for the n-th mode. The coefficient h_n describes the strength of frequency pulling, in explicit terms it reads

$$h_n = \begin{cases} \dfrac{1}{\pi^{1/2} \varepsilon k} \left\{ 1 + \dfrac{1}{3}\left(\dfrac{\Omega - \omega_n}{\varepsilon}\right)^2 \right\} & \text{in case (i)} \\ \dfrac{1}{2\gamma k} & \text{in case (ii).} \end{cases} \qquad (7)$$

Finally, the abbreviation

$$\Psi_{\mu\rho\sigma n} \equiv \phi_\mu - \phi_\rho + \phi_\sigma - \phi_n \qquad (8)$$

has been introduced. The expressions for the coupling constants F and G are rather lengthy. The reader will find them in [3, 4] for case (i) and in [5] for case (ii).
The time is in units of k^{-1}.

We examined numerically the evolution of the field amplitudes and the phases in all modes under consideration, starting from small amplitudes and random phases. 13 modes labeled $n = 0, \pm 1,..., \pm 6$ were taken into account in our computer program. For simplicity, we generally put $\gamma = \frac{1}{2}(\gamma_a + \gamma_b)$.

4. Free-Running Approximation

In the so-called free-running approximation only those terms are retained in Eq. (1) for which $\Psi_{\mu\rho\sigma n}$ vanishes identically. Then Eq. (1) reduces to the simple form

$$\dot{\tilde{E}}_n = \frac{1}{2}\left(\frac{g_n}{k} - 1\right)\tilde{E}_n - \tilde{E}_n \sum_i \tilde{E}_i^2 \, \theta_{ni}, \qquad (9)$$

where $\theta_{nn} = G(nnn)$ and $\theta_{ni} = \theta_{in} = G(nii) + G(iin)$ for $i \neq n$.

Since the amplitude equations are now decoupled from the phase equations, the latter need no longer be considered, and thus the problem becomes much easier to handle. However, the range of validity of this approximation is a rather restricted one. In fact, our numerical study shows that the free-running approximation applies only when the homogeneous linewidth γ is smaller than, or comparable with, the mode spacing $\Delta\omega$ ($= \Omega_{n+1} - \Omega_n$). However, it breaks down in the case of strong mode coupling, $\gamma \gg \Delta\omega$.

5. Both Homogeneous and Inhomogeneous Line Broadening

We consider multimode gas lasers where the inhomogeneous line broadening (due to the Doppler effect) dominates the homogeneous one. The output characteristics of the laser come about as the result of mode competition originating from homogeneous line broadening, counteracted by the effect of spectral hole burning and the presence of inhomogeneous line broadening.

Our numerical analysis reveals the following general features:

A. Dependence on the strength of mode coupling

Characteristic of that strength is the ratio $\gamma/\Delta\omega$. In case of weak coupling ($\gamma \ll \Delta\omega$) there is complete mode coexistence. The evolution leads to a steady state with all the modes that are above threshold oscillating according to their individual gain. This picture changes with growing coupling strength in the following way: For $\gamma \approx \Delta\omega$, the system still reaches a steady state, but the modes affect one another which results in a significant asymmetry of the output spectrum or in complete quenching of certain modes. Specifically, the latter effect has been observed in the Ar-ion laser, where stable two-mode oscillations have been found to occur [6-8].

Our theoretical study indicates that the approach to the steady state is accompanied by mode locking. This means, the relative phases
$$\Psi_n = (\phi_n - \phi_{n-1}) - (\phi_{n+1} - \phi_n) = 2\phi_n - \phi_{n+1} - \phi_{n-1},$$
after a transient stage, attain constant values. The latter are different, in general, from both o (mod 2π) and π (mod 2π), so that the locking phenomenon differs from the familiar types of amplitude modulation (AM) phase locking ($\Psi_n = 0$) and frequency modulation (FM) phase locking ($\Psi_n = \pm\pi$). When the coupling becomes strong ($\gamma \gg \Delta\omega$), the regular behaviour changes into an irregular one. The amplitudes display violent fluctuations in all the modes, and the relative phases vary strongly. There is no indication that the evolution tends to any steady state.

The following numerical examples, depicted in Figs. 1 and 2, may serve to illustrate what has been said before. (More numerical results are presented in [3, 4, 9]. Especially in [4] the case $\gamma_a/\gamma_b \neq 1$ has been studied). Varying only the mode spacing $\Delta\omega$ and keeping the remaining parameters fixed, we found a single-pulse regime for $\Delta\omega/\gamma = 1.25$ (Fig. 1a) originating from AM phase locking in 3-mode oscillation, a three-pulse regime for $\Delta\omega/\gamma = 0.62$ (Fig. 1b) brought about by phase locking of the general type mentioned above in a 7-mode oscillation, and irregular behaviour for $\Delta\omega/\gamma = 0.3125$ (Figs. 2a, b).

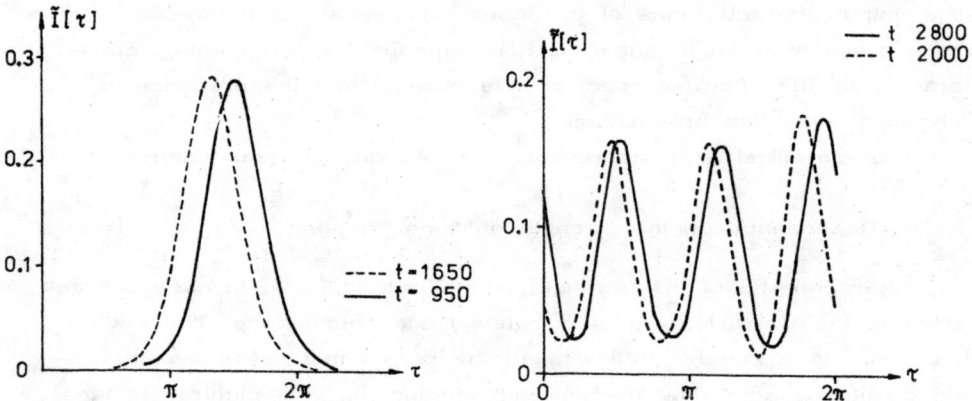

Fig. 1a, b. Intensity modulation of multimode gas laser radiation in the regular regime. The parameters are $g/K = 1.3$, $\Gamma/\gamma = 1.39$, $\gamma_a/\gamma_b = 1$, $\Delta\omega/\gamma = 1.25$ (a) and $\Delta\omega/\gamma = 0.62$ (b). The time τ is in units of $\Delta\omega^{-1}$. The number of oscillating modes is 3(a) and 7(b)

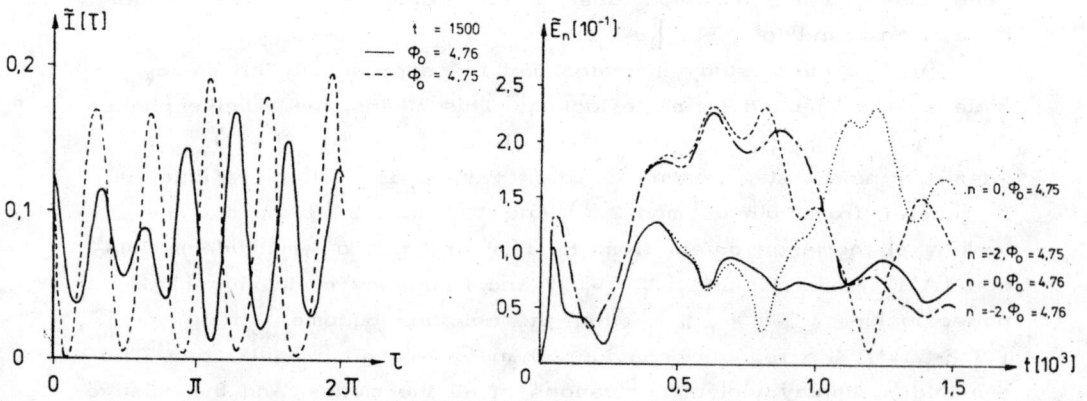

Fig. 2a, b. Irregular output characteristics of a multimode gas laser: (a) intensity modulation and (b) evolution of the field amplitudes in the modes $n = 0$ and $n = -2$ for two sets of initial conditions differing in ϕ_0 only. The parameters are $g/k = 1.3$, $\Gamma/\gamma = 1.39$, $\gamma_a/\gamma_b = 1$, $\Delta\omega/\gamma = 0.3125$. t is in units of k^{-1}, τ in units of $\Delta\omega^{-1}$. All 13 modes are oscillating

B. Dependence on inhomogeneous line broadening

Large inhomogeneous line broadening, in comparison to homogeneous line broadening, is a mechanism that favours mode coexistence. Hence, a reduction of the ratio Γ/γ gives preponderance to mode competition. This leads, in case of strong and intermediate mode competition, to the effect that some modes suppress the remaining ones, where by a phase-locked steady state is attained. In particular, single-mode operation can result in this way.

C. Dependence on the exitation level

At low excitation level, regular few-mode operation is favoured (even single-mode operation occurs) also in the case of strong and intermediate mode competition. With growing excitation level the number of oscillating modes increases in the latter case, while the systems passes to an irregular regime in the former case. Such transitions have been observed, e.g., in the Ar-ion laser, where an increase of the excitation level leads to jumps from one 2-mode regime to a different one (with increased frequency difference), until a state with violently fluctuating mode amplitudes is reached [6-8]. It is interesting to note that a hysteresis effect is present in the transition between different 2-mode oscillations, as has been predicted in [10] and established experimentally in [8].

6. Purely Homogeneous Line Broadening

Such a situation will be found, in some idealization, in solid state, dye, and high-pressure gas lasers (e.g. CO_2 lasers).

Evidently, for $\gamma \ll \Delta\omega$ only one mode can oscillate. For $\gamma \gtrsim \Delta\omega$ mode competition plays an important role. Actually, it has a stronger effect than in the presence of inhomogeneous line broadening, since the latter counterbalances it to some degree.

Our numerical study showed the existence of phase-locked steady-state regimes for intermediate and strong coupling. While the amplitudes grow up to constant values in time intervals $\Delta t \approx 1000$ (in units of k^{-1}), the time intervals needed for the relative phases to approach their final values, varies over a wide range, dependent on the parameters chosen. Both AM and FM type phase locking has been found, also mixtures of them and locking to relative phases different from both 0 and $\pm\pi$.

An example of simultaneous AM and FM locking is presented in Fig. 3. Interestingly, the intensity varies only slightly in this 9-mode oscillation (Fig. 3a), as a consequence of mode locking. It should

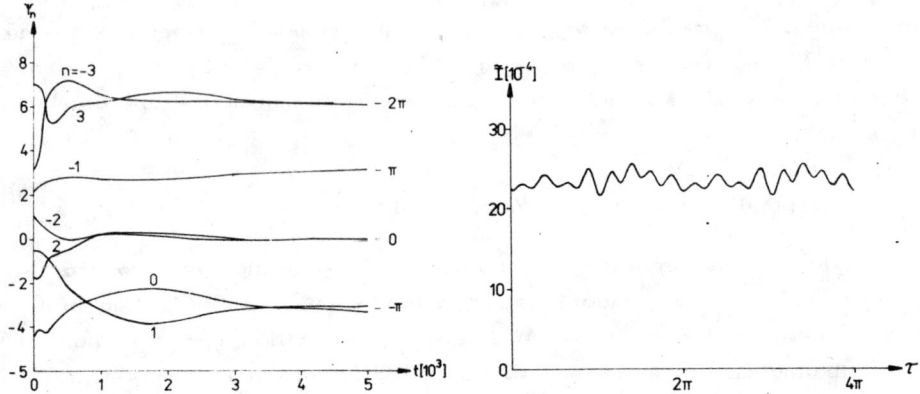

Fig. 3a, b. Output characteristics of a laser with a purely homogeneously broadened emission line: (a) evolution of the relative phases, (b) intensity modulation. The parameters are $g/k = 1.2$, $\Delta\omega/\gamma = 0.035$, $\gamma_a/\gamma_b = 0.1$. τ is in units of $\Delta\omega^{-1}$. 9 modes are oscillating

be emphasized that we did not find indications of irregular behaviour. (For more numerical results see [5, 9]).

7. Deterministic Chaos?

One of the main results of our analysis is the prediction of irregular output characteristics of a multimode laser. Such a behaviour will be found in case of strong mode competition, however, the latter must be counterbalanced, to some degree, by strong inhomogeneous line broadening. Now, the question arises whether this irregular behaviour is of that type which has been termed "deterministic chaos". This question is difficult to answer, in principle, due to the limitations in computation time. In particular, the determination of the intensity spectrum or the Lyapunov exponents suffers from this deficiency. A strong hint at a confirmative answer is provided by the sensitive dependence of the evolution on the initial conditions we found. This is exemplified in Fig. 2.

References

1. W. E. Lamb, Jr., Phys. Rev. 134A, 1429-1450 (1964).
2. M. Sargent III, M. O. Scully and W. E. Lamb, Jr., Laser Physics (Addison-Wesley, Reading, Mass., 1974).
3. W. Brunner and H. Paul, Opt. Quantum Electron. 15, 87-94 (1983).

4. W. Brunner, R. Fischer, H. Paul and Dinh van Hoang, Kvant. Electron. (Moscow) 10, 103-111 (1983); Sov. J. Quantum Electron. 13, 58-62 (1983).
5. W. Brunner, R. Fischer and H. Paul, Appl. Phys., B33, 187-193 (1984).
6. T. J. Bridges and W. W. Rigrod, IEEE J. Quant. Elect., QE-1, 303--308 (1965).
7. M. Bass, G. deMars, and H. Statz, Appl. Phys. Lett., 12, 17-20 (1968).
8. K. Berndt and E. Klose, Opt. Commun., 35, 417-420 (1980).
9. W. Brunner, R. Fischer and H. Paul, J. Opt. Soc. Am., B2, 202-210 (1985).
10. W. Brunner and H. Paul, Opt. Commun., 35, 421-426 (1980).

Gy. FARKAS

Central Research Institute for Physics,
H - 1525 Budapest 114,
P. O. B. 49, Hungary

EXPERIMENTAL INVESTIGATIONS OF THE LASER INDUCED MULTIPHOTON AND TUNNELING PROCESSES IN ATOMS AND SOLIDS

I. Introduction

Since the early paper of Keldysh, theoretical works dealing with the fundamental laws of the intense field QED predict that the general processes governing the laser induced electron emission from atoms or solids traditionally may be interpreted as two complementary limiting interaction processes of the same phenomenon. The first is the multiphoton type process, when the electron interacts only with several well determined small number of photons (quantum limit), the second is the tunneling type, when the number of the interacting photons is increasingly high (classical limit).

While practically all research activity was concentrated both theoretically and experimentally to the multiphoton questions, less attention was paid to the tunneling case, in spite of the fact that the early beginning of the intense field QED started with the pronounciation of this latter.

After a brief summarizing formulation of the topics we describe several characteristic multiphoton experiments performed both for atoms and metal surfaces, whose results have lead to the conclusion that the laser induced optical tunneling process may be also investigated experimentally.

At last we describe our recent experiments carried out for the study of the intense field optical tunneling induced by strong CO_2 laser pulses in the case of noble gas atoms and a gold surface.

II. Summarizing formulation of the topics

In the following part we briefly summarize the facts which are necessary for the description of the emission process both for atoms (ionization) and metal surfaces (photoelectric effect).

The characteristic parameters describing the ionization in the case of atoms are widely used and well known: analytic Coulomb potential, ground or excited (sometimes resonant) levels, well determined ionization potential value "A", etc. As for the photoelectric effect in the case of metals, several addition remarks should be made. The metal system consists of an ensemble of electrons, the potential is not analytic, but is approached by an idealized simplified rectangular potential well. In this case the ionization potential "A" is called "work function" defined as the energy difference between the Fermi-level and the vacuum level and has a statistical meaning only; no resonance effects can be expected.

The simultaneous fulfilment of energy and momentum conservation laws requires the presence of a "third body" which is the nucleus for the gas atom. In the case of metals this third body may be: a) the abrupt potential jump on the surface, considering that the conduction electrons (Sommerfeld-model) are moving freely only in direction parallel to the surface: in this case only that $|E_\perp|$ light electric field component induces emission, which is perpendicular to the surface, this interaction is called "surface photoeffect". b) the third body is the crystal lattice: the emission then depends only on the whole $|E|$ light electric field, regardless to its direction, which process is called "volume photoeffect".

We have also to keep in mind, that using strong laser beams the metal may be heated by two ways: when the whole crystal is heated due its Dulong-Petit heat capacity, or when only the conduction electron cloud (of which the own quantum statistical heat capacity is 3 orders of magnitude less) is heated very fastly, independently of the crystal lattice. Both process may lead to a temperature rise, which in turn, causes the distortion of the Fermi-distribution with possible Richardson-type thermionic emission.

After these remarks, we may outline qualitatively the types of the expected processes.

1) <u>Photoionization (atoms) and photoeffect (metals)</u>. These processes manifest themselves no longer in the classical Lenard-Einstein single photon sense, but in <u>multiphoton</u> form, the $\hbar\omega$ laser quantum being less than A for the present lasers.

2) When the $E(t)$ oscillating electric field of the laser light is high enough, the potential barrier both of atoms and metals may be broken down periodically by the laser field; in each second positive half period electron tunneling may occur through the narrowed barrier which process is called "<u>optical tunneling</u>".

In addition due to thermally induced effects further processes can be imagined only for metals:

3) Pure, <u>laser induced thermionic Richardson-electron emission</u> both in a slow (crystal lattice heating) and ultrafast (heating the conduction eletron cloud only) form,

4) when the distortion of the Fermi-distribution is moderate, the modified occurrence of the 1) and 2) processes may be expected: "<u>thermally assisted multiphoton photoeffect</u>" and "<u>thermally assisted dynamic optical tunneling</u>".

5) Finally a very important question remains: after the ionisation processes what happens with the already "<u>free</u>" <u>electrons in the strong laser field</u>?

III. Theories and their most important statements

The detailed theories relating to the atoms are extensively elaborated and are treated by prominent specialists this School. Therefore we repeat here only the mean results presenting at the same time the problems connected with the metal surfaces in a little more pronounced way.

a) <u>Multiphoton theories</u>

The preliminary theories are based on the simple higher order perturbation calculation. The procedure used here is to calculate the transition probability from a given initial state to a final state, which latter is a free electron state. The result is:

$$N \propto |E|^{2n_o} \propto I^{n_o} \text{ for atoms and } j \propto |E|^{2n_o} \propto I^{n_o} \text{ for volume photoeffect;}$$

$j \propto |E_\perp|^{2n_o} \propto I^{n_o}$ for surface photoeffect, where N is the number of created ions or electrons and j is the n_o-th order photocurrent, E is the laser electric field strength, $E_\perp$ is its component perpendicular to the metal surface, $I \propto |E|^2$, $I \propto |E_\perp|^2$ are the respective laser intensities and $n_o = [A/\hbar\omega + 1]_{integer}$ is the multiphoton order of the effect. This perturbative method is extensively elaborated and generalised developing it in many modern directions (see in the References).

b) <u>Tunneling theories</u>

The other class of the theories is based on the semiclassical theory of Keldysh. The procedure is to determine N or j from the transition probability between a given intial (fundamental) state and a final state which latter is now essentially the Volkov's "bound" state in the strong laser field. The result is a closed analytical formula with two possible approaches depending on the "Keldysh-parameter" defined as $\gamma \equiv \omega\sqrt{2mA}/Ee = \sqrt{A/E_{osc}} = \tau/T$, where m and e are the electron mass and charge, respectively, $E_{osc} = e^2 E^2/2m\omega^2$ is the electron oscillation energy, T is the laser oscillation half-period and τ is the tunneling time through the laser-field-narrowed potential barrier.

For $\gamma \gg 1$ the results coincide with the lowest order perturbative multiphoton approach outlined previously in a) (= quantum limit). For $\gamma \ll 1$, however, the other approach (classical limit) predicts the occurrence of the optical tunnel emission, when $N \propto j \propto E^2 \exp(-\frac{const}{E})$. This formula is similar to that of cold emission in a strong static electric field.

We focus now our attention to the case of metals for a while. For $\gamma \gg 1$, the direct application the previously mentioned general pure quantum mechanical method for the case of surface photoeffect was made by Bunkin and Fedorov, Silin, Brodsky, etc., who predicted a departure from the $j \propto I^{n_o}$ perturbation theory resulting in a decrease of the $n \equiv d \log j/d \log I$ slope value or effective order of nonlinearity, with increasing I laser intensity, i.e. $n(I)$ decreases with I and is less than n_o.

Other theories elaborated by Bloembergen, Anisimov, etc., took into account the influence of the distortion of Fermi-distribution due to laser heating ("thermally assisted multiphoton photoeffect"). They predicted that in contrast with the pure QED calculations $n(I)$ exhibits an increasing character with I. (We note here that, in addition, Anisimov's theory predicted, that a "multiphoton resonance" is expected even in the case of metals by changing the laser wavelength λ, similarly to the case of atoms, which manifests itself with a maximum in the $n(I, \lambda)$ dependence).

When the experimental parameters are adjusted to provoke predominantly high laser induced Richardson-emission, the calculated $n(I)$ dependence shows first an increase and after a maximum value a decrease.

At last at the extreme high I values, when $\gamma \ll 1$, either the strongly modified perturbative formulae, or the Keldysh's (Bunkin-Fedorov's) simple tunneling formula may be studied. This latter furnishes again an $n(I)$ dependence decreasing with I.

In the case of metals we may speak again on the "thermally assisted" or "dynamic" tunneling process, too, when the electron is raised first due to heating (or also to inverse Bremsstrahlung) to a higher energy level. This level presents an $A'(+) < A$ dynamic work function, and a dynamic $\gamma'(t) \ll \gamma$ value, so the electron may tunnel easily. The $n(I)$ dependence is decreasing, but the emission appears at already much lower I range.

In conclusion we may identify experimentally all mentioned processes by measuring the $n(I)$ dependence (among many other experimental possibilities, of course).

We have to stress the importance of the statistics of the laser light (Lambopoulos, Zoller): in the lower intensity range (perturbative approach) passing from single mode to multimode case the well known constant $n_o!$ enhancement enters, whereas in the high (tunneling) intensity range this dependence is somewhat weaker and intensity dependent (Delone). However, each multiphoton process, being extremely sensitive to the higher order coherence properties, furnishes a unique higher order quantum detector.

A very important inherent problem of each kind laser induced electron emission is the energy and angular distribution (Prof. Walther) of the emitted electron in the strong laser field. One side of this problem is investigated in recent times both experimentally (Agostini) and theoretically in the form of the so called "above threshold ionization; process, about which the best theoreticians (many Polish scientists, e.g., Prof. Z. Białynicka-Birula; Eberly) speak here in details. Their new results are exciting. The other side is the "free-free" transition of electron in presence of a scattering potential (inverse Bremsstrahlung) predicted by Bunkin-Fedorov and Kroll-Watson, when the electron absorbs $n\hbar\omega$ integer number of laser photons.

It is interesting, however, that the question was raised from a quite different context since nearly 20 years by a lot of physicists (Eberly, Körmendi, Fedorov, Bergou, Varro, etc.), namely: can at all the completely free electron absorb integer number of photons from a strong laser field? The question is fundamental from the side of QED and the answer seems to be yes (e.g., Eberly's publication at the 1968. O.P. A.L. Conf. in Warsaw) and presents an exciting challenge for the experiment.

IV. Typical experimental results

In the following we enumerate briefly those most characteristic experimental results, by which the theoretical predictions listed in the previous parts have been proven. Most of these results are widely known and can be found in authentic and detailed forms in a number of relevant monographs (see References).

a) <u>Multiphoton experiments</u>
Atoms:

Most of the preliminary experiments were performed on the $\gamma > 1$ region. For gas atoms the experimental results exactly coincided with the perturbative law dependence: $N \propto I^{n_o}$, with its modification in the resonance case, its polarization and coherence dependence etc. Later on correct results have been obtained for the "above threshold ionization" electron energy distribution and for the angular distribution. The correspondence between the theory and experiment is quite satisfactory.

Metals:

Again, the first experiments carried out at $\gamma > 1$ have proven the validity of the perturbative approach here, i.e., the $j \propto |E_\perp|^{2n_o} \propto I^{n_o}$ relation for surface multiphoton photoeffect and the $j \propto |E|^{2n_o} \propto I^{n_o}$ relation for the volume photoeffect, depending on the metals used. Polarization, intensity, work function dependence, time and coherence properties were investigated. Distinction between surface and volume effect succeeded, too.

The most important point was here, however, the problem of the predicted departures from the perturbation law. The measurements (Budapest, Saclay) using mode-locked picosec pulses revealed very clearly, that for the surface multiphoton photoeffect this departure does exist in the $n(I) < n_o$ form, as predicted by the pure QED theories. At the same time an opposite deviation $n(I) > n_o$ was observed for the volume emission only (Harvard, Bloembergen). The mentioned "thermally influenced" photoeffect was not observed at all with the surface photoeffect in contrast with the predictions, whereas it appeared with the volume one. Further confirmation was found observing a very sharp "multiphoton resonance" in the $n(I, \lambda)$ dependence, where the half width of the resonance curve was two orders of magnitude narrower than predicted. These results suggest that electrons behave in the strong laser field as if they were at nearly 0 K$^\circ$ temperature. The explanation of these facts is still lacking.

These experiments determined the critical laser intensity value at which the perturbation approach begins to loose its validity for metal surfaces. This value for gold surface is $I = 10$ GW/cm^2 which coincides with the relevant theory (Silin). Considering, that for both the surface multiphoton photoeffect and the optical tunneling the $E_\perp$ component is effective, the mentioned observations $(n(I) < n_o)$ may present the first possible weak trace of the optical tunnel emission.

It should be noted that all these results could be observed with transform limited $(\Delta \nu \Delta t \sim 1)$ picosecond laser pulses only; otherwise very scattering confusing results were detected: the surface emission process is a unique detection method which is extremely sensitive to the higher order coherence properties of the laser.

With these experiments we have shown that the surface photoeffect may be as useful experimental tool as the ionization of atoms for the studies of the optical tunneling at $\gamma \ll 1$ in the extreme high laser intensity range.

Free electrons:

Among the multiphoton processes we have to mention our experiment performed for low (meV) energy free electrons, which absorbed $n\hbar\omega$ integer number of Nd laser photons $(n = 0, 1, 2,...)$ with I^n dependence in a strong laser beam according to the expectations (Eberly, Körmendi, etc.). As for the interpretation of this QED process, many attempts can be found in the literature (see e.g. Varro's poster).

b) <u>Optical tunneling experiments</u>

Atoms:

After several unsuccessful attempts in the first systematic experiment at the $\gamma < 1$ condition with Nd laser the perturbation power law $N \propto I^{n_o}$ was observed even at $I = 10^{15}$ W/cm^2. The possible explanation may be the strong background coming from the predominance and saturation of multiphoton ionization.

Using, however, a long wavelength ($\lambda = 10.6\,\mu$) CO_2 laser, this multiphoton background, due to its extremely high order ($n_o \sim 120$), completely should be neglected in principle: therefore only the tunneling process may be observed, if it exists at all. We really succeeded (with Chin at Laval University) to observe ionization of noble gas atoms in 1983. Since that time all noble gases have been ionized there with $I \sim 10^{14}\,W/cm^2$ at $\gamma \sim 0.1$ value, creating even double and triple ions. The most important result is that the perturbation power law $N \propto I^{n_o}$, or $n(I) = n_o$ <u>was no longer observed, always $n(I) < n_o$ relation was found</u>. From only this fact, however, cannot be drown direct unambiguous decision whether the process is of tunneling or multiphoton origin: the ionization probability is increasing so extremely strong way with I in both cases causing immediate ionization saturation. Therefore a very careful numerical evaluation is performed just in these days, which takes into account this aboupt saturation expected for both processes in a given differential time and space domain. Integrating for the whole interaction time a focal volume we obtain the total number of ions N as a function of I. The numerical fit with the experimental points shows a somewhat better agreement with the Keldysh type tunneling. The final conclusions, however, will certainly furnish very interesting new results in the near future.

Metals:

The first experiment investigating the possibility of optical tunneling was also performed at the Laval university in 1983 using CO_2 laser and a gold surface. The decisive experimental parameter was again the $n(I)$ slope dependence as we summarized it previously for the different processes. Here $n_o = 40$, which is extremely high. The experimental results showed an $n(I)$ dependence <u>decreasing</u> with I, supporting a tunneling character in contrast with the other two processes for which $n(I)$ is increasing (Richardson-emission) or $n(I) = n_o = 40 =$ const. (pure multiphoton perturbative process), therefore these two latter ones may be exluded. The previous arguments made for the abrupt saturation in the case of gases cannot be applied directly here: the concentration of gas atoms was of $10^{12}/cm^3$, whereas the density of the conduction electrons in metals capable for emission is of $10^{22}/cm^3$, i.e. 10 orders of magnitude higher. The electron emission, however, occurs already at much lower I values, than predicted by Keldysh. The possible explanation is the "dynamic tunneling", when the electron is raised first to a higher energy level inside the potential well either by thermally or via inverse Bremsstrahlung. This level presents an $A'(t) < A$ dynamic work function and a $\gamma'(t) < \gamma$ parameter. Performing a numerical fit with the Keldysh formula $A' \sim 0.112$ eV and $\gamma' \sim 0.08 \ll \gamma = 6.68$ was obtained. The results were confirmed in the course of two independent experiments: both with 4nsec single pulses and with mode-locked trains containing 2 nsec pulses.

Finally, it can be concluded, that as in the case of atoms, tunneling type process can be observed with metal surfaces either. To be able to draw final decisive conclusions however, the use of much more higher intensities (~ 100 GW/cm^2) and shorter (picosecond) duration pulses in needed.

References

Detailed discussions and further reading can be found in the following references:

1. Multiphoton Ionization of Atoms, Eds. S. L. Chin and P. Lambropoulos (Academic, 1984).
2. N. B. Delone and V. P. Krainov, Atoms in Strong Light Fields (Springer Series, Berlin, Heidelberg, 1985).
3. L. V. Kedysh, Sov. Phys. JETP, 1307 (1965).
4. F. V. Bunkin, M. V. Fedorov, Sov. Phys. JETP 21, 896 (1965).
5. S. I. Anisimov, V. A. Bendersky, Gy. Farkas, Usp. Fiz. Nauk., 122, 185 (1977).
6. Gy Farkas, in Multiphoton Processes, Eds. J. H. Eberly and P. Lambropoulos (Wiley, New York, 1978) pp. 81-100.
7. Gy. Farkas, S. L. Chin, in Multiphoton Processes, Eds. P. Lambropoulos and S. J. Smith, Springer Series on Atoms and Plasmas (Springer, Berlin, Heidelberg, New York, Tokyo, 1984) pp. 191-199.
8. Progress in Optics, VII. p. 361. (North Holland 1969).

J. H. EBERLY[*]

Max-Planck-Institut für Quantenoptik
Theoretical Container 3
D-8046 Garching,
West Germany

ESSENTIAL STATES IN MULTIPHOTON IONIZATION AND ELECTRON SCATTERING

Abstract: A set of states labelled Essential States is defined for atomic electron continuum-continuum transitions. These have been used in a partial theory of above-threshold photo-absorption in non-resonant multiphoton ionization. An example is given showing that the use of the Essential States method does not depend on the use of separable potentials.

1. Above-Threshold Ionization, Background

Experimental evidence has accumulated [1] showing that atomic electrons excited above their ionization threshold can continue to absorb photons from a sufficiently intense laser field before they are detected. The characterisitics of the above-threshold ionization (ATI) phenomenon are summarized as follows (see fig. 1):

(a) The photo-electron energy spectrum can exhibit one or two or many peaks, and the peaks are separated by $\hbar\omega$, where ω is the laser frequency.

(b) At low intensities (fig. 1a) the first peak is the strongest and the second and later peaks get successively rapidly smaller. At intermediate intensities (fig. 1b) the second or third peak may be the highest, and at still higher intensities (fig. 1c) this trend continues to

---

[*] Permanent address: Department of Physics and Astronomy, University of Rochester, Rochester NY 14627 USA.

the extent that the first, or the first several, peaks may disappear altogether. This process can be called "peak switching". The complete switching transition from low-intensity to high-intensity behavior can occur with an increase of intensity by a factor of only 3 or 4.

(c) If the strength of a peak is written proportional to I^k, then k is called the multiphoton index for the peak. Measurements indicate that, at low intensities and for the mth spectral peak, $k \approx N - 1 + m$, where N is the number of photons required to exceed the normal ionization threshold. But at higher intensitites a kind of saturation of the index occurs and $k \approx N - 1$, independent of m.

ABOVE-THRESHOLD IONIZATION PEAKS FOR THREE VALUES
OF LASER INTENSITY

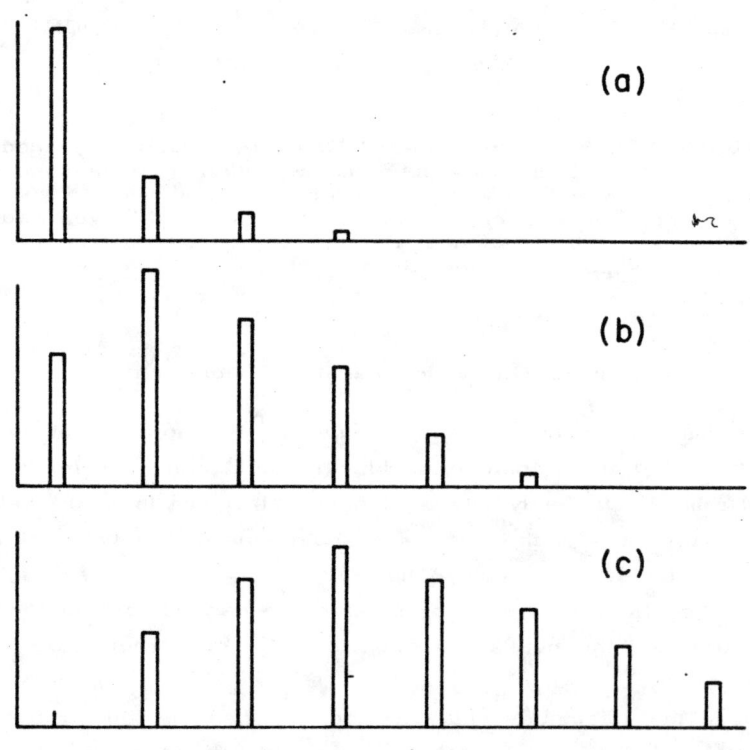

IONIZED e ENERGY →

Fig. 1. Schematic representation of ATI energy spectra showing electron energy peaks separated by $\hbar\omega$ in three cases: (a) low intensity laser, only the first peak is significant; (b) intermediate intensity, several peaks are significant and the first is no longer the strongest; (c) high intensity, many peaks are significant and the first peak has vanished

(d) The strength of the electron spectral peaks is linearly, not quadratically, proportional to atomic gas pressure, indicating that ATI is a single-atom process.

These ATI features have been observed in a variety of atoms, at laser intensities between 100 and 100,000 GW/cm^2, for laser pulses of about 10 psec, and for at least two laser frequencies, corresponding to the wavelengths 1.06 μm and 1/2 x 1.06 μm. The ATI process is evidently both coherent and non-perturbative. It is collision-free and not due, for example, to inverse bremsstrahlung. No shifts in peak positions have been reported and there is no evidence of autoionizing or other quasi-bound-state resonances in the spectra. There is no indication that a peak that has disappeared can reappear at a higher intensity level. There have been no specific measurements of peak widths, but observed widths are significantly greater than instrumental widths.

2. Basic Elements of a Theory

Assuming the absence of collisions and other statistical effects, it should be satisfactory to work with a state-vector theory and avoid the complications of the density matrix. Assuming the possible presence of saturation or other non-perturbative effects, it is essential to have a theory simple enough to solve non-perturbatively, at least approximately. The main complication arises from the need to take account of many transitions above the ionization threshold, i.e., continuum-continuum transitions. A model theory of this type involving several continua was formulated and solved in a study of radiationless relaxation in molecules in 1972 by Lefebvre and Beswick [2]. Similar models have been proposed for ATI processes by Aymar and Crance [3] and Białynicka-Birula [4]. It was the work of Białynicka-Birula that first showed that such a simple model could predict peak-switching.

Deng and I have further developed [5-7] this model to permit inclusion of the infinitely many continua that can contribute to ATI. We have obtained expressions for (i) the peak widths and lineshapes, (ii) the depletion time (the time after which electronic probability is essentially all in the kontinuum), (iii) a dimensionless parameter Z_{EE}, that determines when two successive peaks have approximately equal strength, (iv) a formula for the multiphoton index k (see also refs. 8 and 9), and (v) we have shown how the degree of saturation of the mth photo-peak is governed by the m-dependence of the continuum-continuum matrix elements.

All of this prior theoretical work with soluble models applied to ATI has relied on the flat-continuum approximation. This means that the C-C matrix elements $V_{mm'}(E, E') = \langle E_m | V | E'_{m'} \rangle$ depend on E and E' very weakly or are constant as a function of E and E'. This approximation is in goog agreement with experimental observations, but it can also be replaced by a slightly less restrictive factorization approximation, which uses the ansatz $V_{mm'}(E, E') = u^*_m(E) u_{m'}(E')$. This less restricted version of the method [10] allows one to retain Principal Value contributions to continuum integrations but still permits exact solution for the Schrödinger state amplitudes.

It has been claimed that even the most basic features of agreement between experiment and the predictions of these theoretical models (e.g., the existence of photo-peaks separated by $\hbar\omega$) rests essentially on the flat-continuum or factorization hypothesis. It is the purpose of this note to show that this claim is based on an insufficient understanding of the Essential States method. We present an exactly soluble example within the Essential States framework for which the matrix elements are not smooth and are not factorable. The example nevertheless predicts peaks in the electron spectrum separated by $\hbar\omega$.

3. The Essential States Formulation

There are many different possible ways to label electron continuum states (via energy, angular momentum, linear momentum, quantum numbers associated with parabolic coordinates, etc.). Let us take a complete set of labels associated with a state and abbreviate this set by the symbol m. The initial state will be denoted $|0\rangle$. Transitions can of course be made from this initial state to other states $|m\rangle$ if an appropriate interaction operator exists. We assume, for convenience in orienting the present discussion, that for reasons of selection rules of various kinds the state $|0\rangle$ is a continuum state many transitions away from relevant bound states.

The Hamiltonian can be written:

$$H = H_{atom} - (e/\mu c)\mathbf{A}\cdot\mathbf{p} + (e^2/2\mu c^2)\mathbf{A}\cdot\mathbf{A}, \qquad (1)$$

where H_{atom} refers to the electron's kinetic and potential energy in the absence of radiation. The expansion of the state function $|\psi\rangle = \sum_m E_m |m\rangle$ allows the Schrödinger equation to be written as a set of equations for the coefficients C_m:

$$dC_{-1}/dt = -i\Delta_{-1}C_{-1} - i\sum_{-2}V(-1,-2)C_{-2} - i\sum_{0}V(-1,0)C_{0}$$

$$dC_{0}/dt = -i\Delta_{0}C_{0} - i\sum_{-1}V(0,-1)C_{1}$$

$$dC_{1}/dt = -i\Delta_{1}C_{1} - i\sum_{0}V(1,0)C_{0} - i\sum_{2}V(1,2)C_{2} \qquad (2)$$

....

$$dC_{m}/dt = -i\Delta_{m}C_{m} - i\sum_{m-1}V(m,m-1)C_{m-1} - i\sum_{m+1}V(m,m+1)C_{m+1}$$

....

Here the symbol $\sum_m$ means a sum (or integration) over the quantum numbers associated with state $|m\rangle$, and $V(m, m+1) = \langle m|V|m+1\rangle$, where V is the interaction Hamiltonian: $V = H - H_{atom}$, and $m\pm 1$ designates the states corresponding to transitions accomplished via the absorption and emission of one photon from the state m. The states that are connected in this way to the state $|0\rangle$ are the <u>Essential States</u>.

The detuning Δ_m is identified as follows. Let E_I be some fixed energy which we can associate with the energy of the initial electron, and let $E_{mI} = E_m - E_I$ be the transition energy from the initial energy to the energy of the mth state. This is a continuum variable because E_m is a continuous parameter. Then we define $\Delta_m = E_{mI} - m\hbar\omega$, which is obviously also a continuum variable.

The specific example to be discussed here is free electron scattering, so it is most convenient to identify the basis states with electron plane-waves. We assume circularly polarized plane-wave radiation for convenience:

$$\mathbf{A}(r, t) = \text{Re}\left[\hat{\mathbf{e}} A e^{-i(\omega t - \mathbf{k}\cdot\mathbf{r})}\right] \qquad (3)$$

in order to set $\mathbf{A}\cdot\mathbf{A} = A^2 =$ constant, which means its effect will be only a trivial shift of the state energy. The state amplitudes are just

$$a_p(t) = \int d^3 r\, e^{-i\mathbf{p}\cdot\mathbf{r}} \psi(r, t) \qquad (4)$$

and the general equation of motion is:

$$i\hbar\, da_p/dt = (2\pi)^{-3} \int d^3 q\, a_q(t) \int d^3 r\, e^{-i\mathbf{p}\cdot\mathbf{r}} \left[p^2/2\mu - (e/\mu c)\mathbf{p}\cdot\mathbf{A}(r,t) + e^2 A^2/2\mu c^2\right] e^{i\mathbf{q}\cdot\mathbf{r}}. \qquad (5)$$

It is clear that the r integral can be carried out in (5). The matrix element of the **p · A** interaction is in fact the delta function $\delta^3(p-q\pm k)$ times $\exp[\pm i\omega t]$. We point out explicitly that the delta function cannot be factorized into a factor depending on p and another depending on q. Because of the delta function the q integration can also be performed and we can write the general equation more simply:

$$da_p/dt = -iW_p a_p + i(eA/2\mu\hbar c)\left[(\hat{e}\cdot p)a_{p-k}e^{-i\omega t} + (\hat{e}\cdot p)^* a_{p+k}e^{i\omega t}\right], \quad (6)$$

where $W_p = (1/\hbar)(p^2/2\mu + e^2A^2/2\mu c^2)$. Now we designate by $\mathbf{p}_0$ the initial momentum, and by a_0 the corresponding amplitude, and by a_m the amplitude corresponding to the Essential State with momentum $\mathbf{p}_m = \mathbf{p}_0 + m\mathbf{k}$. The general equation becomes

$$da_m/dt = -iW_m a_m + i(eA/2\mu\hbar c)\left[(\mathbf{p}_0+m\mathbf{k})\cdot\hat{e}\, a_{m-1}e^{i\omega t} + (\mathbf{p}_0-m\mathbf{k})\cdot\hat{e}^*\, a_{m+1}e^{i\omega t}\right]. \quad (7)$$

Of course, since the radiation field is transverse, $\mathbf{k}\cdot\hat{e} = 0$ and only $\mathbf{p}_0\cdot\hat{e}$ is significant.

It is convenient to shift the energy origin by the constant amount W_0, and to invoke the rotating wave approximation. We also define a constant phase ϕ by $\mathbf{p}_0\cdot\hat{e} = |\mathbf{p}_0\cdot\hat{e}|e^{-i\phi}$. Then the relevant amplitudes are

$$C_m(t) = a_m(t)\exp[iW_0 t]\exp[im(\omega t + \phi)] \quad (8)$$

and they satisfy

$$dC_m/dt = -i\Delta_m C_m + i(eA|\mathbf{p}_0\cdot\hat{e}|/2\mu\hbar c)\left[C_{m-1} + C_{m+1}\right], \quad (9)$$

where the generalized detuning Δ_m is given by

$$\Delta_m = W_m - W_0 - m\omega =$$
$$= \left[p_m^2/2\mu + e^2A^2/2\mu c^2\right] - \left[p_0^2/2\mu + e^2A^2/2\mu c^2\right] - m\omega = \quad (10)$$
$$= m\left[\mathbf{p}_0\cdot\mathbf{k}/\mu - \omega\right] + m^2 k^2/2\mu.$$

The $k^2/2\mu$ term is negligible in the nonrelativistic limit. In fact, we can recall that the fully relativistic propagator (indicating 4-vectors by removing bold-face) is just $1/\left[(p_0 + mk)^2 - \mu^2\right]$, which reduces in the nonrelativistic limit to $1/2\mu\Delta_m$.

The consequence is that one sees that eqn. (8) already contains the full time dependence explicitly, and (9) has the simple steady-state form:

$$2mC_m = -2(\mu/p_0 \cdot k)(eA|\mathbf{p}_0 \cdot \hat{\mathbf{e}}|/2\mu\hbar c)\left[C_{m-1} + C_{m+1}\right], \quad (11)$$

with the solution:

$$C_m = J_m(\beta), \quad (12)$$

where J_m is the ordinary cylindrical Bessel function and

$$\beta = -(eA/\hbar c)(|\mathbf{p}_0 \cdot \hat{\mathbf{e}}|/p_0 \cdot k). \quad (13)$$

Now the whole wave function can be reconstructed by inverting the phase transformations made on the amplitudes. One finds:

$$\Psi(r,t) = (2\pi)^{-3}\int d^3q\, e^{iq\cdot r}\, a_q(t) =$$
$$= \sum_m e^{ip_0\cdot r}\, e^{imk\cdot r}\, a_m(t) = \quad (14)$$
$$= e^{-i(W_0 t - p_0\cdot r)}\sum_m e^{-im(\omega t - k\cdot r + \phi)} J_m(\beta).$$

This is easily recognized to be a semi-nonrelativistic form of the well-known Volkow wave function for a free electron in a plane wave radiation field. It has been obtained here by an application of the Essential States method. Other simpler methods of derivation of the Volkow wave function are available, of course. The purpose of this derivation has been to demonstrate that the Essential States method is not restricted to situations in which the continuum-continuum matrix elements are smooth.

Acknowledgements: It is a pleasure to acknowledge conversations and correspondence about ATI physics during the past eighteen months with many colleagues, initially with Z. Białynicka-Birula and Principally with Z. Deng, but also with P. Agostini, L. Armstrong, I. Berson, M.

Crance, A. Dulčić, F. H. M. Faisal, M. V. Fedorov, M. Gavrila, Y. Gontier, G. Leuchs, E. Kyrölä, M. Lewenstein, G. Mainfray, C. Manus, J. Mostowski, G. Petite, K. Rzążewski, Y. R. Shen, B. W. Shore, A. Szöke and H. Walther. The scientific atmosphere of the Max-Planck Institut für Quantenoptik, and particularly the warm hospitality of P. Meystre and H. Walther during the past seven months, has been greatly appreciated. This work has been financially supported by the U. S. Department of Energy and by receipt of a Senior Alexander von Humboldt Award.

References

1. The initial experimental observation was reported in: P. Agostini, F. Fabre, G. Mainfray, and N. K. Rahman, Phys. Rev. Letters $\underline{42}$, 1127 (1979). A more recent experimental report, which shows evidently non-perturbative behavior for the first time is P. Kruit, J. Kimman, H. G. Muller, and M. J. van der Wiel, Phys. Rev. A $\underline{28}$, 248 (1983). See also L. A. Lompre, et al., J. Opt. Soc. Am. B (in press).
2. R. Lefebvre and J. A. Beswick, Mol. Phys. $\underline{23}$, 1223 (1972).
3. M. Crance and M. Aymar, J. Phys., B $\underline{13}$, L421 (1980).
4. Z. Białynicka-Birula, J. Phys., B $\underline{17}$, 3019 (1984).
5. Z. Deng and J. H. Eberly, Phys. Rev. Letters, $\underline{53}$, 1810 (1984).
6. Z. Deng and J. H. Eberly, J. Opt. Soc. Am., B $\underline{2}$, 486 (1985).
7. Z. Deng and J. H. Eberly, J. Phys., B $\underline{18}$, L287 (1985).
8. M. Edwards, L. Pan and L. Armstrong, Jr., J. Phys., B $\underline{17}$, L515 (1984).
9. M. Edwards, L. Pan and L. Armstrong, Jr., J. Phys., B $\underline{18}$, 1927 (1985).
10. Z. Deng and J. M. Eberly, unpublished. See also ref. 9.

S. M. BARNETT and P. L. KNIGHT

Optics Section, Blackett Laboratory,
Imperial College,
London SW7 2BZ, England

and

P. M. RADMORE

Physics Department,
Queen Mary College,
London E1 4NS, England

DEPHASING AND DECAY OF QUANTUM OPTICAL RESONANCES

Abstract: We discuss two examples of intense-field dressing of atomic resonances and show how they dephase or decay through environmental influences or through the influence of other states.

1. Introduction

Dressed states of strongly interacting resonant atomic states and coherent radiation have been used widely in quantum optics [1-2]. The essence of the technique is to isolate those few levels or states participating in the resonant dynamics and solve the interaction problem for these states exactly. Here we analyse the effects of states or perturbations extraneous to the dressing interaction which cause dephasing or decay in two fundamental models which have attracted much attention in the past. The Jaynes-Cummings model of a two-level atom interacting with a single quantised field mode is generalised to include cavity dissipation. The quasicontinuum model of photoexcitation is generalised to describe the excitation of a Rydberg series of highly excited atomic states.

2. Field Dissipation of Dressed Atom Coherence

The first example of a dressed atom perturbed by extrinsic perturbations that we discuss concerns the effect of resistive cavity field damping in the Jaynes-Cummings model [3]. In this model of a single two-level atom interacting with a quantized single cavity field mode, the time-evolution of the strongly-coupled atom-field system is dictated by the discrete nature of the field energy and its statistical distribution. The atom and field exchange energy in an oscillatory, Rabi fashion whose frequency is determined by the photon occupation number. A distribution of photon numbers causes a dephasing or "collapse" of the coherent time-evolution as the distribution of dressed state eigenvalues beat against each other. Eberly and coworkers have demonstrated [4] that the discrete nature of the photon number distribution leads to a further purely quantum effect as the dephased Rabi oscillations partially rephase or "revive". The revival time is governed by the single-photon Rabi frequency and its "granularity". We have shown [5] that the revivals of the coherent evolution which are the signature of the quantum nature of the field are strongly affected by field dissipation even when the damping hardly affects the underlying Rabi oscillations.

We describe the field damping in terms of a bath of harmonic oscillators representing the resistive cavity-wall degrees of freedom which couple directly in rotating-wave approximation to the electromagnetic field mode. This field mode is coupled in turn to the two-level atom. We imagine the range of frequencies present in the dissipative bath to be so large as to justify the Born-Markov approximation in the field-bath coupling. The reduced density-matrix for the atom-field mode system can then be obtained by standard master equation techniques. In the simplest case of a cavity at $0°$ K and a field mode exactly resonant with the two-level atomic transition the reduced density matrix obeys the equation of motion

$$\frac{d}{dt}\rho = -ig\left[(a^+\sigma_- + \sigma_+ a), \rho\right] + \gamma a \rho a^+ - \frac{\gamma}{2} a^+ a \rho - \frac{\gamma}{2} \rho a^+ a. \tag{1}$$

The atom-field mode coupling is g, the field represented by creation and annihilation operators a^+ and a and the atom by the usual Pauli spin-operators [6]. Instead of solving this by numerical methods, we transform to a new "dissipation picture" which stresses the departure of our solution from semiclassical Rabi behaviour.

The density-matrix in the dissipation picture is $\chi = \exp[-(J+L)t]\rho$ where J and L act to the right [7] on the density operator as

$$J\rho = \gamma a \rho a^+, \qquad L\rho = -\frac{\gamma}{2} a^+ a \rho - \frac{\gamma}{2} \rho a^+ a$$

so that

$$\frac{d}{dt}\chi = -ig\left\{ e^{\gamma t/2} a^+ \sigma_- \chi - 2\sinh(\frac{\gamma t}{2}) \sigma_- \chi a^+ \right.$$
$$+ e^{-\gamma t/2} \sigma_+ a \chi - e^{\gamma t/2} \chi \sigma_+ a +$$
$$\left. + 2\sinh(\frac{\gamma t}{2}) a \chi \sigma_+ - e^{-\gamma t/2} \chi a^+ \sigma_- \right\}. \qquad (2)$$

If the field is treated as a classical variable entirely unaffected by the atomic evolution, eq. (2) simplifies to

$$\frac{d}{dt}\rho = -i\Omega e^{-\gamma t/2}\left[(\sigma_+ + \sigma_-), \rho\right], \qquad (3)$$

where $\Omega = ga$ is the semiclassical Rabi frequency and $\rho = \chi$ in this semiclassical approximation. So in semiclassical approximation the cavity dissipation merely damps out the atom-field perturbation exponentially. We retain the full quantum nature of the field and solve for components $\chi(n,m; +.-)$ etc. of the interaction picture density matrix, where n, m label field occupation numbers and $|\pm\rangle$ are atomic excited and ground states. We find the solution of eq. (2) for these components is considerably simplified in the underdamped limit $n^2 \gamma \ll (n+1)^{1/2} g$. This limit is precisely that which interests us: we want the dissipation not so strong as to damp out Rabi oscillations but to damp out revivals. After lengthy algebra we find

$$\chi(n, n; +, +) = \frac{1}{2} e^{-\gamma t/2} \left\{ p(n) + \right.$$
$$\left. + \sum_{l=0}^{\infty} p(l) \cos\left[2g(l+1)^{1/2} t\right] (e^{-\gamma t} - 1)^{l-n} \frac{l!}{n!(l-n)!} \right\}, \qquad (4)$$

where $p(n)$ is the photon number distribution. We find from (4) that the probability of being excited at time t is given by

$$P_+(t) = \frac{1}{2} e^{-\gamma t/2} \left\{ 1 + \sum_{m=0}^{\infty} p(m) e^{-m\gamma t} \cos 2g(m+1)^{1/2} t \right\}. \qquad (5)$$

The atomic inversion is given by $w(t) = 2P_+(t) - 1$.

In Fig. 1-3 we plot the atomic inversion w(t) for an initially excited atom in a coherent field with a mean photon number of five for (1) $\gamma = 0$, (2) $\gamma = 0.01$ g and (3) $\gamma = 0.03$ g. We note the collapse of the atomic Rabi oscillations and their revival. As the damping in-

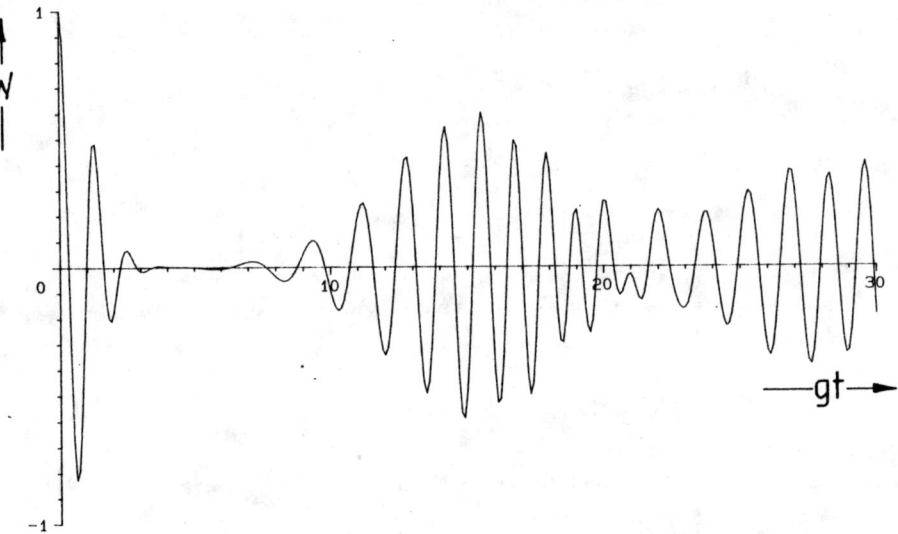

Fig. 1. Time-evolution of inversion w for atom interacting with initially coherent field with a mean photon number of 5 for $\gamma = 0$

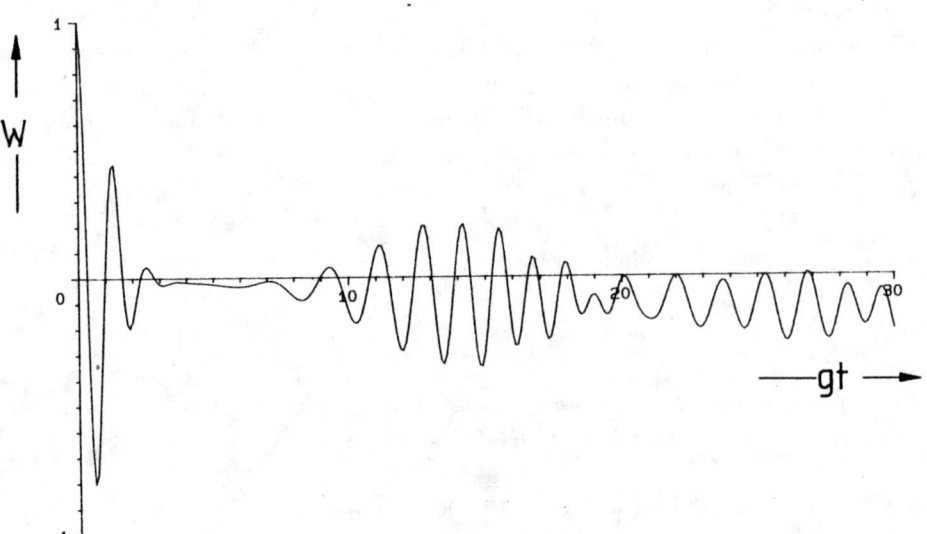

Fig. 2. As in Fig. 1 but for $\gamma = 0.01$ g

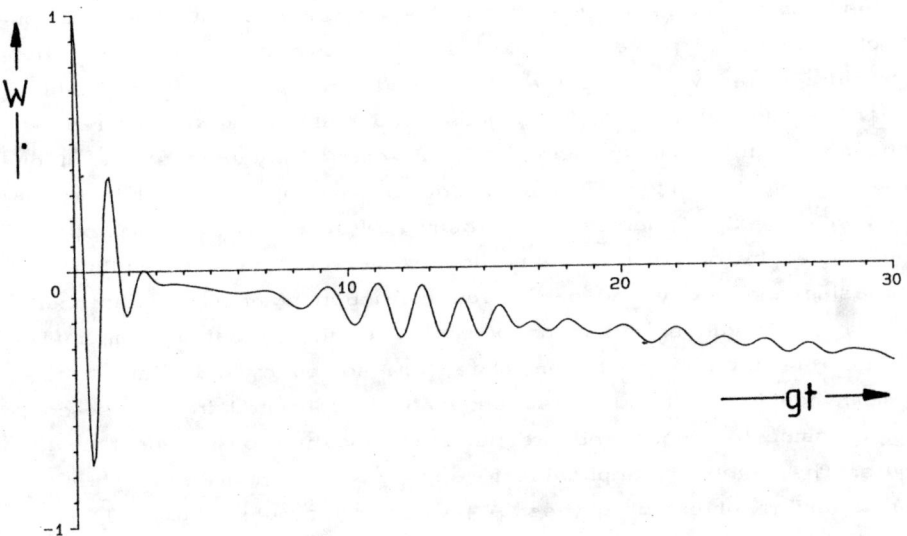

Fig. 3. As in Fig. 1 but for $\gamma = 0.03\,g$

creases, the revivals rapidly diminish in amplitude even though little dissipation or decay is evident in the initial collapse. The revivals depend on the discrete nature of the photon number distribution and this granularity is translated into a discrete spectrum of Rabi frequencies. The field dissipation broadens these spectral components (the dressed-atom eigenenergies) so that the Fourier sum in eq. (5) becomes more effectively a Fourier integral (with, therefore, no revivals or repeats). Current experimental work on Rydberg atoms in high Q cavities are certainly within this underdamped limit [8] and should be able to resolve these effects.

3. Dephasing of Dressed Atom Coherences by a Quasicontinuum of Nearby States

Conventional studies of Rabi oscillations ensure that only two resonant levels participate in the dynamics [6]. The Rabi oscillations are a quantum beat phenomenon depending on the coherent excitation of superposition states (the two dressed states). The excitation of many atomic states in a coherent superposition has been of particular interest recently with particular attention being paid to quasicontinuum resonances [9]. Indeed, atomic superposition states excited by short optical pulses form the basis of quantum beat spectroscopy. With the availability of femtosecond pulses for laser spectroscopy, superpositions of fine-structure states widely separated in frequency space are accessible [10]. We suggest [11] that a femtosecond pulse of ultraviolet radiation can create a superposition of highly-excited Rydberg states of different principal quantum number n. The time-evolution of such a superposition will exhibit many of the features of a quasicontinuum excitation studied by others [9]. In particular, we would expect oscillation, dephasing and decay and partial reconstruction of atomic coherence.

The Schrödinger equations of motion for the probability amplitudes for the relevant quasicontinuum states are solved by a Laplace-transform method in which near-resonant states are treated in a close-coupling dressing technique and are perturbed by their off-resonant neighbours. The method is applicable to any quasi-continuum which has convergent resolvent properties. A particularly simple example [9] assumes equal level spacings in the (usually infinite) quasicontinuum with constant (and equal) Rabi frequencies between the ground state 0 and those continuum states α. We will show how our general method reproduces earlier results for this case, where Rabi oscillations between 0 and the resonant state in α are periodically interrupted by a "kick" from the rephased non-resonant levels and a sequence of coarse-grained Golden Rule decay rates generated. For the Rydberg series quasicontinuum we use hydrogenic matrix elements and energy level splittings to evaluate the probability of photoexcitation. Again the result appears as a perturbation around the Rabi oscillatory solution by the background states (Fig. 4).

We consider the level scheme shown in Fig. 4 in which $|0\rangle$ is excited to states $|\alpha\rangle$ with detuning $\Delta_{\alpha'}$, and matrix elements of the excitation $V_{0\alpha}$. The equations of motion for the Schrödinger amplitude are

$$\dot{C}_0 = -i \sum_{\alpha} V_{0\alpha} C_{\alpha} \qquad (6a)$$

$$\dot{C}_{\alpha} = -i \Delta_{\alpha} C_{\alpha} + i V_{\alpha 0} C_0. \qquad (6b)$$

We use Laplace transform equations (6) using the initial condition $C_0(t=0) = 1$, to give

$$\tilde{C}_0 = \frac{1}{Z + \sum_{\alpha} \dfrac{V_{\alpha}^2}{Z + i\Delta_{\alpha}}}, \qquad (7)$$

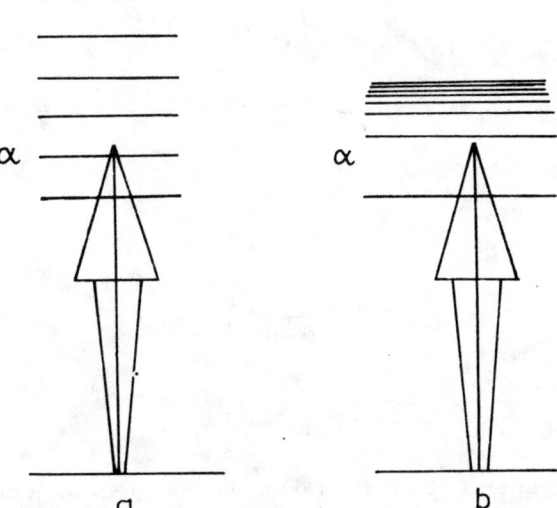

Fig. 4. Photoexcitation of atomic quasicontinuum (a: equal spacing, b: hydrogenic Rydberg series)

where $V_\alpha = |V_{0\alpha}|$. The inversion of eq. (7) gives

$$C_0(t) = \frac{1}{2\pi i} \int_{Br} \frac{e^{Zt}}{Z + S(Z)} dz, \qquad (8)$$

where the resolvent or self-energy is

$$S(Z) = \sum_\alpha V_\alpha^2 / (Z + i\Delta_\alpha), \qquad (9)$$

so that $C_0(t)$ is given by the sum of residues in the complex Z-plane of the integrand of eq. (8) at places where $Z + S(Z) = 0$. It is easy to see that substitution of $Z = u + iv$ and taking real and imaginary parts of $Z + S(Z)$ shows that $u = 0$. We therefore put $Z = ix$ and consider the zeros of $f(x)$, where

$$f(x) = x - \sum_\alpha V_\alpha^2 / (x + \Delta_\alpha). \qquad (10)$$

If one state P in $|\alpha\rangle$ is resonant with $|0\rangle$, then we can assume the excitation is weak (i.e. $V_\alpha \ll \Delta_\alpha$) for all α except that resonant state. To second order in this background coupling we find after extensive algebra, the general resonant result

$$C_0(t) = \tfrac{1}{2}(1 + \tfrac{1}{2}S_1 - S_2)\exp\left\{iV_p t(1 + \tfrac{1}{2}S_1 - \tfrac{1}{2}S_2 + \tfrac{1}{8}S_1^2)\right\} +$$
$$+ \tfrac{1}{2}(1 - \tfrac{1}{2}S_1 - S_2)\exp\left\{-iV_p t(1 - \tfrac{1}{2}S_1 - \tfrac{1}{2}S_2 + \tfrac{1}{8}S_2^2)\right\} + \quad (11)$$
$$+ \sum_{m \neq p} \frac{V_m^2}{\Delta_m^2} \exp(-i\Delta_m t),$$

$$S_1 = \frac{1}{V_p} \sum_{\alpha \neq p} V_\alpha^2 / \Delta_\alpha, \quad (12)$$

$$S_2 = \sum_{\alpha \neq p} V_\alpha^2 / \Delta_\alpha^2. \quad (13)$$

From this general resonant result we can extract various examples. If the quasicontinuum levels are equally spaced, infinite in both directions from p, all of which are coupled equally to 0 with matrix elements $V_\alpha = V$ then $\Delta_\alpha = \alpha\Delta$, $|\alpha| = 1, 2, \ldots$ with $\alpha = 0$ for the resonant state. We find from eq. (12), $S_1 = 0$ and from eq. (13), $S_2 = (\pi^2 V^2/3\Delta^2)$ so that

$$C_0(t) = (1 - \frac{\pi^2 V^2}{3\Delta^2})\cos\left[Vt(1 - \frac{\pi^2 V^2}{6\Delta^2})\right] +$$
$$+ \frac{2V^2}{\Delta^2} \sum_{n=1}^{\infty} \frac{\cos(n\Delta t)}{n^2}, \quad (14)$$

as in earlier work [9]. The first term represents a perturbed Rabi solution for the strong coupling between 0 and p, and the second term is a sum of contributions from the background levels. This sum may be performed to give

$$C_0(t) = (1 - \frac{\pi^2 V^2}{3\Delta^2})\cos\left[Vt(1 - \frac{\pi^2 V^2}{6\Delta^2})\right] +$$
$$+ \frac{V^2}{\Delta^2}(\tfrac{1}{2}\Delta^2 t^2 - \pi\Delta t + \frac{\pi^2}{3})\Big|_{\frac{2\pi}{\Delta}, t}, \quad (15)$$

where $|\frac{2\pi}{\Delta}, x$ means the expression holds for $0 \leq x \leq \frac{2\pi}{\Delta}$ and then **repeats** periodically. We have shown the initial state population $|C_0|^2$ as a function of time in Fig. 5 for $\alpha = \pi V/\Delta = 0.5$ so the regular disruptions of the Rabi oscillations are visible. For small α, and small distance x from the last (nth) interruption, we find

$$|c_0(t)|^2 = A^2(1 - 2\Gamma x + \ldots),\tag{16}$$

where for small n, $A \simeq 1 + \mathbf{0}(n^2 a^2)$ and

$$\Gamma \cong (2n+1)\alpha = \alpha, 3\alpha, 5\alpha \ldots \tag{17}$$

is the cascade of interrupted "coarse-graining" rates found previously [9].

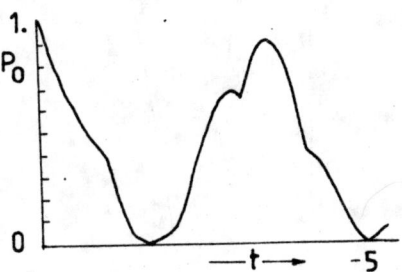

Fig. 5. Time-evolution of initial state population $P_0 = |C_0|^2$ resonantly excited by a laser from 0 to an equally spaced quasicontinuum

Our general result eq. (11) for resonant excitation can also be specialised to a Rydberg series photoexcitation. If we take the lowest level to be resonant, and the energy distance from that lowest state to the ionisation limit to be D, then

$$\Delta_n = D(1 - \frac{1}{n^2})\tag{18}$$

and the hydrogenic matrix elements from 0 to n are

$$V_n = V \cdot n^{-3/2},\tag{19}$$

where $\Delta_1 = 0$ and $V_1 = V$. Again the sums may be performed to give $S_1 = V/4D$ and $S_2 = 5V^2/16D^2$. We find, therefore

$$\begin{aligned}C_0(t) = &\tfrac{1}{2}(1 + \tfrac{V}{4D} - \tfrac{5V^2}{16D^2})\exp\left\{iVt(1 + \tfrac{V}{8D} - \tfrac{19V^2}{128D^2})\right\} + \\ &+ \tfrac{1}{2}(1 - \tfrac{V}{4D} - \tfrac{5V^2}{16D^2})\exp\left\{-iVt(1 - \tfrac{V}{8D} - \tfrac{19V^2}{128D^2})\right\} + \\ &+ e^{-iDt}\tfrac{V^2}{D^2}\sum_{n=2}^{\infty}\frac{n}{(n^2-1)^2}e^{iDt/n^2}.\end{aligned}\tag{20}$$

In Fig. 6 we plot the initial state population $P_0(t) = |C_0(t)|^2$ as a function of time for the same choice of parameters as Fig. 5 (basically $\alpha = \frac{\pi V}{D} = 0.5$). The periodic disruption due to the precise rephasing of the equally spaced background is visible in Fig. 5 is now

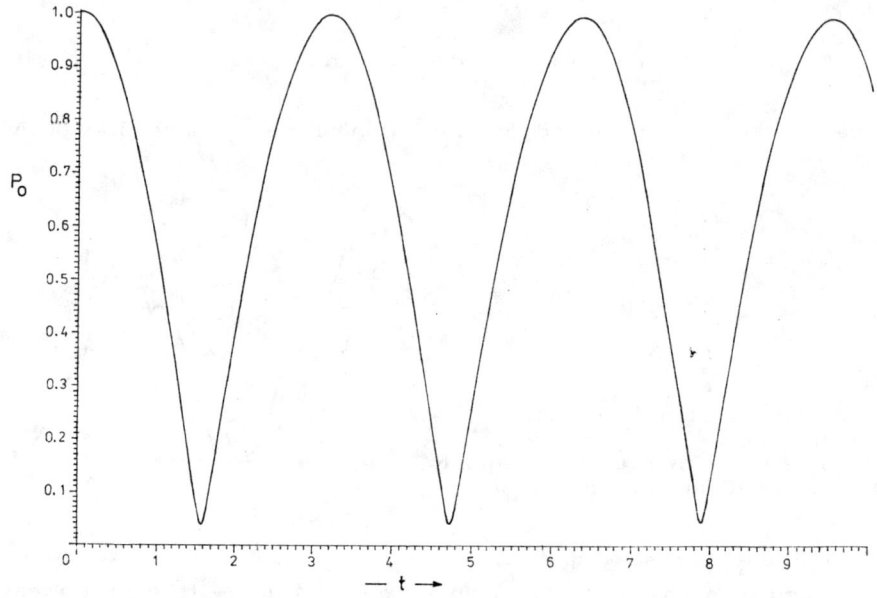

Fig. 6. As in Fig. 5 but for a Rydberg series

absent; what is visible is a sharpening of Rabi "troughs" compared with peaks (as if a harmonic component were present). We have also examined the non-resonant case and included further continua and will report our results elsewhere [11].

In both of our examples we have shown that extrinsic coupling of other states or perturbations to a strongly-dressed resonance disrupts or dephases simple Rabi oscillations. In the Jaynes-Cummings example, bath interactions resistively damp the field component of the strongly--interacting atom-field system. In the quasicontinuum example, other levels act as a dephasing (not as a bath, unless their density becomes so large as to act as a continuum).

Acknowledgements: We are grateful to the UK Science and Engineering Research Council for their support of this work and the award of a research studentship to SMB. We would like to acknowledge discussions with J. H. Eberly and E. Kyrölä, and M. A. Lauder for the computer-generated graphs.

References

1. P. L. Knight and P. W. Milonni, Phys. Rep., 66, 21 (1980).
2. C. Cohen-Tannoudji, B. Diu and F. Laloë, "Quantum Mechanics", (New York: Wiley), 1977.
3. E. T. Jaynes and F. W. Cummings, Proc. IEEE, 51, 89 (1963).
4. J. H. Eberly, N. B. Narozhny and J. J. Sanchez-Mondragon, Phys. Rev. Lett., 44, 1323 (1980); N. B. Narozhny, J. J. Sanchez-Mondragon and J. H. Eberly, Phys. Rev., A23, 236 (1981); J. J. Sanchez-Mondragon, N. B. Narozhny and J. H. Eberly, Phys. Rev. Lett., 51, 550 (1983); H. I. Yoo, J. J. Sanchez-Mondragon and J. H. Eberly, J. Phys., A14, 1383 (1981) and refs therein.
5. S. M. Barnett and P. L. Knight, to be published 1985.
6. L. Allen and J. H. Eberly, "Optical Resonance and Two Level Atoms", (New York: Wiley), 1975.
7. E. B. Davies, "Quantum Theory of Open Systems", (London: Academic Press, 1976).
8. D. Meschede, H. Walther and G. Muller, Phys. Rev. Lett., 54, 551 (1985).
9. eg R. Lefebvre and J. Savolainen, J. Chem. Phys., 60, 2509 (1974); J. J. Yeh, C. M. Bowden and J. H. Eberly, J. Chem. Phys., 76, 5936 (1982); P. W. Milonni, J. R. Ackerhalt, H. W. Galbraith and M.-L. Shih, Phys. Rev., A28, 32 (1983); E. Kyrölä, JOSA B1, 737 (1984).
10. J. E. Rothenberg and D. Grischkowsky, Opt. Lett., 10, 22 (1985).
11. P. M. Radmore and P. L. Knight, to be published 1985.

G. REMPE, H. WALTHER, und P. DOBIASCH

Max-Planck-Institut für Quantenoptik and
Sektion Physik der Universität München,
D-8046 Garching, Fed. Rep. of Germany

THE ONE-ATOM MASER - A TEST SYSTEM FOR SIMPLE
QUANTUMELECTRODYNAMICAL EFFECTS

I. Introduction

When a valence electron of an atom is excited in an orbit with high principal quantum number and therefore far from the ionic core, the energy levels of the atom can simply be described by the Rydberg formula. This is the reason why atoms in these highly excited states are often called Rydberg atoms.

The existence of hydrogen Rydberg atoms has been known from radio astronomy for many years. In space, they are formed when protons capture electrons in high orbits; radio waves are then emitted when the electron jumps down to lower lying levels. Until recently, however, it was impossible to generate such atoms in the laboratory. In particular, it is not possible to populate highly excited states in discharges, owing to their large collisional cross sections. It is the availability of tunable lasers of sufficient intensity and narrow bandwidth that first made the selective excitation of these states possible. Because of their high principal quantum numbers, and also because the relative energy difference between neighbouring levels is small, Rydberg atoms are expected to exhibit a number of classical properties. In particular, according to Bohr's correspondence principle, the transition frequency between neighbouring levels approaches the orbital frequency of the electron.

Rydberg atoms represent an ideal testing gruond for some of the most fundamental models and predictions of low energy quantum electrodynamics (QED). The reasons are the following:

a) The matrix elements for electric dipole transitions between neighbouring Rydberg states scale as n^2, where n is the principal quantum number. For high enough n, stimulated effects overcome spontaneous emission already for very small photon numbers. As a consequence, Rydberg atoms are very sensitive e.g. to blackbody radiation (see Ref. [1] and [2] for recent reviews).

b) The transitions to neighbouring levels are in the region of millimeter waves, therefore it is possible to physically modify the nature of the environment into which they decay using for example conducting walls. Introducing conductors imposes boundary conditions on the electromagnetic field and leads back to a discrete spectrum in the case of a finite volume enclosed in a cavity. In principle, there are essentially two distinct cases to be discussed. First, the situation of an atom in close proximity to a conducting plate [3-8]. The induced image charges give rise to extra contributions of a van der Waals type force to the inner atomic forces, thus leading to position dependent level shifts. Second, there are effects from a discrete mode structure of the electromagnetic field inside a cavity due to its geometry. Of course, it is not possible to consider one of these phenomena without the other, but in most cases only one of the two produces the major influence. Consequences of the discrete mode structure of a cavity for Rydberg atoms are: the rate of the spontaneous emission is enhanced or diminished depending upon the cavity being tuned on or off resonance with a transition frequency [9-13], as well as modifying the Lamb shift of Rydberg levels [14].

c) For cavities with high quality factors, the photon emitted by an atom in a Rydberg state remains stored inside the resonator long enough to be reabsorbed by the same atom with a finite probability. In this way, it is possible to realize a single-atom maser [15]. A single Rydberg atom inside a low loss, single mode resonator is an experimental realization of the Jaynes-Cummings model [16], describing the interaction between a single two-level atom and a single mode of the electromagnetic field. This model has been the object of considerable attention in the past, and a number of purely quantum mechanical predictions on the dynamics of this system have been made. These include the collapses and revivals in the dynamics of the atomic population. Rydberg atoms will for the first time offer the possibility to test these predictions [16-18] (see also the contributions of P. Meystre and J. H. Eberly in this volume).

In the following we will briefly review the most important properties of Rydberg atoms and then discuss the QED in a cavity and in the last chapter we will describe the present status of our one-atom maser experiment.

II. Properties of Rydberg Atoms

The scaling laws for the properties of Rydberg atoms are compiled in Table I. The energy can be simply described by the Rydberg formula with n replaced by an effective principal quantum number n^*. In general, n^* depends on the phenomenological quantum defect δ_ℓ of the state of angular momentum ℓ. For low-ℓ states, where the orbits of the classical Bohr-Sommerfeld theory are ellipses of high eccentricity, the penetration and polarization of the electron core by the valence electron lead to large quantum defects and strong departures from hydrogenic behaviour. As ℓ increases, the orbits become more circular and the atom more hydrogenic: as a result δ_ℓ scales as ℓ^5. Table I compiles the scaling laws for further properties of Rydberg atoms: The radius of the charge distribution of the valence electron scales as n^{*2}, and for $n^* = 50$ the linear dimension of the atom is already comparable with the wavelength of light in the visible region and competes with the size of large biomolecules.

Table I: Scaling laws for properties of Rydberg atoms

Energy: $\quad E_n = R/(n - \delta_\ell)^2 = R/n^{*2}$
δ_ℓ quantum defect, n^* effective quantum number
R Rydberg constant

Radius: $\quad \langle r \rangle \sim n^{*2}$

Lifetimes: $\quad \tau \sim n^{*3}$ (low angular momentum states)
$\tau \sim n^{*5}$ (high angular momentum states)

Fine-structure interval: $\quad \Delta E \sim 1/n^{*3}$

The electric polarizability for the quadratic Stark effect increases as n^{*7} and the diamagnetic interaction as n^{*4}. This allows one to perform experiments at field strengths high enough to make the interaction energy in the external electric or magnetic field comparable with or larger than the Coulomb energy of the atom. For practical reasons the corresponding field strengths for ground state atoms cannot be reached in the laboratory. The study of highly excited atoms in external electric and magnetic fields is therefore interesting in itself. (For reviews see References [19-22]). The sensitivity of Rydberg atoms to external electric fields also means that the atoms already ionize in rather weak fields. This opens the possibility of a very effective detection, as will be discussed later.

The large Rydberg atom orbitals are characterized by natural lifetimes much longer than the ones of less excited atoms. In the case of hydrogen Rydberg states, the dependence of the lifetime on n can be obtained by fully quantum mechanical radiation rate calculations involving hydrogenic coulombic wave-functions. For Rydberg states of other species the lifetimes (and the other radiative parameters) scale not exactly as a power of n but rather as a power of n^*. The n^* scaling law can be determined using Bates and Damgaard type of calculations [23]. The lifetimes scale either with n^{*3} (when ℓ is small) or with n^{*5} (when $\ell \cong n$).

The rate of spontaneous emission of radiation for a transition from a state n to n' is given by the Einstein A coefficient:

$$A_{n \to n'} = 16\pi^3 \upsilon^3 \langle r_{nn'} \rangle^2 / 3\varepsilon_0 hc^3,$$

where υ is the transition frequency and $\langle r_{nn'} \rangle$ the matrix element of the electric dipole operator between the initial n and the final state n'. For the case $n' \ll n$ one has a small matrix element for dipole transitions $\langle r_{nn'} \rangle$ owing to the small overlap of the radial wave functions for n and n', $A_{n \to n'} \sim n^{-3}$. If n' is close to n, the energy difference $E_n - E_{n'} \sim n^{-3}$ and $\langle r_{nn'} \rangle^2 \sim n^4$, and so $A_{n \to n'}$ becomes proportional to n^{-5}.

Since the matrix element $\langle r_{nn'} \rangle$ ($n \cong n'$) scales with n^2, this leads to rather high transition probabilities for induced transitions. Rydberg atoms therefore strongly absorb microwave or far-infrared radiation. As a consequence, blackbody radiation may cause strong mixing of the states. This is especially the case for states with high angular momenta since the spontaneous lifetimes for these are much larger than for low ℓ states and the induced transitions can therefore be saturated more easily.

We now wish to discuss the scaling laws related to blackbody--induced effects. The induced transition rate due to blackbody radiation is proportional to $\langle r_{nn'} \rangle^2 S_\nu$, where S_ν is the energy flux of the blackbody radiation per unit bandwidth and unit surface area. At low frequencies (Rayleigh-Jeans limit) S_ν changes as ν^2. Considering the distance between the Rydberg states to scale as n^{-3} (here again we perform the discussion with n instead of n^*), it is therefore found that S_ν is proportional to n^{-6}. Since $\langle r_{nn'} \rangle^2 \sim n^4$, it follows that the induced transition rate behaves as n^{-2}. Important in experiments is the ratio between the induced transition rate and the spontaneous rate, which changes as n^{-3} for low ℓ and as n^{-5} for high ℓ. This means that for a given atom and a given temperature there exists an n, above which the blackbody-induced rate overcomes the spontaneous rate.

The influence of blackbody radiation on Rydberg atoms was first demonstrated in lifetime measurements. For instance, Gallagher et al. [24] observed that the measured lifetimes of the 16p and 17p states of Na were three times shorter than expected; the shorter lifetime was supposed to be due to blackbody interaction. Haroche et al. [25] found a population transfer to nearby levels which could not be explained by spontaneous decay. More direct evidence of interaction with blackbody radiation was observed later [26-28].

A consequence of the long radiative decay time of Rydberg levels and the very large value of the electric dipole matrix elements is that the saturating power for transitions between closely lying Rydberg levels is very small. The corresponding saturating power fluxes are proportional to n^{-10} for low and to n^{-14} for high angular momentum states. A very vivid way of describing the behaviour is to express the saturation power flux in terms of number of photons per surface of the size λ^2 and per lifetime, (the size λ^2 corresponds to the resonant cross--section). For $n \cong 30$ one obtains 10^2 and 1 for low and high angular momentum states, respectively. This means that for high angular momentum states a single photon is required (in the chosen units) to saturate the transition to a neighbouring Rydberg level [29].

There are many applications of wide ranging importance for detectors in the submillimeter region, e.g. infrared and radio-astronomy, diagnostics of plasmas for nuclear fusion, stratospheric monitoring and materials research. The investigation of new principles for detectors is therefore as important as the further development and improvement of known detector principles. Ducas et al. [30] demonstrated the very sensitive detection of low-power far infrared laser radiation at 600 GHz by inducing transitions between Rydberg states of sodium atoms. To check the ultimate sensitivity of the Rydberg detector to microwave or far infrared radiation, two improvements of the previous experiments have to be effected. First, the population of the Rydberg atoms must be performed by continuous wave lasers in order to increase the duty cycle and, in addition, the walls of the surrounding chamber have to be cooled to a low temperature, so that the influence of the thermal background radiation is minimized. Such an experiment has been performed by Figger et al. [27] with sodium Rydberg atoms.

In a more recent experiment [34] on rubidium Rydberg atoms the environment was cooled to liquid helium temperature. Fig. 1 shows the experimental set-up. Two Rb atomic beams were excited stepwise by the light of three semiconductor diode lasers. The corresponding level scheme is shown in Fig. 2. The atomic beam on the left side of Fig. 2 was used to stabilize the frequency of the diode lasers. Lasers 1 and 2 were stabilized using the fluorescence signal of the beam, whereas laser 3, which populated the Rydberg level, was stabilized by means of the electron current produced in the field ionization region.

The Rb beam on the right side of Fig. 1 was used to study the interaction with the blackbody radiation. Fig. 3 shows the signal obtained in the field ionization region of this beam when the $40P_{3/2}$ level was populated. The blackbody radiation populates the higher levels as indicated by the arrows on the left-hand side of the Figure. The strongest contribution results from the 181 GHz transition. The $p \rightarrow d$ transitions have a transition probability about 100 times smaller than the $p \rightarrow s$ ones.

Changing the field strength of the ionizing field allowed the population of either the $43S_{1/2}$ or $42S_{1/2}$ level to be monitored. (The contribution of the 40D and 41D levels can be neglected.) The two signals are shown on the right-hand side of Fig. 3, upper and lower trace respectively. Since the induced transition rate of the $40P_{3/2} \rightarrow 42S_{1/2}$ transition is about a factor of two larger than that of $40P_{3/2} \rightarrow 43S_{1/2}$, the field ionization signal of the $42S_{1/2}$ is correspondingly larger.

The signals on the right-hand side of Fig. 3 were obtained for 77K and 300K blackbody radiation, respectively. In the first case, it was emitted by the walls of the liquid nitrogen cooled surface, and injected inside the apparatus through the waveguide shown in Fig. 1 (liquid helium cooled flap open), while in the second case both flaps

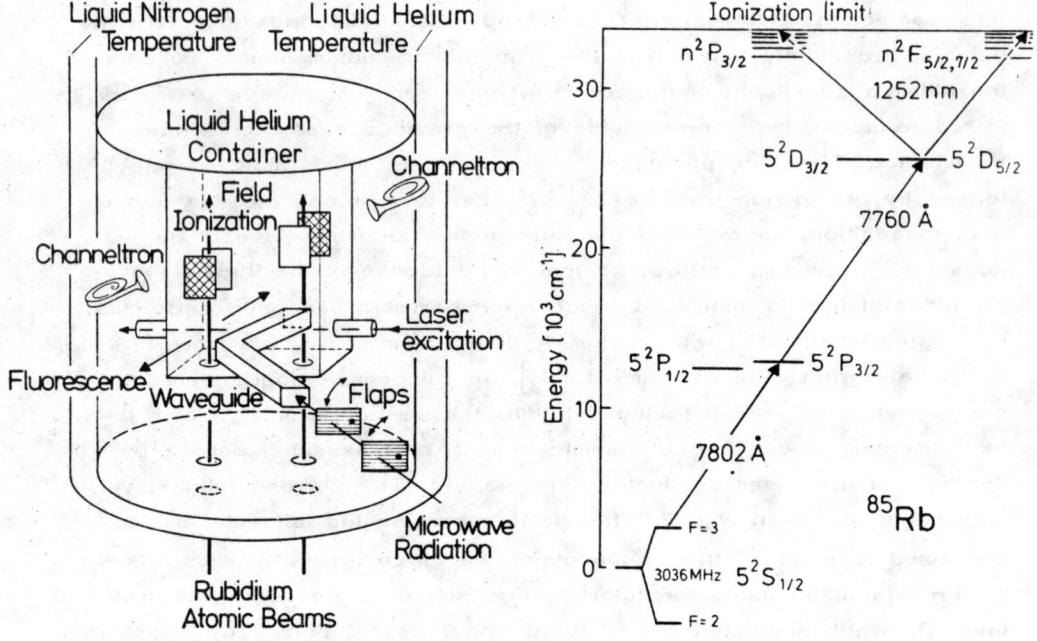

Fig. 1. Experimental setup for demonstrating the interaction of blackbody radiation with Rydberg atoms. The two atomic beams are surrounded by shieldings which are cooled to liquid nitrogen and liquid helium temperatures. The left atomic beam is used to stabilize the frequency of the semiconductor diode lasers; the right beam is used to study the interaction with blackbody radiation (see also Reference [35])

Fig. 2. Level scheme of the rubidium atom. The stepwise excitation with the semiconductor diode lasers is shown (see also Reference [35])

were open and the radiation was due to the 300K walls of the vacuum chamber. When comparing the results, it is of course important to take into account the different solid angles in both situations.

The evaluation of the data of Fig. 3 leads to a noise equivalent power (NEP) of the "Rydberg detector" of 10^{-17} W/$\sqrt{Hz}$, a value comparable to the result of our earlier experiment on sodium atoms [27]. But this new set-up clearly demonstrates that semiconductor lasers can be used to populate efficiently Rydberg levels, making the Rydberg detector a more practicable device.

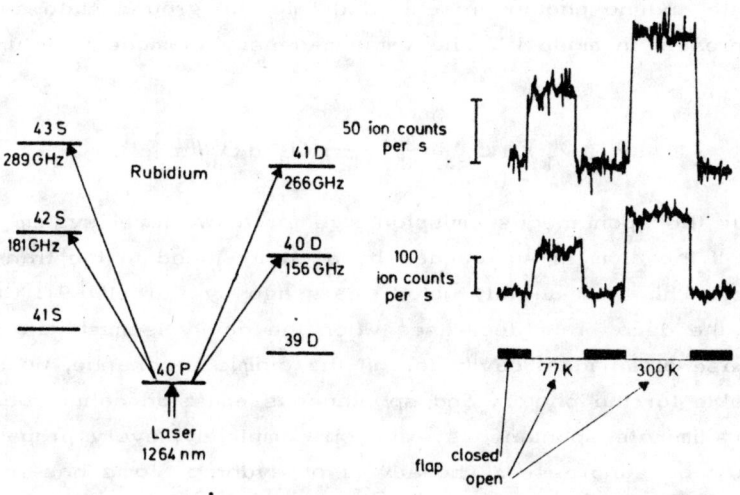

Fig. 3. Left: Rubidium Rydberg states relevant for the interaction with the blackbody radiation. The $40\,P_{3/2} - 42\,S_{1/2}$ transition has the highest transition probability. The transition probabilities for the p → d transitions are about hundred times smaller than those for p → s. The right side of the figure shows the field ionization of the $43\,S_{1/2}$ level (upper trace) and for $42\,S_{1/2}$ (lower trace), respectively (see also Reference [35])

In addition to population changes induced by the blackbody radiation, energy shifts of the atomic levels also occur. Their magnitudes depend on the match of the atomic frequencies with the blackbody frequencies and the strength of the coupling of the Rydberg atom to the blackbody radiation [31, 32, 33].

III. Cavity Quantum Electrodynamics

It is well-known that the spontaneous lifetime of an excited atom is proportional to the density of modes of the electromagnetic field $\rho(\omega_k)$ about the atomic transition frequency ω_0. Specifically, the Weisskopf-Wigner spontaneous lifetime is given by [29]

$$\gamma = 2\pi \int d\Omega_k \left[|V_{12}^k|^2 \rho(\omega_k)\right] \omega_k = \omega_0, \qquad (1)$$

where $\rho(\omega_k) d\omega_k \, d\Omega_k$ represents the number of field modes in the frequency range ω_k to $\omega_k + d\omega_k$, $d\Omega_k$ is the differential solid angle, and V_{12}^k is the dipole matrix element between the states "atom in excited state and no photon present" and "atom in ground state and one photon present in mode k". The vacuum density of modes per unit volume is

$$\rho_f(\omega_k) d\omega_k = \tfrac{1}{2}(\tfrac{1}{4\pi}) \omega_k^2 d\omega_k / \pi^2 c^3. \qquad (2)$$

Thus the spontaneous emission rate for a two-level system is increased if the atom is surrounded by a cavity tuned to the transition frequency. This was already noted years ago by Purcell [9]. Conversely, the decay rate decreases when the cavity is mistuned [10]. In the case of an ideal cavity far off the atomic resonance, no mode is available for the photon, and spontaneous emission cannot occur. In order to eliminate spontaneous emission completely, every propagating mode must be suppressed. The advent of Rydberg atoms has rendered the realisation of such experiments possible.

Consider a cubic cavity of length L and quality factor Q. Owing to the finite Q, the cavity exhibits some losses which yield a cavity linewidth $\Delta\omega = \omega/Q$. For a Lorentzian lineshape, the density of modes around a cavity mode of frequency ω_c and mode volume V_c is given by

$$\rho_c(\omega) = \frac{1}{2\pi V_c Q} \cdot \frac{\omega_c}{(\omega - \omega_c)^2 + (\omega_c/2Q)^2}. \qquad (3)$$

At resonance $\omega_0 = \omega_c$, this gives

$$\gamma_c = \gamma_f (Q/4\pi^2) \cdot (\lambda_0^3/V_c), \qquad (4)$$

where γ_c and γ_f are the spontaneous emission rates in the resonator and in vacuum, respectively.

From this equation, we draw two important conclusions: At optical frequencies, the size of the resonator is large compared to the wavelength, resulting in a small ratio λ_0^3/V_c. Since even the best available resonators, of $Q \sim 10^5$, are unable to compensate for this factor, one cannot hope to observe cavity-enhanced spontaneous decay in this regime. However, for cavities operating near their fundamental frequencies, the ratio $\lambda_0^3/V_c \sim 1$, which results typically in high enhancements in the case of high Q resonators. We conclude that when an atom is placed inside a cavity with a single mode at its transition frequency, it radiates about Q/π^2 faster than in free space. (Here, we have taken $V_c \sim \lambda_0^3$.) This enhancement of the radiation rate was first pointed out by Bloembergen and Pound [37] and is the rationale for using resonant cavities in lasers and masers.

Since a resonant cavity can enhance spontaneous emission, it is not surprising that a nonresonant cavity can depress it. Consider for instance a cavity whose fundamental frequency is at twice the resonant frequency of the atomic transition. In this case, the radiation rate becomes

$$\gamma_c = \gamma_f \cdot \rho_c(\omega_0 = 2\omega)/\rho_f(\omega) = \gamma_f/4\pi^2 Q. \quad (5)$$

In principle, γ_c can be made arbitrarily small by making Q sufficiently large.

It is important to realize that a cavity is not absolutely necessary to modify the spontaneous decay rate of an atom. Any conducting surface placed near it will affect the mode density and hence its decay rate. For instance, parallel conducting plates can somewhat alter the emission rate, but at most reduce it by a factor of 2 because of the existence of TEM modes, which are independent of the separation. The effect of conducting surfaces on the radiation rate has been studied theoretically in a number of investigations (for details see Reference [1]).

To demonstrate experimentally the modification of the spontaneous decay rate, it is not always necessary to go to single-atom densities. The experiments where the spontaneous emission is inhibited can also be performed with higher densities. However, in the opposite case where an increase of the spontaneous rate is observed, a large number of excited atoms increases the field strength in the cavity and the induced transitions disturb the experiment.

The first experimental work on inhibited spontaneous emission is due to Drexhage [6]. The fluorescence of a thin dye film near a mirror was investigated. Drexhage observed an alteration of the fluorescence lifetime arising from the interference of the molecular radiation with its surface image. An experiment with Rydberg atoms was recently performed by Vaidyanathan, Spencer and Kleppner [11]. They observed a wavelength dependent cutoff in the absorption of blackbody radiation by Rydberg atoms arising from a discontinuity in the density of modes between parallel-conducting plates. Absorption at a wavelength of 2/3 cm by atoms between planes 1/3 cm apart was measured at a temperature of 180 K. The discontinuity in the absorption rate occurred when

the absorption wavelength was varied across the cutoff of the parallel-
-plate modes. The experiment was performed with Na atoms and the
transition employed was 29d → 30p. For the tuning of the atomic resonance across the cutoff frequency a small electric field was applied to
the parallel plates. In a new experiment the inhibited spontaneous emission has been demonstrated for Rydberg atoms with electrons in circular orbits [38].

Inhibited spontaneous emission was observed clearly for the first
time by Gabrielse and Dehmelt [12]. In these experiments on a single
electron stored in a Penning trap, they observed that the cyclotron excitation shows a lifetime up to ten times larger than that calculated for
a cyclotron orbit in free space. The electrodes of the trap form a cavity which decouples the cyclotron motion from the vacuum radiation field,
leading to a longer lifetime.

The first observation of enhanced atomic spontaneous emission of
Rydberg atoms in a resonant cavity was published by Goy, Raimond,
Gross and Haroche [13]. Their experiment was performed with Rydberg
atoms of Na excited in the 23s state in a niobium superconducting cavity resonant at 340 GHz. Cavity-tuning dependent shortening of the lifetime was observed taking advantage of the very strong electric dipole of
these atoms and of the high Q value of the superconducting resonator.
This cooling, necessary for superconducting operation, also had the
advantage of totally suppressing the blackbody field contributions, a
necessary requirement to test purely spontaneous emission effects in
the cavity.

Of course, the cavity affects only the partial decay rate from the
23S to the 22P state, which in vacuum is $(\gamma_{23S \rightarrow 22P})_f = 150$ sec^{-1},
this being only a small fraction of the total decay rate $(\gamma_{23S})_f \sim 6.9 \cdot 10^4$ sec^{-1}. To make the effect observable, the probability of a cavity-
-enhanced spontaneous transition during the transit time $\Delta t \sim 2 \cdot 10^{-6}$ sec
of the atom through the cavity has to be near one, thus calling for a
quality factor of about 10^6 in view of $(\gamma_{23S \rightarrow 22P})_c \cdot \Delta t \sim .22 \cdot 10^{-6}$ Q.
This has actually almost been achieved and a spontaneous emission
rate of single atoms of $(\gamma_{23S \rightarrow 22P})_c \sim 8 \cdot 10^4$ sec^{-1} has been observed.

In spite of this enhancement of spontaneous emission, the cavity
damping rate of the radiation was still faster by a factor of 35 in this
experiment. The probability of reabsorbing the emitted photon either by
the same atom, or by a new atom entering the cavity could be neglected.

Improving further on the cavity Q, however, leads to a regime where these effects become important. This allows in particular the experimental realisation of a single-atom maser [15], and brings us for the first time to experimental conditions close to those required to test the Jaynes-Cummings model [16], which describes the simplest and at the same time most fundamental interaction between a single mode of the electromagnetic field and a single two-level atom. Experiments of this type will be described in the next section.

Now we are going to discuss the question to what extent radiation corrections like the Lamb shift can be altered if they are calculated under the restriction of a certain mode structure due to a cavity [14]. For the calculations we assume that cavity only affects the propagation of real and virtual photons leaving the atomic system unchanged. This can be justified by comparing the extension of the atomic wave function ($\cong$ 1 μm even for highly excited Rydberg states) to the size of the cavity. Under this assumption the QED corrections have to be calculated as usual but the conventional propagation function of the photon has to be replaced by that of the photon in the cavity. It is obvious that this procedure causes tiny "apparatus-dependent" deviations from the high-precision predictions of QED, as recently pointed out by Fischbach and Nakagawa [39] for the anomalous magnetic moment of the electron.

These deviations are expected to depend on both the shape and size of the surrounding cavity. It has been shown for rectangular cavities [40] that the photon propagator in ordinary space has the same form as for an unaltered vacuum; however, changes appear in the momentum space representation because of the discrete structure of the modes. Essentially, the integral over the wave vector is replaced by a sum over the mode wave vectors. For a certain high-frequency cutoff Λ, the cavity is assumed to become transparent to that the mode structure only enters for $\omega < \Lambda$, whereas beyond Λ the usual integration is performed [39]. In the case of the Lamb shift, this argument excludes cavity effects on the vacuum polarization and on the high-frequency part of the self-energy diagram for one virtual photon emission.

In the calculation of the low-frequency part of the Lamb shift only those intermediate states will contribute that accompany the emission of a virtual photon with momentum appropriate to the cavity. Since there are broadening mechanisms both for the atomic levels of the atomic system due to the natural line width and for the cavity modes due to a finite quality factor Q, the coincidences between atomic transition frequencies and eigenfrequencies of the cavity are appreciably larger than it might be expected. A special case is the waveguide for which the mode density approaches that of the undisturbed vacuum just above its fundamental cutoff. Therefore the calculation of the modification of the Lamb shift is especially simple for this case and our calculation was restricted to this situation. That means that in the Lamb shift calculation those terms were omitted corresponding to energies $\hbar\omega$ of the virtual photons below the cutoff frequency ω_c of the waveguide. It should be noted that the mass renormalization counterterm is also changed owing to the cavity, however, this correction is the same for all levels with the same main quantum number.

The results for the change of the Lamb shift of s-states $\delta E_{n,0}$ due to the model waveguide outlined above are plotted in Fig. 4 versus n and the cutoff wavelength $\lambda_c = 2\pi c/\omega_c$. Some points are worth men-

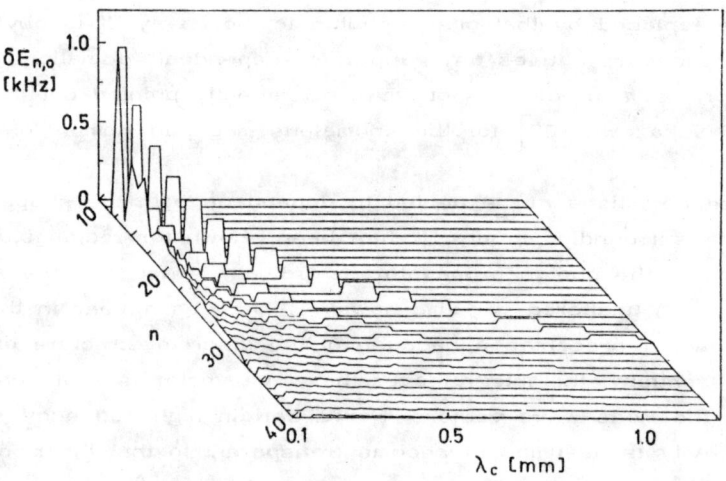

Fig. 4. Absolute change of the Lamb shift of hydrogen s-states as a function of principal quantum number n and cut-off wavelength λ_c [14]

tioning. The overall effect is extremely small, even for very short cutoff wavelengths. For a fixed λ_c, the largest $\delta E_{n,0}$ occurs for the state with the smallest quantum number n^* that is just above the threshold for a suppressed transition to the $(n^*+1)p$ state. As λ_c is decreased, also n^* gets smaller. For values $n > n^*$, two facts tend to decrease $\delta E_{n,0}$: First, the general n^{-3} behaviour of the Lamb shift itself; second, there may be virtual cancellation of contributions to $\delta E_{n,0}$ resulting from transition to neighbouring states as e.g. $n^*s \rightarrow (n^* + 1)p$ and $n^*s \rightarrow (n^* - 1)p$ since they enter with opposite signs.

Since the change of the Lamb shift of an hydrogen atom in a waveguide shows a strong variation with the main quantum number n, it is not necessary to perform a measurement in the optical region in order to demonstrate the phenomenon experimentally. It has been shown in ref. [14] that the change of the Lamb shift can be measured by comparing transitions of the kind $n^*s \rightarrow (n^* + 1)p$ with $(n^* + 2)s \rightarrow (n^* + 1)p$. To achieve the necessary precision the Ramsey method with two spatially separated microwave fields [41] has to be used. In the space between the microwave fields the atoms are moving between the conducting plates of a waveguide (for details see ref. [14]).

As mentioned above Rydberg atom experiments in high Q superconducting cavities approach the conditions required to test the Jaynes--Cummings 16-18, 42-46 model of interaction between a single two--level atom and a single mode of the electromagnetic field. This model is amenable to an exact analytical solution, and has been the object of extensive theoretical analysis. A number of purely quantum-mechanical effects, whose experimental verification would be of considerable interest, have been predicted. These include the so-called Cummings collapse in the evolution of the population inversion [16, 17, 42], as well as its partial revival for longer interaction times [18]. The Jaynes-Cummings theory will not be discussed in this contribution since it is the subject of two other reviews in this volume [45, 46]. In the following the experiments performed with the single atom maser will be discussed.

IV. The Single-Atom Maser

For the Rydberg maser experiment [15] an atomic beam of highly excited Rydberg atoms was used. The beam oven is carefully heat--shielded from the cavity by copper plates cooled by water, liquid nitrogen and liquid helium; the atoms pass through small apertures into the liquid helium cooled part of the apparatus. There they are excited to the upper maser level and enter the cavity. Behind the cavity the atoms in Rydberg states were monitored by field ionization (see Fig. 5).

The Rydberg states were populated using the frequency doubled radiation of a continuous wave ring dye laser. The constant stream of Rydberg atoms is ionized in an inhomogeneous dc electric field of a

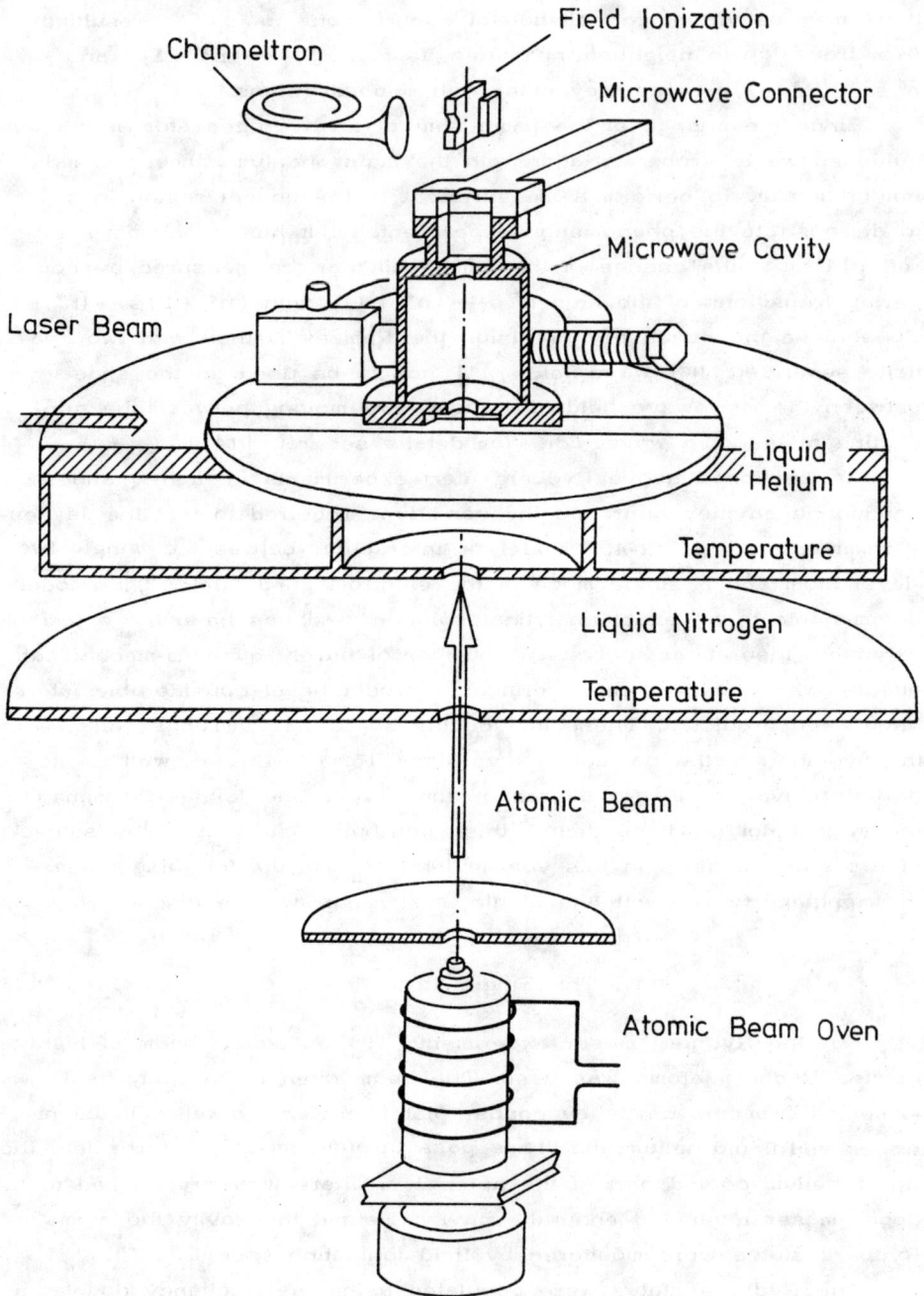

Fig. 5. Vacuum chamber with the atomic-beam arrangement and the microwave cavity. The upper part is cooled to liquid-helium temperature (see also Reference [15])

plate capacitor. The atoms attain the point of maximum field strength in front of a hole in the anode through which the ejected electrons pass and reach a channeltron multiplier. If the field strength is adjusted properly mainly the atoms in the upper maser level are monitored. Transitions from the initially prepared state to the lower maser level are thus detected by a reduction in the electron count rate. The cylindrical cavity (diameter 24,8 mm, length 24 mm) was manufactured from pure niobium rods. It is enclosed in a cryoperm shield to reduce the influence of ambient magnetic fields on the Q value due to frozen-in flux. The temperature of the cavity could be varied from 4.3 to 2.0 K, corresponding to Q factors of 1.7×10^7 and 8×10^8, respectively. The atomic beam passes through the cylindrical cavity along its axis, where only the TE_{1np} and TM_{1np} modes possess a nonvanishing transversal electric field. For our experiment the TE_{121} mode was used. This mode has a plane field distribution and is doubly degenerate in an ideal cylindrical cavity. The degeneracy is removed by a slight deformation of the circular cross section into an oval shape, which then determines the direction of polarization of the field mode. The deformation is achieved by squeezing the cylinder with a screw and a piezoelectric transducer for fine tuning (0.5 MHz/1500 V). The upper maser level was the $63p_{3/2}$ level of ^{85}Rb. The fine structure splitting between $63p_{3/2}$ and $63p_{1/2}$ amounts to 396 MHz (see Fig. 6). It is, therefore no problem to excite a single fine structure level with the narrow-band ultraviolet radiation ($\Delta\nu \cong 2$ MHz).

To demonstrate maser operation, the cavity was tuned over the $63p_{3/2}$ - $61d_{3/2}$ transition by changing the voltage of the piezoelectric transducer; the field ionization was recorded simultaneously. Transitions from the initially prepared $63p_{3/2}$ state to the $61d_{3/2}$ level (21.50658 GHz) are detected by reduction of the electron count rate.

In a recent experiment also the transition $63p_{5/2}$ - $61d_{5/2}$ at 21.456 GHz (Fig. 6) was observed in a single atom maser. The transition matrix element for this transition is

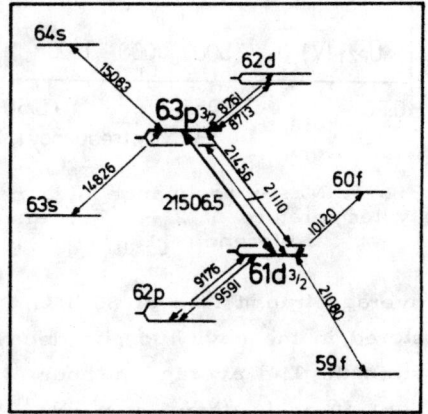

Fig. 6. Rubidium level scheme with the maser transition

about 7 times larger than that for $63p_{3/2} - 61d_{3/2}$. For this transition the quantum revival was observed. The results will be published elsewhere.

In the case of measurements at a cavity temperature of 2K [15], a reduction of the $63p_{3/2}$ signal could be clearly seen for atomic fluxes as small as 800 atoms/s. An increase in flux caused power broadening and finally an asymmetry and a small shift (Fig. 7). This shift is attributed to the ac Stark effect, caused predominantly by virtual transitions to the $61d_{5/2}$ level, which is only 50 MHz away from the maser transition (Fig. 6). The fact that the field ionization signal at resonance is independent of the particle flux (between 800 and 22×10^3 atoms/s) indicates that the transition is saturated. This, and the observed power broadening show that there is a multiple exchange of photons between Rydberg atoms and the cavity field.

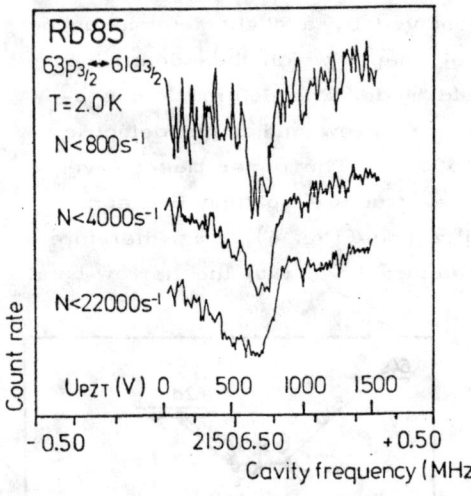

Fig. 7. Maser resonance at a cavity temperature of 2 K (see also Reference [15])

With an average transit time of the Rydberg atoms through the cavity of 80 μs one calculates for a flux of 800 atoms/s a probability of 0.06 of finding a Rydberg atom in the cavity. According to Poisson statistics this implies that more than 99% of the events are single atoms. This clearly demonstrates that single atoms are able to maintain continuous oscillation of the cavity.

Since the transition is saturated, half of the atoms initially excited in the $63p_{3/2}$ state leave the cavity in the lower $61d_{3/2}$ maser level. The decay to other levels can be neglected for the average transit time of 80 μs. The energy radiated by those atoms is stored in the cavity for its decay time, increasing the average field strength. The average number of photons left in the cavity by the Rydberg atoms is given by $n = \tau_d N/2$ where τ_d is the characteristic decay time of the cavity and N the number of Rydberg atoms entering the cavity in the upper maser level per unit time. For the highest particle flux used in our experiment $N = 22 \times 10^3$ atoms/s, one finds $n \cong 55$

photons at 2 K ($\tau_d \cong$ 5 ms) and n $\cong$ 1.4 photons at 4.3 K ($\tau_d \cong$ .13 ms). This last value is smaller than the average number of blackbody photons $n_{b\ell} \cong$ 4 at 4.3 K. For N $\cong$ 800 atoms/s one obtains n $\cong$ 2 at 2 K, which means that the radiation generated by the Rydberg atoms in the cavity has about the same energy as the blackbody radiation ($n_{b\ell} \cong$ = 1.5).

When the squares of the halfwidth $\Delta\nu$ of the signal curves are plotted versus the Rydberg atom flux, a straight line is obtained as expected. This line intersects the $(\Delta\nu)^2$ axis at a finite value, from which the number of blackbody photons originally in the cavity can be evaluated. The result (3 $\pm$ 1) is in reasonable agreement with the value given above. It follows that as the atomic flux decreases the thermal radiation becomes the dominant part of the field [15].

The experimental setup described above is suitable to test the Jaynes-Cummings model describing the dynamics of the interaction of a single atom with a single cavity mode. An important requirement is, however, that the atoms of the beam have a homogeneous velocity so that it is possible to observe the Rabi nutation in the cavity directly. In a modified setup a Fizeau type velocity selector is inserted between atomic beam oven and cavity, so that a fixed atom-field interaction time is obtained. Changing the selected velocity leads to a different interaction time and leaves the atom in another phase of the Rabi cycle when it reaches the detector.

Since the cavity contains thermal photons the transient behaviour has to be described by a sum of elementary Rabi oscillations in a field in which the number of photons is a random quantity following the Bose--Einstein statistics. The distribution of Rabi frequencies results in an apparent random oscillation which for larger n values collapses very quickly and then revives again. This behaviour is typical of a chaotic quantum field; a semiclassical description of a random field does not give this result. Fig. 8 shows the Rabi nutation for a simple atom entering the cavity in the excited state for T = 0 K and T = 3 K. The single photon Rabi frequency assumed is 10 kHz. The cavity contains zero thermal photons at T = 0 K and 2.5 thermal photons on the average at T = 3 K. Plotted is the probability P for the population of the upper (P = 1) and the lower maser level (P = 0). In a thermal field the time dependence of P is given by

$$P(t) = \tfrac{1}{2}(1 - \exp(-h\nu/kT))\sum_n \exp(-\tfrac{h\nu n}{kT})(1 + \cos(2\Omega\sqrt{n+1}\,t)).$$

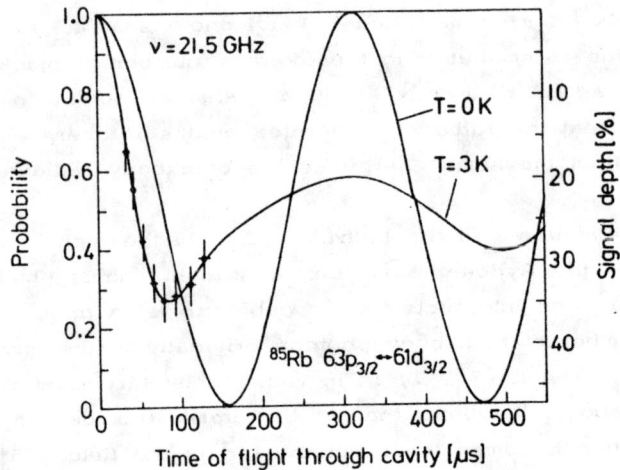

Fig. 8. Rabi-nutation with and without a thermal field. The crosses on the T = 3 K curve have been obtained with a velocity selected beam. The velocity distribution of the atomic beam only allows to measure for interaction times in the cavity between 30 and 140 μs

The crosses on the T = 3 K curve are obtained in the experiment. For these measurements the cavity temperature was raised to 3 K in order to have more thermal photons (2.5) in the cavity. The agreement between theory and experiment is excellent.

For the transition $63p_{3/2}$ - $61d_{5/2}$ which has a larger transition matrix element as $63p_{3/2}$ - $61d_{3/2}$ and therefore a larger Rabi frequency also a revival could be observed so that it has been demonstrated that the described apparatus is suitable to test the QED in a cavity in detail [47].

References

1. S. Haroche, J. M. Raimond: Advances in Atomic and Molecular Physics, Vol. 20, Eds. D. Bates and B. Bederson, pp. 350-411 (Academic Press, New York 1985).
2. J. A. Gallas, G. Leuchs, H. Walther, H. Figger: Advances in Atomic and Molecular Physics, Vol. 20, Eds. D. Bates and B. Bederson, pp. 413-466 (Academic Press, New York 1985).
3. V. B. Berestetskii, E. M. Lifshitz, and L. P. Pitaevskii: "Quantum Electrodynamics" (Pergamon Press, Oxford 1982).

4. G. Barton: Proc. Roy. Soc. London A320, 251 (1970).
5. P. Stehle: Phys. Rev., A2, 102 (1970).
6. K. H. Drexhage: Progress in Optics, Vol. 12, Ed. by E. Wolf (North Holland, Amsterdam 1974).
7. P. W. Milonni, and P. L. Knight: Opt. Commun., 9, 119 (1973).
8. E. A. Power, and T. Thirunamachandran: Phys. Rev., A25, 2473 (1982).
9. E. M. Purcell: Phys. Rev., 69, 681 (1946).
10. D. Kleppner: Phys. Rev. Lett., 47, 233 (1981).
11. A. G. Vaidyanathan, W. P. Spencer, and D. Kleppner: Phys. Rev. Lett., 47, 1592 (1981).
12. G. Gabrielse, H. Dehmelt: (to be published).
    G. Gabrielse, R. van Dyck, Jr., J. Schwinberg, H. Dehmelt: Bull. Am. Phys. Soc. 29, 926 (1984).
13. P. Goy, J. D. Raimond, M. Gross, S. Haroche: Phys. Rev. Lett., 50, 1903 (1983).
14. P. Dobiasch, and H. Walther: Annales No6 - Alfred Kastler Symposium (Editions de Physique 1985), in print.
15. D. Meschede, H. Walther, G. Müller: Phys. Rev. Lett., 54, 551 (1985).
16. E. T. Jaynes, and F. W. Cummings: Proc. IEEE 51, 89 (1963).
17. P. Meystre: PhD Thesis, Ecole Polytechnique Fédérale Lausanne (1974); P. Meystre, E. Geneux, A. Quattropani, A. Faist, Nuovo Cimento, 25B, 521 (1975).
18. J. H. Eberly, N. B. Narozhny, J. J. Sanchez-Mondragon: Phys. Rev. Lett., 44, 1323 (1980), and references therein.
19. S. Feneuille, P. Jacquinot, Adv. Atom. Mol. Phys., 17, 99 (1981).
20. D. Kleppner, in "Laser-Plasma Interactions", Les Houches XXXVI, edited by R. Balian, pp. 733-784, North-Holland, Amsterdam 1982.
21. D. Kleppner, M. G. Littman, M. L. Zimmerman, in "Rydberg States of Atoms and Molecules" edited by R. F. Stebbings and F. B. Dunning, pp. 73-116, Cambridge University Press, Cambridge 1983.
22. D. Delade, J. C. Gay, Comments At. Mol. Phys., 13, 275 (1983).
23. D. R. Bates, A. Damgaard, Philos. Trans. Roy. Soc. London, 242, 101 (1949).
24. T.F. Gallagher, and W. E. Cooke, Phys. Rev. Lett., 42, 835 (1979).
25. S. Haroche, C. Fabre, P. Goy, M. Gross, and J. M. Raimond, in "Laser Spectroscopy IV", edited by H. Walther, and K. W. Rothe, Springer Series in Optical Sciences, Vol. 21, Springer Verlag, Berlin, Heidelberg, New York 1979.

26. E. J. Beiting, G. F. Hildebrandt, F. G. Kellert, G. W. Foltz, K. A. Smith, F. B. Dunning, and R. F. Stebbings, J. Chem. Phys., 70, 3551 (1971).
27. H. Figger, G. Leuchs, R. Straubinger, H. Walther, Opt. Comm., 33, 37 (1980).
28. P. R. Koch, H. Hieronymus, A. F. J. Van Raan, W. Raith, Physics Letters, 75A, 273 (1980).
29. S. Haroche, in "Atomic Physics 7", edited by D. Kleppner, and F. M. Pipkin, Plenum Press, New York, London 1981, p. 141.
30. T. W. Ducas, W. P. Spencer, A. G. Vaidyanathan, W. H. Hamilton, and D. Kleppner, Appl. Phys. Lett., 35, 382 (1979).
31. W. E. Cooke, T. F. Gallagher, Phys. Rev., A21, 588 (1980).
32. T. F. Gallagher, W. Sandner, K. A. Safinya, W. E. Cooke, Phys. Rev., A23, 2065 (1981).
33. J. W. Farley, W. H. Wing, Phys. Rev., A23, 2397 (1981).
34. G. Rempe, H. Walther 1985 (unpublished).
35. P. Filipowicz, P. Meystre, G. Rempe, H. Walther, Optica Acta (in print).
36. V. Weisskopf and E. Wigner, Z. für Phys., 63, 54 (1930).
37. N. Bloembergen and R. V. Pound, Phys. Rev., 95, 8 (1954).
38. R. G. Hulet and D. Kleppner 1985 (to be published).
39. E. Fischbach, and N. Nakagawa: Phys. Lett., 149B, 504 (1984).
40. E. Ledinegg, Acta Phys. Austr., 51, 85 (1979).
41. N. F. Ramsey: Phys. Rev., 76, 996 (1949).
42. T. von Foerster: J. Phys., A8, 95 (1975).
43. S. Stenholm: Phys. Rep., 6, 1 (1975).
44. P. L. Knight, and P. W. Milonni: Phys. Rev., C66, 21 (1980).
45. P. Filipowicz, J. Javanainen, P. Meystre (to be published).
46. J. H. Eberly, this volume.
47. G. Rempe, H. Walther, N. Klein, (publication in preparation).

Z. BIAŁYNICKA-BIRULA

Institute of Physics,
Polish Academy of Sciences,
al. Lotników 46, 02-668 Warsaw,
Poland

IONIZATION OF ATOMS BY HIGH-INTENSITY LASERS: COLLECTIVE EFFECTS

Recent experiments on the multiple ionization of atoms by high-intensity lasers have yielded unexpected results. They have clearly demonstrated that the standard theory of laser spectroscopy was incapable, with discrepancy as high as several orders of magnitude, of describing the behavior of many-electron atoms irradiated by intense laser light. The ≫ intense light ≪ means here the light whose electric field is of the order of the atomic field E_A ($E_A = e/a_0^2 = 5 \times 10^9$ V/cm). It corresponds to the light intensity of the order of 10^{17} W/cm^2. Most experiments on multiple ionization in this regime were performed in Chicago by the group headed by C. K. Rhodes [1].

In these experiments, the following class of physical processes was studied:

$$n\gamma + A \rightarrow A^{q+} + qe^-. \qquad (1)$$

Here, n is the number of absorbed photons, which for heavy atoms is of the order of 30-80, up to 99 for uranium. The excimer ArF* laser was used that gives ∼5psec, ∼3 GW pulses of 193 nm radiation. The corresponding single photon energy is $\hbar\omega_L$ ∼6 eV. The laser pulses were focused to give the intensities $10^{16} - 10^{17}$ W/cm^2. These laser pulses were used to irradiate dilute atomic* vapours of different ele-

---

* In the case of uranium the gas of UF_6 molecules was used.

ments (He, Ne, Ar, Kr, I, Xe, Eu, Yb, Hg, and U) under the collision-
-free conditions. Ions with different electric charge multiplicity q were
observed. Charge multiplicities detected for different atoms varied from
1 to 10.

The experiments have shown an unusually strong coupling between
heavier atoms and the radiation that leads to a large energy transfer.
High charge multiplicities were observed for heavier atoms. For example, such ions as Ar^{6+}, Kr^{6+}, Xe^{8+} and U^{10+} were detected, whose
production requires the absorption of the energy from 500 eV for Ar^{6+}
to 633 eV for U^{10+}. The average energy transferred from the light to
an atom shows strong dependence on the atomic number Z. As a general rule, it grows with growing Z, but it also shows drastic differences
for adjacent elements. For example, the average energy transfer per
atom for xenon (Z = 54) is about 4 times larger than that for iodine
(Z = 53). In conclusions of their paper [1], the authors of the experiments suggested that "the experimental evidence points to a collective
coherent atomic motion involving several electrons, possibly an entire
shell, as the main physical mechanism enabling the scale of the energy
transfers to be seen".

The standard theory used until now to describe multiphoton ionization of atoms by laser light is based on the assumption of the single-
-electron coupling to the light field. Collective excitations in atoms were
studied over the last 20 years by many authors [2], but only in relation to high-energy spectroscopy that uses short waves (from the X-ray
region), but not very intense sources. The biggest success of this
approach was the prediction, confirmed by experiments, of the "giant
dipole resonance" in the cross-section for ionization. For example, such
a resonance for Xe lies at about 100 eV. However, this theory is restricted to the weak-field (linear response) regime and in its present
form is uncapable of describing excitations induced by strong lasers.

The exact theory of the many-electron excitations induced by intense laser field would require solving the many-body Schrödinger
equation without resorting to perturbation theory. Therefore, it has not
yet been used to give any quantitative results. Under these circumstances, simple models were proposed to explain the experimental data.
The fact, that many electrons in the atom are involved in the process
of an excitation by a laser was accounted for in statistical models of
M. Crance [3] and S. Geltman [4]. In these models, electrons from
a given atomic shell are treated as independent, undistinguishable par-

ticles. Geltman used simple combinatorics to express all probabilities of observed multiple ionizations by phenomenological probabilities to eject one electron from any given shell. He then fitted his formulas to the experimental data. His fit to the Rhodes' data for relative abundances of xenon charge states Xe^{q+} produced by laser of various intensities is qualitatively good, but shows significant deviations from experimental points. In view of a preliminary character of the experimental data, it is difficult to judge the success of this fit and to decide whether the correlations and interactions between electrons play a significant role. The exact machanism of many-electron multiphoton atomic excitations is still unknown. In particular, the role of the repulsive forces between the electrons has not been clarifield.

To gain at least a partial understanding of collective excitations in atoms we [5] developed an exactly soluble model that has all the features of the realistic situation (such as the electron-nucleus attraction as well as the electron-electron repulsion, Fermi-Dirac statistics and the minimal electromagnetic coupling to radiation) with the only exception that all coulombic forces are replaced by harmonic forces.

The model is formulated in the language of second quantization which is best suited for handling collective phenomena. In this formulation, the basic objects are the operators $\psi_\sigma(\vec{r}, t)$ and $\psi_\sigma^\dagger(\vec{r}, t)$ that annihilate and create an electron at the point $(\vec{r}, t)$ with the spin projection σ ($\sigma = 1/2, -1/2$). These operators obey the following anticommutation relations (because of the Fermi-Dirac statistics),

$$\left[\psi_\sigma(\vec{r}, t), \psi_{\sigma'}^\dagger(\vec{r}', t) \right]_+ = \delta_{\sigma\sigma'} \delta(\vec{r} - \vec{r}'). \tag{2}$$

The operator $\psi_\sigma(\vec{r}, t)$ obeys the Heisenberg equation of motion ($\hbar = 1$),

$$i\frac{\partial}{\partial t} \psi_\sigma(\vec{r}, t) = \left[\psi_\sigma(\vec{r}, t), H_A + H_I \right], \tag{3}$$

where the atomic Hamiltonian H_A has the following form:

$$H_A = 1/2 \sum_\sigma \int d^3r \, \psi_\sigma^\dagger(\vec{r}, t) \left[\vec{p}^2/m + m\Omega^2 \vec{r}^2 \right] \psi_\sigma(\vec{r}, t)$$
$$-1/4 m\Omega_1^2 \sum_{\sigma\sigma'} \int d^3r\, d^3r' \, \psi_\sigma^\dagger(\vec{r}, t) \psi_\sigma(\vec{r}, t) (\vec{r} - \vec{r}')^2 \psi_{\sigma'}^\dagger(\vec{r}', t) \psi_{\sigma'}(\vec{r}', t). \tag{4}$$

Here, m is the electron mass and $\vec{p} = -i\vec{\nabla}$.

The first part of this Hamiltonian describes the motion of electrons in an external potential of an isotropic harmonic oscillator which simulates the Coulomb potential of the nucleus. The second part describes repulsive forces between the electrons, replaced also by harmonic forces. Owing to these modifications, the Hamiltonian H_A can be exactly expressed in terms of a few collective operators. To this end we define:

$$N = \int \psi_\sigma^\dagger (\vec{r}, t) \psi_\sigma (\vec{r}, t), \qquad (5a)$$

$$\vec{R}(t) = \int \psi_\sigma^\dagger (\vec{r}, t) \vec{r} \psi_\sigma (\vec{r}, t), \qquad (5b)$$

$$\vec{P}(t) = \int \psi_\sigma^\dagger (\vec{r}, t) \vec{p} \psi_\sigma (\vec{r}, t), \qquad (5c)$$

$$U_{ij}(t) = 1/2 \int \psi_\sigma^\dagger (\vec{r}, t) r_i r_j \psi_\sigma (\vec{r}, t) - \frac{1}{2N} R_i R_j, \qquad (5d)$$

$$T_{ij}(t) = 1/2 \int \psi_\sigma^\dagger (\vec{r}, t) p_i p_j \psi_\sigma (\vec{r}, t) - \frac{1}{2N} P_i P_j, \qquad (5e)$$

$$W_{ij}(t) = 1/4 \int \psi_\sigma^\dagger (\vec{r}, t) (r_i p_j + p_j r_i) \psi_\sigma (\vec{r}, t) - \frac{1}{4N} (R_i P_j + P_j R_i). \qquad (5f)$$

Here, the integral sign means both the summation over σ and the integration over $\vec{r}$.

The operator N is the operator of the total number of the electrons. It is a constant of motion. Operators $\vec{R}(t)$ and $\vec{P}(t)$ represent the position of the center of mass and the total momentum of the atomic electrons. The remaining operators represent the quadrupole degrees of freedom of the electron cloud.

The Hamiltonian H_A has the following form when expressed in terms of collective operators,

$$H_A = \frac{1}{2mN} \left[P^2 + (m\Omega R)^2 \right] + \frac{1}{m}(T + \kappa U) = H_D + H_Q, \qquad (6)$$

where

$$\kappa = (m\omega/2)^2$$

and $\qquad (7)$

$$\omega = 2\sqrt{\Omega^2 - N\Omega_1^2}.$$

The traces of operators U_{ij}, T_{ij} and W_{ij} have been denoted by U, T and W, respectively.

Since $\vec{R}$ and $\vec{P}$ commute with the remaining collective operators, two parts of the atomic system, described by H_D and H_Q respectively, in the absence of an interaction with an external field evolve independently. The "giant dipole" represented by $\vec{R}$ and $\vec{P}$ oscillates with the frequency Ω, while the quadrupole operators U, T and W oscillate with a much smaller frequency ω. The softening of the frequency of the quadrupole oscillations is á result of the presence of the electron-electron repulsion. As seen in Eq. (7) the quadrupole frequency ω results from substracting two large quantities. In the absence of electronic repulsion it would be twice the dipole frequency Ω.

The interaction of the model atom with the laser light, represented by a monochromatic, linearly polarized plane wave, is assumed in the form of the minimal electromagnetic coupling.

$$H_A + H_I = 1/2 \sum_\sigma \int d^3 r\, \psi_\sigma^\dagger(\vec{r}, t) \left[(\vec{p} - \frac{ea\vec{\varepsilon}}{c} \sin(\omega_L t - \vec{k}\cdot\vec{r}))^2/m + m\Omega^2 \vec{r}^2 \right] \psi_\sigma(\vec{r}, t)$$
$$- 1/4 m\Omega_1^2 \sum_{\sigma\sigma'} \int d^3 r d^3 r'\, \psi_\sigma^\dagger(\vec{r}, t)\, \psi_\sigma(\vec{r}, t)\,(\vec{r}-\vec{r}')^2\, \psi_{\sigma'}^\dagger(\vec{r}', t)\, \psi_{\sigma'}(\vec{r}', t), \quad (8)$$

where ω_L, $\vec{k}$ and $\vec{\varepsilon}$ are the frequency, the wave vector and the polarization vector of the wave.

In the dipole approximation,

$$H_I = -\frac{ea}{mc} \vec{\varepsilon}\cdot\vec{P} \sin\omega_L t + \frac{e^2 a^2}{2mc^2} N \sin^2\omega_L t. \quad (9)$$

This interaction excites only the "giant dipole" of the atom, leaving other collective degrees of freedom unaffected. Since the "giant dipole resonances" in real atoms occur at much higher frequencies than the optical or UV frequencies used in the experiments described at the beginning of this lecture, the interaction in the dipole approximation will be of no interest to us.

The interaction Hamiltonian H_I up to the quadrupole approximation has the form:

$$H_I = -\frac{ea}{mc}\vec{\varepsilon}\cdot\vec{P}\sin\omega_L t + \frac{e^2 a^2}{2mc^2} N\sin^2\omega_L t - \frac{e^2 a^2}{2mc^2}\vec{k}\cdot\vec{R}\sin 2\omega_L t +$$
$$+ \frac{ea}{mc}\frac{1}{N}(\vec{k}\cdot\vec{R})(\vec{\varepsilon}\cdot\vec{P})\cos\omega_L t + \frac{2ea}{mc}\vec{k}\cdot\hat{W}\cdot\vec{\varepsilon}\cos\omega_L t. \quad (10)$$

Only the last term of this Hamiltonian affects the quadrupole degrees of freedom, and is of interest to us. We will fix the coordinate system by assuming that the wave propagates in the z-direction and its polarization vector is in the y-direction. Then, the last term in the Hamiltonian H_I can be written as $2g(t)W_{zy}$, where

$$g(t) = \frac{e\mathcal{E}}{mc} \cos \omega_L t \tag{11}$$

and $\mathcal{E}$ is an amplitude of the electric field of the wave.

The quadrupole degrees of freedom still evolve independently of the dipole. Under the influence of the light wave, the atomic energy H_Q associated with the quadrupole degrees of freedom will change in time. To study this time-evolution, one has to solve the set of coupled equations of motion for various components of the quadrupole operators U_{ij}, T_{ij} and W_{ij}. Fortunately, this large set splits into two sets of three equations for the following linear combinations,

$$K_{\pm} = \frac{2}{m\omega}(-T_{zy} + \kappa U_{zy}) \pm (W_{zz} - W_{yy}), \tag{12a}$$

$$L_{\pm} = \frac{2}{m\omega}(T_{yy} + \kappa U_{zz}) \mp 2W_{zy}, \tag{12b}$$

$$M_{\pm} = \frac{1}{m\omega}\left[-(T_{yy} - T_{zz}) + \kappa(U_{yy} - U_{zz})\right] \pm (W_{zy} + W_{yz}). \tag{12c}$$

Equations for K_+, L_+ and M_+ have the similar form as equations for K_-, L_- and M_-; we shall write them together in a compact form.

$$\dot{K}_{\pm} = \pm \omega M_{\pm} + g(t)L_{\pm}, \tag{13a}$$

$$\dot{L}_{\pm} = \pm \omega K_{\pm}, \tag{13b}$$

$$\dot{M}_{\pm} = \mp \omega K_{\pm} + g(t)K_{\pm}. \tag{13c}$$

The rate of change of the atomic quadrupole energy is:

$$\frac{dH_Q}{dt} = \frac{\omega}{2} g(t)(K_+ + K_-). \tag{14}$$

Equations (13) exhibit a parametric coupling between the atom and the wave field. This parametric coupling introduces field dependent modulations of the quadrupole oscillations. The amplitude and the period of the modulation depend on the field intensity and the detuning. However, there is a region in the wave-frequency and wave-amplitude plane, in the vicinity of the quadrupole frequency ω, where parametric resonance occurs. In this region oscillatory motions change their character and lead to an exponential growth. When parametric resonance occurs, the atomic energy grown exponentially. The figure shows the time dependence of the atomic quadrupole energy, obtained by a numerical integration of Eqs. (13). We assumed that the light field was strong, i.e. that the analog of the Rabi frequency ω_R,

$$\omega_R = \frac{e\mathcal{E}}{mc}, \qquad (15)$$

was 10% of the laser frequency

$$\omega_R/\omega_L = 0.1 \qquad (16)$$

The initial state of the atom was assumed to be spherically symmetric

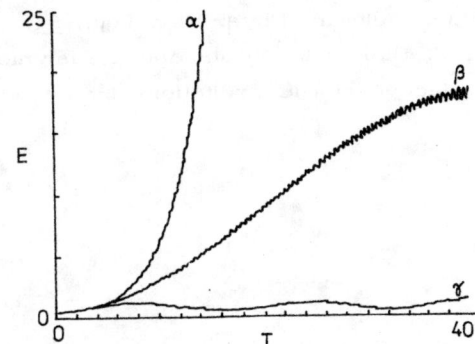

Atomic quadrupole energy E in units of single-photon energy $\hbar\omega_L$ as function of time T. The wave period is used as the time unit.
$\alpha: \Delta = 0, \quad \beta: \Delta = 0.52\,\omega_R,$
$\gamma: \Delta = 0.8\,\omega_R$

$(K_\pm(0) = 0 = M_\pm(0)$ and $L_+(0) = L_-(0))$. The three curves differ in the value of the detuning Δ,

$$\Delta = \omega_L - \omega. \qquad (17)$$

The curve γ represents the behavior far from the resonance region. The energy oscillates rather mildly; few photons are exchanged between the wave and the atom. As ω_L approaches the region of the parametric resonance, as is the case for the curve β, both the period and the amplitude of modulation increase dramatically. Finally, at the resonance, the energy grows exponentially with time. After only 15 wave periods the atom absorbs as many as 25 photons.

Thus, we see that the energy transferred from the light wave to the atom through the collective motion of many electrons can be very large for those atoms whose characteristic quadrupole frequencies are close to the laser frequency. That can happen for real atoms irradiated by UV laser, due to the softening of the quadrupole frequency owing to the electronic repulsion.

References

1. T. S. Luk, U. Johann, H. Egger, H. Pummer and C. K. Rhodes, Phys. Rev., A32, 214 (1985).
2. For a recent review and a bibliography on this subject see G. Wendin, Application of many-body problems to atomic physics in: New trends in atomic physics, edited by G. Grynberg and R. Stora (North-Holland, Amsterdam 1984).
3. M. Crance, J. Phys. B: At. Mol. Phys., 17, L355 and 3503 (1984), 18, L155 (1985).
4. S. Geltman, Phys. Rev. Lett., 54, 1909 (1985).
5. Z. Białynicka-Birula and I. Białynicki-Birula, Model studies of collective atomic excitations by intense laser fields, (to be published).

E. G. PESTOV

P. N. Lebedev Physical Institute,
USSR Academy of Sciences Moscow,
Leninsky Prospect 53

RELAXATION OF QUANTUM SYSTEMS IN STRONG ELECTROMAGNETIC FIELD-NEW NONLINEAR EFFECTS

I. Introduction

The strong electromagnetic field (EMF) has a substantial influence on the relaxation of quantum systems (QS) in the interaction with a thermostat and electromagnetic vacuum. The relaxation parameters become dependent on the intensity of the external EMF. Besides, the relaxation scheme changes, namely, there appears a new correlation between the density matrix elements in the quantum system, which is absent in a standard Bloch equation. This correlation is proportional to the EMF amplitude of intensity.

Argyres and Kelley [1] were the first who paid attention to this fact, when they investigated the relaxation of spin systems in 1964. We also treated this fact in [2], in frames of the atomic collisions theory, where we predicted the effect of the field narrowing of spectral lines in the radiation of gaseous atoms in strong resonant EMF.

In 1975 Lisitsa and Yakovlenko [3] predicted the decreasing of absorption of the gaseous medium in the quasistatical region in the strong field, which follows from the effect of the field narrowing. These effects were later confirmed experimentally by Bonch-Bruevich and the coworkers [4]. Note that the first experimental observation of the dependence of atomic scattering cross-section on the EMF intensity has been made by Szöke et al. [5] in 1977 when they investigated the spectra of the resonant fluorescence.

Much attention has been paid of late to the influence of the strong EMF on the relaxation processes. Serious investigations by Cooper et al., Apanasevich et al., Eberly et al., Cohen-Tannoudji and others were devoted to this problem. For the lack of time we are unable to discuss them now in detail. At present, the Keldysh diagram method [6] finds an extensive application in studying the QS relaxation in the strong EMF. Interesting results have been obtained in this field by An. Vinogradov [7] and Fanchenko [8].

The dependence of the "thermostat" relaxation parameters on the EMF intensity has been investigated most thoroughly. However, the variation of the spontaneous relaxation characteristics under the action of the strong resonant EMF has been still insufficiently studied. We, therefore, consider here in detail the spontaneous relaxation of the quantum systems.

The influence of EMF on the spontaneous relaxation is revealed in the frequency shift of disperse characteristics of the quantum systems, an order higher than the well-known Bloch-Siegert shift. And as a result, the saturation of QS in the strong EMF is not complete [8]. The shape of the absorption line induced emission turns out to be asymmetrical.

II. Quantum-Kinetic Equation for the Density Matrix

A quantum-kinetic equation for a single-particle density matrix $\rho(t)$ of and atom, which interacts with a thermostat and an electromagnetic vacuum in the strong (classical) electromagnetic field, in frames of the interaction representation, has the following form ($\hbar = 1$):

$$\frac{\partial \rho}{\partial t} = i\left[V(t), \rho(t)\right] + i\sigma_B + i\sigma_R, \qquad (1)$$

where $V(t)$ is the operator of the atom-EMF interaction; $\sigma_B(t)$ and $\sigma_R(t)$ are the "collision" integrals determined by the influence of a thermostat (σ_B) and zero fluctuations (σ_R) on the density matrix evolution $\rho(t)$.

Expressions for the collision integrals $\sigma(t)$ are derived from the Neumann equation for a three-particle density matrix

$$F_3(t) = \rho(t)\rho_B\rho_R + g(t),$$

where the thermostat is independent of the atom. As a result we obtain:

$$\sigma_B(t) = \frac{1}{i}\left\{ Sp_B\left[W(t), \rho(t)\rho_B\right] - Sp_B\left[W(t), i\int_{-\infty}^{t} G^+(t,t')\left[W(t'), \rho(t')\rho_B\right]G(t,t')dt'\right]\right\}; \quad (2)$$

$$\sigma_R(t) = -\left\langle\left[V(t), \int_{-\infty}^{t} G^+(t,t')\left[V(t'), \rho(t')\right]G(t,t')dt'\right]\right\rangle. \quad (3)$$

Here $W(t)$ and $V(t)$ are the energy operators of the atomic interaction with a thermostat and electromagnetic vacuum, respectively; ρ_B is the matrix of the thermostat density and $G^+(t,t')$ is the evolution operator satysfying the differential equation:

$$i\frac{\partial G^+(t,t_0)}{\partial t} = \left[V(t) + W(t)\right]G^+(t,t_0); \quad G^+(t,t) = I; \quad (4)$$

the angular brackets denote averaging over the EMF-vacuum parameters.

The definition of the evolution operator $G^+(t,t_0)$, in the general case of multi-level quantum systems and a nonmonochromatic EMF is the most complicated task. In the analysis that follows we restrict ourselves to a consideration of a two-level quantum system and a monochromatic quasiresonant EMF. As the condition is fulfilled which is analogous with the adiabaticity condition in the atomic collisions theory:

$$|\Delta\dot{w}| \equiv \left|\frac{d}{dt}\Delta w\right| \ll \left[(\omega - \Delta w)^2 + 4|V|^2\right]^{1/2};$$

$$\Delta w = w_{mm} - w_{nn}; \quad \omega = \omega_0 - \omega_{mn}; \quad |V| \equiv |V_{mn}| \quad (5)$$

the operator of the evolution $G^+(t, t-\tau)$ for the elastic collisions ($W_{mn} = 0$) can be presented in the form:

$$G^+(t, t-\tau) = \frac{1}{2}\left\|\begin{array}{cc} a(t,\tau) & b(t,\tau) \\ -b^*(t,\tau) & a^*(t,\tau) \end{array}\right\|, \quad (6)$$

where

$$a(t,\tau) = (C_+ e^{i\varphi/2} + C_- e^{-i\varphi/2})e^{-i\omega\tau/2};$$

$$b(t,\tau) = (2|V|/\dot\varphi)e^{i\omega\tau/2}e^{-i\omega t}(e^{-i\varphi/2} - e^{i\varphi/2}); \quad (7)$$

$$\varphi(t,t-\tau) = \int_{t-\tau}^{t} \sqrt{\left[\omega - \Delta W(\tau_1)\right]^2 + 4|V|^2} \, d\tau_1,$$

$$C_{\pm}(t) = 1 \pm \tilde{\omega}(t)/\dot{\varphi}(t); \quad \tilde{\omega}(t) = \omega - \Delta W(t). \tag{7}$$

After substituting (6) and (7) in Eqs. (2) and (3) for the collision integrals, we obtain the following system of the differential equations:

$$\frac{\partial \rho_{mn}}{\partial t} = -(\gamma + i\Delta)\rho_{mn} + iVe^{-i\omega t}\left[\rho_{mm} - \rho_{nn} - i(C_m \rho_{mm} - C_n \rho_{nn})\right] - De^{-2i\omega t}\rho_{nm},$$

$$\frac{\partial \rho_{ss}}{\partial t} = -\gamma \rho_{ss} \pm 2\text{Re}\left[iV^* e^{i\omega t}(1+iK_s)\rho_{mn}(t)\right] + \gamma_s q_s + \Gamma \rho_{mm} \delta_{ns}. \tag{8}$$

Here the components $\gamma_s q_s$ $(s = m, n)$ characterize the noncoherent pumping; Γ denotes a cascade transition from the upper level (m) to the lower one, and the relaxation parameters are of the form:

$$\gamma = \gamma_B(|V|,\omega) + \gamma_R(|V|,\omega); \quad \Delta = \Delta_B(|V|,\omega) + \Delta_R(|V|,\omega);$$

$$D = D_B(|V|,\omega); \quad C_s = C_{sR}(|V|,\omega) + C_{sB}(|V|,\omega); \quad K_s = K_{sR}(|V|,\omega). \tag{9}$$

The system of equations (8) is distinct from the Bloch equations. First, it has additional correlations between the diagonal and off-diagonal matrix elements, and second, is characterized by the dependence of the relaxation parameters on the intensity and frequency of the external EMF. Hence, the effect of EMF changed the scheme of QS relaxation, including the spontaneous relaxation. The scheme includes entirely new components that are proportional to the coefficients D and C_s, D being determined by the EMF effect on the "thermostat" relaxation only (in respect with the elastic collisions).

The analysis of D, C_s and K_s shows that they stem from the off-diagonal elements of the evolution operator $G^+(t,t-\tau)$ and are determined by the splitting of energy levels m and n under the action of external EMF (Stark effect). Here we give evident expressions for these coefficients without their derivation:

$$\gamma_B(|v|,\omega) + i\Delta_B(|v|,\omega) = iSp_B\left[\Delta W \rho_B\right] +$$

$$+ Sp_B \int_0^\infty d\tau \left\{\Delta W(t-\tau)\Delta W(t)b_1(t,\tau)\rho_B + \right.$$

$$\left. + W_{mm}(t-\tau)\left[\Delta W(t)b_1(t,\tau),\rho_B\right]\right\} ;$$

$$D(|v|,\omega) = Sp_B \int_0^\infty d\tau \left\{\Delta W(t-\tau)\Delta W(t)b_2(t,\tau)\rho_B + \right. \quad (10)$$

$$\left. + W_{mm}(t-\tau)\left[\Delta W(t)b_2(t,\tau),\rho_B\right]\right\} ;$$

$$C_{sB}(|v|,\omega) = Sp_B \int_0^\infty d\tau W_{ss}(t-\tau)\left[\Delta W(t)b_3(t,\tau),\rho_B\right] ,$$

where

$$b_1(t,\tau) = \frac{2|v|^2}{\dot\varphi(t)\dot\varphi(t-\tau)} + \left[C_+(t)C_+(t-\tau)e^{i\varphi} + C_-(t)C_-(t-\tau)e^{-i\varphi}\right]/4$$

$$b_2(t,\tau) = -\frac{2|v|^2}{\dot\varphi(t)\dot\varphi(t-\tau)} + \left[C_-(t)C_+(t-\tau)e^{i\varphi} + C_+(t)C_-(t-\tau)e^{-i\varphi}\right]/4 \quad (11)$$

$$b_3(t,\tau) = \frac{1}{\dot\varphi(t)}\left[\frac{\tilde\omega(t-\tau)}{\dot\varphi(t-\tau)} + \frac{1}{2}C_-(t-\tau)e^{-i\varphi} - \frac{1}{2}C_+(t-\tau)e^{i\varphi}\right]$$

$$\gamma_s(|v|) = \frac{1}{2}\sum_{i<s} A_{si}\left(1 + 3\frac{|v|^2}{\omega_{si}^2}\right) + \frac{3}{2}A_{mn}\frac{|v|^2}{\omega_{mn}^2}\delta_{sm};$$

$$\gamma_R = \frac{1}{2}(\gamma_m + \gamma_n) + \frac{3}{2}A_{mn}\frac{|v|^2}{\omega_{mn}^2}; \quad \omega_{si} = (E_s - E_i)/\hbar;$$

$$C_{sR} = \sum_{i<s} R_{si} + R_{mn}\delta_{sm}; \quad K_{sR} = \sum_{i<s} R_{si} + R_{mn}\delta_{sn}; \quad (12)$$

$$R_{si} = \frac{3}{4}A_{si}\frac{1}{\omega_{si}}\left(1 + \frac{\omega}{\omega_{si}}\right);$$

$$\Gamma = A_{mn}/2$$

A_{si} is the Einstein coefficient.

From Eqs. (10) and (11) it follows that the relaxation parameters determined by the atom-thermostat interaction, have a complicated dependence on the intensity and frequency of the external EMF. The effect of EMF is basically connected with the components of $b_k(t,\tau)$ proportional to $\exp[\pm i\varphi(t,t-\tau)]$ and which turn to zero in the strong EMF ($|v|^2 \gg \tau_c^2$, τ_c is the time of the correlation, collision).

A so called field narrowing effect follows from this, which was predicted by us in 1973. It is interesting to note that γ_B and D turn out to be equal in value and opposite in sign, in the strong EMF. This leads to a considerable decrease in the absorption coefficient and narrowing of spectral lines in the Rayleigh scattering.

From Eq. (12) it is seen that the spontaneous relaxation parameters γ_S and γ_R are practically independent of the external EMF, because normally, the condition for the resonant approximation applicability $|V|^2 \ll \omega_{mn}^2$ is fulfilled in the optical wavelength region. However, the evolution of the atomic density matrix may be essentially affected by new components proportional to C_{sR} and K_{sR} which determine an unusual correlation between the elements of the density matrix even when there is no interaction with a thermostat. It should be noted that the influence of EMF on γ_S and γ_R would be noticeable if a two-level open system has the energy levels lying close to the resonance levels, so that the condition $|V|^2 > \omega_{si}^2$, where $i \neq m, n$ is fulfilled.

To investigate new physical processes that follow from the modified differential equations for the density matrix, we consider two simple cases, namely, quasiclassical collisions of gas atoms in the strong EMF and the spontaneous relaxation without interaction with a thermostat.

III. Atomic collisions

To simplify our analysis, we shall neglect the EMF effect on the spontaneous relaxation, taking $C_{sR} = K_{sR} = 0$. Then the equation system (8) yields the following expression for the shape of absorption spectral lines induced emission:

$$I(|V|,\omega) \sim \frac{|V|^2 (q_m - q_n)\left[\gamma_R + \gamma_\alpha(|V|,\omega)\right]}{\left[\omega - \Delta_\alpha(|V|,\omega)\right]^2 + \left[\gamma_R + \gamma_{0C}(|V|,\omega)\right]^2 + 2|V|^2(\gamma_R + \gamma_\alpha)(\frac{1}{\gamma_m} + \frac{1}{\gamma_n}) + pq} \tag{13}$$

where

$$\gamma_\alpha(|V|,\omega) = \gamma_B(|V|,\omega) + \mathrm{Re}D(|V|,\omega) = \\ = n_b \ll \int_{-\infty}^{+\infty} \Delta w(t)\,dt \int_0^{+\infty} \Delta w(t-\tau)\cos\varphi(t,t-\tau)\,d\tau \gg, \tag{14}$$

$$\Delta_\alpha(|V|,\omega) = \overline{\Delta W} + n_b \ll \int_{-\infty}^{+\infty} \Delta W(t)\,dt \int_0^{+\infty} \Delta W(t-\tau)\sin\varphi(t,t-\tau)\,d\tau \gg ;$$

$$\overline{\Delta W} = n_b \ll \int_{-\infty}^{+\infty} \Delta W(t)\,dt \gg ; \quad pq = 2(\gamma_R + \gamma_\alpha)\,\mathrm{Re}\,D. \tag{15}$$

Double angular brackets denote averaging over the relative rates and the impact parameter.

From Eq. (14) it is seen that $\gamma_\alpha(|V|,\omega)$ tends to zero ($\gamma_\alpha(|V|,\omega) \to 0$) for $|V|^2 \gg \tau_c^{-2}$. We called this effect the field narrowing in 1973 [2]. In the quasistatical spectral region, when $\omega^2 \gg |V|^2$ this effect is most evident. In this case there is observed the decreased absorption of the gaseous medium in the strong EMF predicted in 1975 by Lisitsa and Yakovlenko [3] (see Fig. 1).

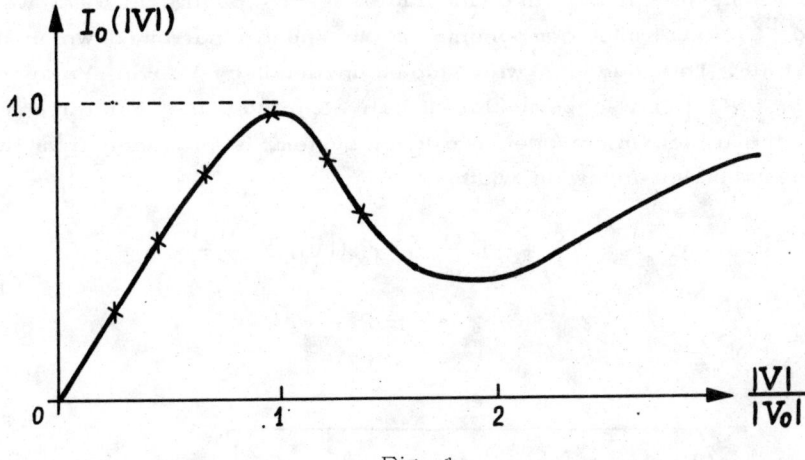

Fig. 1

The dependence of the absorbed light power on $|V|/|V_0|$ where $|V_0|$ is the critical value determined by the properties of the medium, is plotted in Fig. 1. A cross shows the experimental data obtained by Bonch-Bruevich et al. in [4], which confirmed our predictions for the decreased absorption of the medium and the field narrowing effect.

In 1984 we have also predicted the dependence of the spectral line shift on the EMF intensity [9, 10]:

$$\Delta_\alpha(|V|) \approx n_B \ll \int_{-\infty}^{+\infty} dt\,\Delta W(t)\left[1 + \int_0^{+\infty} \Delta W(t-\tau)\sin\varphi(t,t-\tau)\,d\tau\right] \gg . \tag{16}$$

From Eq. (16) a classical expression for the impact shift for

$V \to 0$ follows. An integrand of the nonlinear component (16), in the limiting case of strong EMF ($|V|^2 \gg \tau_c^{-2}$, $|V|^2 \gg (\Delta W)^2$) is an oscillating time function, and its contribution in $\Delta_\alpha(|V|)$ is negligibly small. The shift $\Delta_\alpha(|V|)$ reaches the maximum value:

$$\Delta_\alpha^{max}(|V|) \approx \overline{\Delta W} \equiv n_B \ll \int_{-\infty}^{+\infty} \Delta W(t) dt \gg . \qquad (17)$$

The shift $\Delta_\alpha(|V|)$ increases with the EMF intensity; in general, it is a nonlinear dependence determined by the potential difference $\Delta W(\tau(t))$. The shift $\Delta_\alpha(|V|)$ would grow due to the increase influence of collisions with large impact parameters in the field splitting of energy levels at strong EMF.

The dependence of the relaxation parameters on the EMF intensity is well manifested in the resonant fluorescence spectra. In 1973 we showed [2] that triplet components of the spectra narrow down in the strong EMF. This question was studied in detail by Vdovin, Yakovlenko et al. in 1982 [11]. In particular, it can be proved that central component of the triplet, in an open two-level system, is produced from two components of the following widths:

$$\gamma_c^{(1)} \approx \gamma_R + \frac{1}{2}\Gamma \frac{\omega^2}{\Omega^2} + \gamma_\alpha(|V|, \omega) \frac{4|V|^2}{\Omega^2} + F;$$
$$\gamma_c^{(2)} \approx \gamma_R - \frac{1}{2}\Gamma - F, \qquad (18)$$

where

$$F = \sqrt{B^2 + (\omega^2/4\Omega^2)\left[(\gamma_m - \gamma_n)^2 - \Gamma^2\right]} - B;$$
$$B = 2(|V|^2/\Omega^2)(\gamma_\alpha - \Gamma/2) + \Gamma/2; \quad \Omega^2 = \omega^2 + 4|V|^2.$$

For a closed two-level system with the basic lower level, $\gamma_n = 0$, $\gamma_R = \gamma_m/2 = \Gamma/2$, the F function is zero, and Eqs. (18) turn to the well-known results obtained by Mollow [12]:

$$\gamma_c \approx \gamma_R(1 + \omega^2/\Omega^2) + \gamma_\alpha(|V|, \omega) 4|V|^2/\Omega^2. \qquad (19)$$

It follows that width of the central component, in the strong EMF when $\gamma_\alpha(|V|, \omega) \to 0$, is determined by a spontaneous relaxation γ_R only, and the influence of the thermostat interaction (atomic collisions) is levelled. Lateral components of a triplet depend on EMF in a more intricate way.

So, the consideration of the relaxation parameters versus the intensity and frequency of EMF enabled us to derive, for the first time, the universal equation (13) describing the shape of an absorption line, which is characteristic of the entire spectrum, including impact, intermediate and quasistatic frequency regions, for the arbitrary intensity of EMF. With variation of the relaxation scheme (using the component proportional to D) the well-known Karplus-Schwinger formula was determined more accurately in the region of its applicability (for the therm pq).

IV. Spontaneous Relaxation

Now consider a spontaneous relaxation of an open two-level quantum system in strong monochromatic EMF. The components determined by the thermostat interaction are not included in the system of differential equations (8). The coefficient D is zero in the relaxation scheme. The rest of terms in the relaxation part of the density matrix equations remain the same.

The system (8) takes the form:

$$(\frac{\partial}{\partial t} + \gamma_R) \rho_{mn} = V e^{-i\omega t}\left[(i + C_n)(\rho_{mm} - \rho_{nn}) + C\rho_{mm}\right]$$

$$(\frac{\partial}{\partial t} + \gamma_s) \rho_{ss} = \pm \text{Re}\left[iV^* e^{i\omega t}(1 + iK_s)\rho_{mn}(t)\right] + \quad (20)$$

$$+ \gamma_s q_s + \Gamma \rho_{mm} \delta_{sn}.$$

It follows that the spontaneous relaxation in strong EMF is essentially distinct from the known one. The new physical effects will apparently be determined by an additional component proportional to C, $C = C_m - C_n$.

To determine physical effects which follow from the modified matrix equations, we find a stationary solution for the equation system (20). As a result, we obtain for $\rho_{mn}(t)$:

$$\rho_{mn}(t) = \frac{V e^{-i\omega t}}{D_0}\left[(\omega - i\gamma_R)N + \alpha C|V|^2 + \beta C(\gamma_R + i\omega)\right], \quad (21)$$

where

$$D_0 = \omega^2 + \gamma_R^2 + 2|V|^2 \gamma_R\left(\frac{1}{\gamma_m} + \frac{1}{\gamma_n} - \frac{\Gamma}{\gamma_m \gamma_n}\right); \quad (22)$$

$$N = q_n - q_m(1 - \Gamma/\gamma_n); \quad \alpha = 2(q_m/\gamma_n + q_n/\gamma_m);$$
$$\beta C = C_n q_n - C_m q_m (1 - \Gamma/\gamma_n). \qquad (23)$$

From Eqs. (21) - (23) it follows that the effect of the external EMF on the spontaneous relaxation results in the appearance of two new physical effects.

First, the real part ρ_{mn} for the QS reflection coefficient dispersion has a new frequency shift:

$$\Delta\omega = 2(C/N)(q_m/\gamma_n + q_n/\gamma_m)|V|^2, \qquad (24)$$

which is proportional to the EMF intensity. It is more than an order higher than the well-known Bloch-Siegert frequency shift [13] characterized by the effect of double-frequency oscillations.

The appearance of the shift $\Delta\omega$ violates the effect of the reflection coefficient saturation in the strong EMF. Actually, from Eq. (21) we get for $|V| \to \infty$:

$$\lim_{|V| \to \infty} (\rho_{mn} e^{i\omega t}) = \frac{C|V|(q_m \gamma_m + q_n \gamma_n)}{\gamma_R(\gamma_m + \gamma_n - \Gamma)} \longrightarrow \infty. \qquad (25)$$

Such a violation of the saturation effect, in a closed two-level system, has been noticed for the first time by Fanchenko [8] in 1983 when he investigated the spontaneous relaxation in the strong EMF by Keldysh diagram method [6].

Second, the absorption coefficient (induced emission) in the open two-level system $(\mathrm{Im}(\rho_{mn} e^{i\omega t}))$ is an asymmetric function of EMF frequency, even without the interaction with the thermostat:

$$K(\omega) \sim \mathrm{Im}(\rho_{mn} e^{i\omega t}) = \frac{|V|}{D_0}(\gamma_R N - \beta C \omega). \qquad (26)$$

Equation (26) shows that the spectroscopic effect of the absorption coefficient saturation is not violated, namely, $K(\omega)$ tends to zero when $|V|$ turns to the infinity $(K(\omega) \to 0$ if $|V| \to \infty)$.

Consider a closed two-level system investigated by Fanchenko [8] with the help of another method. For such a system the equations (20) take the form:

$$\left(\frac{\partial}{\partial t} + \gamma_R\right)\rho_{mn} = iVe^{-i\omega t}(\rho_{mm} - \rho_{nn} - iC_m \rho_{mm}), \qquad (27)$$

$$\left(\frac{\partial}{\partial t} + \gamma_m\right)\rho_{mm} = 2\text{Re}\left[iV^* e^{i\omega t}(1 + iK_m)\rho_{mn}(t)\right],$$

$$\rho_{mm} + \rho_{nn} = 1.$$

(27)

An off-diagonal matrix element $\rho_{mn}(t)$ has a stationary solution:

$$\rho_{mn}(t) = V e^{-i\omega t} \frac{\omega - i\gamma_R + 6|V|^2/\omega_0}{\omega^2 + \gamma_R^2 + 2|V|^2}.$$

(28)

To obtain Eq. (28) it has been taken that $\gamma_m = 2\gamma_R$, $C_m = 2K_m$, $C_n = 0$, in accordance with Eq. (12) for the closed quantum system, and that the dipole moment oscillates on EMF frequency $\omega_{mn} \approx \omega_0$.

From Eq. (28) it follows that an additional frequency shift

$$\Delta\omega = 6|V|^2/\omega_0$$

(29)

is 24 times higher than the Block–Siegert shift

$$\Delta\omega_{BS} = |V|^2/4\omega_0$$

(30)

but unlike Eqs. (21), (26) the absorption coefficient is a symmetrical frequency function ω.

Equation (28) indicates that in strong EMF $(|V|^2 \gg \omega_0(\omega^2 + \gamma_R^2)^{1/2})$ the following correlation is fulfilled:

$$\rho_{mn}(t) \approx 3(V/\omega_0) e^{-i\omega t}.$$

(31)

This asymptotic result coincides with the expression obtained in [6] by the Keldysh diagram method. It was quite easy to derive it on the basis of quantum-kinetic equations (8).

Thus, by using simple examples we have shown that the variation of relaxation properties in quantum systems, at strong EMF, leads to some new physical effects. The investigation of atomic interaction with the strong EMF and other subsystems must be performed with the help of modified quantum-kinetic matrix equations as distinct from the well-known Bloch equations.

I wish to express gratitude to N. Basov, P. Apanasevich, An. Vinogradov, A. Oraevsky, S. Rautian, J. Eberly, S. Stenholm and S. Yakovlenko for their valuable advice and interest in my work.

References

1. P. N. Argyres, P. L. Kelley, Phys. Rev., $\underline{A134}$, 98 (1964).
2. E. G. Pestov, S. G. Rautian, JETP, $\underline{64}$, 2032 (1973).
3. V. S. Lisitsa, S. I. Yakovlenko, JETP, $\underline{68}$, 479 (1975).
4. A. M. Bonch-Bruevich, G. A. Vartanian, V. V. Khromov JETP, $\underline{78}$, 538 (1980).
5. J. L. Carlsten, A. Szöke, M. G. Rymer, Phys. Rev., $\underline{A15}$, 1019 (1977).
6. L. V. Keldysh, JETP, $\underline{47}$, 1515 (1964).
7. An. V. Vinogradov, Kvantovaya Elektronika, $\underline{12}$ (1985) to be published.
8. S. S. Fanchenko, JETP, $\underline{85}$, 1936 (1983).
9. E. G. Pestov, Kvantovaya Elektronika, $\underline{11}$, 7 (1984).
10. E. G. Pestov, JETP, $\underline{86}$, 1643 (1984).
11. D. S. Bakaev, Yu. A. Vdovin, V. M. Ermachenko, S. I. Yakovlenko, JETP, $\underline{83}$, 1297 (1982).
12. B. R. Mollow, Phys. Rev., $\underline{A188}$, 1969 (1969).
13. F. Bloch, A. J. F. Siegert, Phys. Rev., $\underline{57}$, 522 (1940).

S. Ya. KILIN

Institute of Physics,
BSSR Academy of Sciences,
Leninsky pr. 70, Minsk 220602, USSR

THE QUANTUM STATISTICS OF LIGHT SCATTERING AND PROPAGATION EFFECTS

I. Resonant light scattering statistics

The statistical fluctuations of resonant scattered light can, under certain circumstances, exhibit a number of interesting non-classical features: 1. photon antibunching; 2. bunching of different frequency photons; 3. squeezing; 4. sub-Poissonial statistics. 1, 2, 4 have been observed for one-atom resonance fluorescence. 3 has not yet been experimentally.

The first problem discussed here is: how these quantum statistical features change during the propagation in resonant medium?

Statistical description of quantum field fluctuations

The statistics of quantum field fluctuation is usually described in the following two ways. The first way is based on the construction of correlation functions of three types: temporal, spectral, and spectral-temporal correlation functions. Temporal functions correspond to correlations between frequency integrated fields. Spectral and spectral-temporal functions describe correlations between spectral components with definite frequencies and correlations between fluxes with finite spectral bandwidth, respectively [1]. The second way is based on the probability distribution $P(n, t, t+\tau)$ for a number of photons n emitted in a given time interval $(t, t+\tau)$.

The first approach is used to describe photon bunching, antibunching and squeezing. The second one is used to investigate sub-Poissonian statistics of light.

In both approaches, the statistical properties of light are connected with the statistical properties of emitted atoms, i.e. with the atomic dipole correlation functions. So, atomic dipole fluctuations are the source of scattered light fluctuations. The quantum nature of light leads only to a definite order of Heisenberg atomic dipole operators in correlation functions. In order to describe the quantum field fluctuations one must be able to describe the atomic fluctuation.

Nonclassical effects for one-atom resonance fluorescence

Let us briefly consider the nonclassical effects for one-atom resonance fluorescence light.

Let atom A be driven by the strong laser field $E_L^+ = E_0^+ e^{-i\omega_0 t}$. The spectrum of the scattered field is represented by the well-known triplet lines first observed by Wu et al. [2].

As mentioned above, the statistical properties of the scattered field are described by correlation functions. The calculation of the second-order correlation function for the whole scattered light done by Carmichael and Walls [3] predicted the phenomenon of photon antibunching. This phenomenon has been experimentally observed by Kimble, Dagenais and Mandel [4].

If filters extracting sidebands of resonance fluorescence are placed before the photon detectors sidebands photon bunching can be observed. This phenomenon was predicted by Apanasevich and Kilin [5] and experimentally observed by Aspect et al. [6]. The essence of this phenomenon is that the photon created in even multiphoton scattering processes whose lowest-order is the four-photon scattering $2\omega_0 \to \omega_+ + \omega_-$ where $\omega_\pm = \omega_0 \pm \Omega$, $\Omega = (\omega_{ab} - \omega_0)\sqrt{4\chi^2/(\omega_{ab}-\omega_0)^2 + 1}$ appears as correlated pairs. The number of pairs where ω_- photons come first is greater than the number of photon pairs with reversed order. This is due to the fact that the sequence $\omega_- \to \omega_+$ corresponds to the processes beginning with the more populated ground state. It results in asymmetry of the spectral time correlation function with respect to the time delay τ.

If the bandwidth of filters is less than the line shape width it is possible to measure the spectral correlations. Apanasevich and Kilin [7] have shown that scattered photons symmetrically located about the laser frequency are correlated. Note that spectral correlations are described by the anomalous correlator $\langle a_{\omega_-} a_{\omega_+} \rangle$. Spectral correlations have not been yet observed experimentally.

Theoretical calculations of the phase-dependent correlation functions done by Walls and Zoller [8] showed that squeezing was present in the total fluorescence light. As spectral correlations, this nonclassical feature of scattered light has not been observed.

The sub-Poissonian statistics for which the photon number variance $\langle (\Delta n)^2 \rangle$ is smaller than the mean $\langle n \rangle$ was predicted for resonance fluorescence by Mandel [9] and has been observed recently by Short and Mandel [10].

Quantum theory of sources of scattered field fluctuations (propagation effects)

Let us now consider how one-atom resonance fluorescence statistics transforms due to the propagation effect.

We are interested in the problem when the whole field $\mathcal{E}^\pm$ consists of two parts: a strong laser field $E_L^\pm$ which can be considered as classical and a scattered field $E_s^\pm(\underline{r}, t) = \int d\omega_s E_{\omega_s}^\pm(\underline{r}) \exp(\mp i \omega_s t)$ which is quantum in nature.

As previously mentioned, the source of scattered field fluctuations is polarization fluctuation. It can be shown [11] that for a number of problems the polarization operator $P^\pm = \sum_i \delta(\underline{r} - \underline{r}_i(t)) P_i^\pm J_i^\pm(t)$ is a steady-state random operator, since its pair correlation functions depend on the spatial and temporal arguments difference. So, the polarization operator can be expanded into a Fourier series with respect to spatial and temporal arguments $P^\mu(\underline{r}, t) = \iint d\omega_s d\underline{K}_s e^{-i\mu(\underline{K}_s \underline{r} - \omega_s t)} P^\mu_{\omega_s \underline{K}_s}$. The statistical properties of the random Fourier components $P^\mu_{\omega_s \underline{K}_s}$ are determined by the mean value and correlators.

Confining ourselves to the problems linear with respect to the scattered field $E_s^\pm$, it is easy to find that the mean value

$$\langle P^+_{\omega_s \underline{K}_s} \rangle = \delta(\omega_s - \omega_0) \sum_i P_i^+ \langle J_i^+ \rangle e^{i \underline{K}_s \underline{r}_i} +$$

$$+ (i/\hbar) \sum_i E^-_{\omega_s}(\underline{r}_i) e^{i\underline{K}_s \underline{r}_i} \int_0^\infty e^{-i\varepsilon_s \tau} |P_i^+|^2 \langle [J_i^+(\tau), J_i^-(0)] \rangle +$$

$$+ (i/\hbar) \sum_i E^+_{2\omega_0 - \omega_s}(\underline{r}_i) e^{i\underline{K}_s \underline{r}_i} \int_0^\infty e^{-i\varepsilon_s \tau} (P_i^+)^2 \langle [J_i^+(\tau), J_i^+(0)] \rangle ;$$

$$\varepsilon_s = \omega_s - \omega_0.$$

and the correlators $\langle P^\mu_{\omega_s \vec{K}_s} P^\nu_{\omega'_s \vec{K}'_s} \rangle$ are defined through the one-atom correlators $\langle J^\mu_i(\tau) J^\nu_i(0) \rangle$ which have been discussed previously.

Using this scheme for concrete laser field geometry one can derive quantum equations for the slowly varying scattered field amplitude operator.

Quantum fluctuations (spontaneous emission) in the scheme of forward four-wave mixing

Consider as an example of the proposed scheme non-coplanar forward FWM in a resonant homogeneously broadened medium consisting of two-level atoms. Let laser wave vectors $\vec{k}_{10}$, $\vec{k}_{20}$ and scattered wave vectors $\vec{K}_1$ and $\vec{K}_2$ satisfy the phase-matching condition in the x-y plane (Fig. 1). Then from the theory under consideration we obtain the following equations for the slowly varying photon creation and annihilation operators of scattered waves

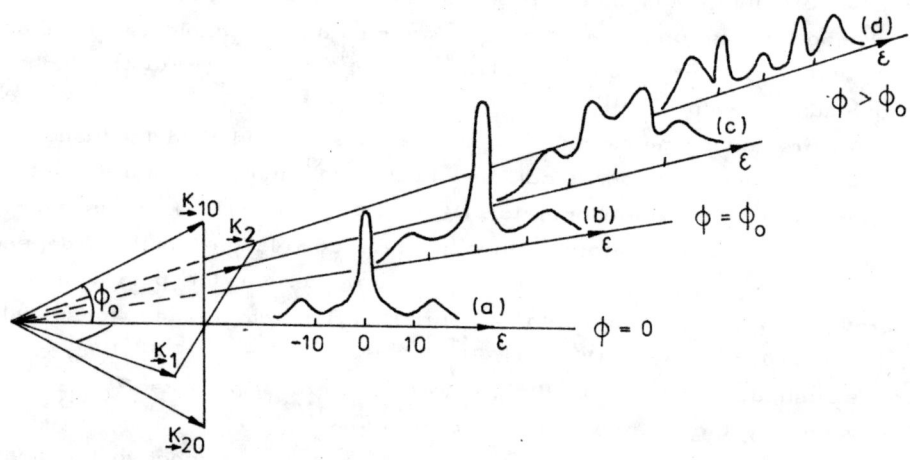

Fig. 1. Spectrum $\ln I(\phi, \varepsilon, z)$ as a function of detuning $\varepsilon = (\omega_1 - \omega_0) T_2$ and scattering angle ϕ at $(\omega_{ab} - \omega_0) T_2 = 10$, $(\chi T_2)^2 = 15$, $\phi_0 = 10^{-2}$, $z \rho_0 = 10^3$, where $\rho_0 = 300$ cm^{-1} is non-saturated absorption at the line centre. $\phi = 0$(a), 10^{-2}(b), $1.1 \cdot 10^{-2}$(c), $2 \cdot 10^{-2}$(d).

$$da_1/dz = -\gamma_1 a_1 - iK_1^* a_2^+ e^{i\Delta z} + G_1(z),$$
$$da_2^+/dz = -\gamma_2 a_2^+ + iK_2 a_1 e^{-i\Delta z} + G_2(z), \quad (1)$$

where loss per unit length γ_i and the coupling coefficients K_i are determined by the atomic correlators

$$\gamma_i = f_i \operatorname{Re} \int_0^\infty e^{i\varepsilon_i \tau} \overline{\langle [J^-(\tau), J^+(0)] \rangle^{\phi_0}} d\tau,$$
$$K_i = -i f_i \int_0^\infty e^{i\varepsilon_i \tau} \overline{\langle [J^+(\tau), J^+(0)] \rangle^{\phi_0}} d\tau, \qquad (2)$$

$\Delta = 2K_{0Z} - K_{1Z} - K_{2Z}$; $\varepsilon_1 = \omega_1 - \omega_0 = -\varepsilon_2$; $\overline{\langle \quad \rangle^{\phi_0}}$ -
- denotes averaging over transverse coordinates.

Noise components G_1 and G_2 are given by the expression

$$\dot{G}_S(Z) = (f_S/2\pi)^{1/2} \int dK_Z e^{i(K_Z - K_{SZ})Z} P^-_{\omega_S \vec{K}_S}. \qquad (3)$$

The random spectral operators $P^\mu_{\omega_S \vec{K}_S}$ are also determined by the atomic correlators. Note that besides the usual correlators

$$\langle P^+_{\omega \underline{K}} P^-_{\omega' \underline{K}'} \rangle = \delta_{\omega,\omega'} \delta_{\underline{K},\underline{K}'} g^{+-}(\varepsilon), \quad \langle P^-_{\omega \underline{K}} P^+_{\omega' \underline{K}'} \rangle = \delta_{\omega,\omega'} \delta_{\underline{K},\underline{K}'} g^{-+}(\varepsilon)$$

corresponding to the autocorrelation of space-time spectral components of fluctuating polarization, the noise sources $P^\mu_{\omega \underline{K}}$ are also determined by the anomalous correlators

$$\langle P^+_{\omega \underline{K}} P^+_{\omega' \underline{K}'} \rangle = \delta_{\omega - \omega_0, \omega_0 - \omega'} \delta_{\underline{K} - \underline{K}_0, \underline{K}_0 - \underline{K}'} g^{++}(\varepsilon),$$

where

$$2g^{\mu\nu}(\varepsilon) = \int_{-\infty}^{+\infty} d\tau \exp(-i\mu\varepsilon\tau) \langle (J^\mu(\tau) - \langle J^\mu(\tau) \rangle)(J^\nu(0) - \langle J^\nu(0) \rangle) \rangle.$$

As a result, the noise components G_1 and G_2 in Eqs. (1) describing quantum atomic fluctuations <u>are not independent</u> $\langle G_1(Z) G_2(Z') \rangle = 2(f_1 f_2)^{1/2} \dot{g}^{--}(\varepsilon_1) \delta(Z - Z')$. Using these Eqs., we calculate the spectrum of scattered light for different scattering angle. The final formula represents the complex function depending on different one-atom correlation functions $g^{\mu\nu}(\varepsilon)$ and rates λ_i:

$$I(\omega, Z, \phi) = \frac{1}{|\lambda_2 - \lambda_1|^2} \left\{ \frac{e^{2\operatorname{Re}\lambda_1 Z} - 1}{2\operatorname{Re}\lambda_1} m_{11} + \frac{e^{2\operatorname{Re}\lambda_2 Z} - 1}{2\operatorname{Re}\lambda_2} m_{22} + \right.$$
$$\left. + 2\operatorname{Re} \frac{e^{(\lambda_1^* + \lambda_2)Z} - 1}{\lambda_1^* + \lambda_2} m_{12} \right\}, \qquad (4)$$

where $\lambda_{1,2} = -(\gamma_1 + \gamma_2 - i\Delta)/2 \pm \sqrt{(\gamma_1 + \gamma_2 - i\Delta)^2/4 + K_1^* K_2}$,

$m_{11} = |C_{21}|^2 g^{+-}(\varepsilon_1) + |K_1|^2 g^{-+}(-\varepsilon_1) + 2\mathrm{Re}\left[C_{21}^* K_1^* g^{++}(\varepsilon_1)\right]$,

$m_{22} = |C_{22}|^2 g^{+-}(\varepsilon_1) + |K_1|^2 g^{-+}(-\varepsilon_1) + 2\mathrm{Re}\left[C_{11}^* K_1^* g^{++}(\varepsilon_1)\right]$,

$m_{12} = -C_{21}^* C_{11} g^{+-}(\varepsilon_1) - |K_1|^2 g^{-+}(-\varepsilon_1) - K_1^* C_{21}^* g^{++}(\varepsilon_1) - K_1 C_{11} g^{++}(-\varepsilon_1)$

$C_{21} = \lambda_2 + \gamma_1; \quad C_{11} = \lambda_1 + \gamma_1.$

It is seen from Fig. 1 that the scattered spectrum differs from the well--known Mollow's triplet. If the scattered angle is smaller than the angle between the pump waves the spectrum is three-peaked. If the scattered angle is near ϕ_0 the spectrum is four-peaked. For $\phi > \phi_0$ the spectrum is five-peaked. The maximum of the scattered light intensity is at an angle near the pump angle. The above results assume perfect radiative damping. The presence of phase-damping processes affects the spectrum. New collisional-induced components appear in the spectrum.

Squeezing in the DFWM scheme

It has been shown recently by Kumar and Shapiro [12] that forward degenerate ($\varepsilon_1 = \varepsilon_2 = 0$, $K_{1,2} = K$, $\gamma_{1,2} = \gamma$) four-wave mixing used to generate squeezed states has a number of advantages over the scheme which uses counterpropagating waves. Their considerations are based on the same equations we have discussed. Kumar and Shapiro assume that the sources of atomic fluctuations $G_1(Z)$ and $G_2(Z)$ are independent. But as I have shown, the noise components $G_1(Z)$ and $G_2(Z)$ are correlated. Now in [11] we show that this correlation leads to atomic noise reduction (see also [13]).

Consider the field mode c which is formed by the 50/50 splitter and the introduction of an additional phase shift into one of the output waves so that $c = (a_1 - i a_2 e^{i\theta})/\sqrt{2}$. The in-phase and out-of-phase quadratures of the mode c are defined by $c_1 = (c + c^+)/\sqrt{2}$, $c_2 = (c - c^+)/\sqrt{2}$. If $a_1(Z)$ and $a_2(Z)$ are in coherent states at the initial point $Z = 0$ we have the following expressions for in-phase and out-of-phase quadrature variances

$$\langle \Delta c_{1,2}^2 \rangle = \langle (c_{1,2} - \langle c_{1,2} \rangle)^2 \rangle = \frac{1}{4}\left(e^{2\lambda_{2,1}Z} + \frac{e^{2\lambda_{2,1}Z} - 1}{\lambda_{2,1}} f R_{2,1}\right). \quad (5)$$

Here $\lambda_{1,2} = -\gamma^{\pm}|K|$; $R_{2,1} = g^{+-}(0) + g^{-+}(0) \mp \text{Re} i e^{-i\theta} g^{++}(0)$ are an in-phase and an out-of-phase component of normally ordered one-atom fluorescence squeezing spectra at zero frequency.

A number of conclusions follow from the obtained relations:

(i) The most significant effect of quantum atomic fluctuations (which are the source of spontaneous emission) on the scattered waves will be observed near the amplification threshold. In this case, the signal--to-noise ratio is

$$\frac{\text{signal}}{\text{noise}} = \frac{\langle a_s^+(0) a_s(0) \rangle}{|KZ|}.$$

For sodium vapors (D_2-line) with the density $N = 10^{14} - 10^{15}$ cm^{-3}, $Z = 1$ cm and the number of photons in the mode $h_s(0) = 10^3$ the ratio reaches unity.

(ii) In the case of exact resonance $\varepsilon = 0$ we have

$$R_1 = 1, \quad R_2 = \frac{1}{\sqrt{2g+1}} \left(1 - \left(1 + \frac{T_2}{T_1}\right) \frac{g(g+2)}{2(2g+1)^2}\right); \quad g = I/I_{\text{sat}}. \tag{6}$$

At saturation $I/I_{\text{sat}} \gg 1$ there is a reduction of the atomic fluctuations component from the in-phase quadrature variance. R_2 approaches zero as $(I/I_{\text{sat}})^{-1/2}$. So, the quantum atomic fluctuations do not limit the possibilities of squeezed state generation via DFWM. The degree of squeezing is equal to

$$r = \frac{\langle \Delta c_1^2 \rangle}{\langle \Delta c_2^2 \rangle} = \begin{cases} (1 + 2Zf)^{-1} & \text{for a thin medium } (|\lambda_2 Z| \ll 1), \\ \alpha (1 + 2Zf)^{-1} & \text{for an intermediate case,} \\ \alpha (2g)^{-3/2} & \text{for a thick medium } (|\lambda_1 Z| \gg 1), \end{cases}$$

where α varies from 5/8 for $T_2/T_1 = 2$ to 7/8 for $T_2/T_1 \to 0$. For example, at $I/I_{\text{sat}} = 100$ (pump power $\sim 1 \frac{W}{cm^2}$ for D_2-line of Na--vapors) we have

$$r = \begin{cases} 3 \cdot 10^{-1} (|\lambda_2 Z| \sim 10^{-1}), \\ 3 \cdot 10^{-4} (|\lambda_1 Z| \sim 10^{-1}), \\ 7 \cdot 10^{-5} (|\lambda_1 Z| \sim 10). \end{cases}$$

(iii) In spite of the fact that in the case of exact resonance and at saturation the mode C state is squeezed, because of $[c_1(Z), c_2(Z)] \approx i/2 e^{-2\gamma Z}$; $\langle \Delta c_1^2(Z) \rangle \langle \Delta c_2^2(Z) \rangle = (1/16) e^{-4\gamma Z}$ and $\langle \Delta c_1^2(Z) \rangle = (1/4) e^{-2(\gamma + |K|)Z} < (1/2) e^{-2\gamma Z}$ the experimental observation of large value of squeezing will be limited by the fact that the intensity of the in-phase component, as well as its variance, decays faster that the out-of-phase component intensity.

II. Statistics of Stokes Pulse Energy Fluctuations for Different Regimes of Stimulated Raman Scattering

Another problem discussed here is: how do statistical properties of spontaneous emission which initiates stimulated Raman scattering transform due to the propagation effects?

It was predicted by Raymer et al. [14] and observed experimentally by Walmsley and Raymer [15] and Fabricius et al. [16] that in the linear regime where neither laser pump nor population are depleted, the statistical properties of the initiating photons are preserved during the amplification. As a result [17] in the linear and transient regimes where the laser pulse duration τ_L is smaller than the scattered field coherence time $\tau_s = g_0 I_L L / \Gamma$ (g_0 is the Raman gain coefficient, I_L is the laser intensity, L is the length of Raman medium, Γ is Raman half linewidth) the probability density function $P(W)$ for the Stokes pulse energy W has the form

$$P(W) = <W>^{-1} \exp(-W/<W>),$$

where $<W>$ is the mean value of the energy.

In the nonlinear regime where the laser field is depleted the statistical properties of the initiating photons are modiefied and stabilization of Stokes pulse energy occurs. The phenomenon of energy stabilization has been studied theoretically by Lewenstein [18], Trippenbach and Rząźewski [19] and observed experimentally by Walmsey et al. [20].

The purpose of this part of the report is to describe the experimental results [21] for the evolution of Stokes pulse energy distribution from the negative exponential to the stabilized form with increasing laser pump.

Raman scattering at λ_s = 683 nm took place in hydrogen cell (H_2 pressure p = 50 atm, cell length L = 120 cm, $1/\Gamma$ = 0.148 ns) under excitation by frequency-doubled Nd:YAG laser pulses (λ_L = 532 nm, τ_L = 10 ns, I_L = $10^7 \div 10^9$ W/cm^2, pulse energy is 10 uJ, repetition rate is 10 Hz). The laser linewidth varied in our experiment from

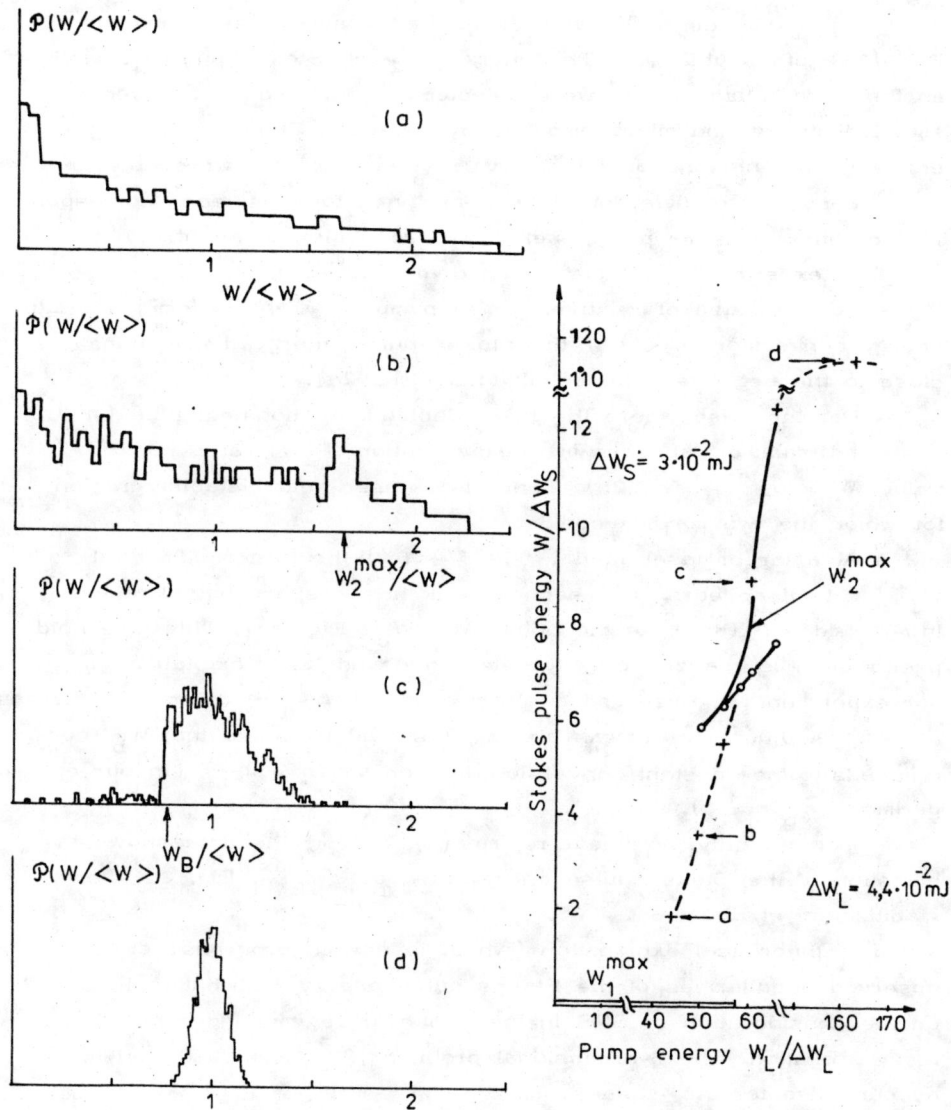

Fig. 2. Dependence of the Stokes pulse energy mean value $\langle W \rangle$ (dashed line, crosses correspond to experimental data), the second maximum W_2^{max} (solid line) and the sharp bound point W_b (open circles) on the laser pulse energy and Stokes pulse energy distribution histograms for different laser fields: η = 1% (a), 4% (b), 6% (c), 46% (d)

0.2 cm^{-1} to 0.06 cm^{-1}. We had an excited volume Fresnel number $F = A/\lambda_s L$ of about $2 \div 5$. The energy of each Stokes pulse as well as the laser pulse energy were detected by photodiodes followed by signal digitizing and microcomputer. By selecting Stokes and laser energies falling into one of 256 intervals (channels) we were able to obtain a number of different Stokes energy histograms, each corresponding to different laser pump energy with resolution of about 0.5%.

The experimental results obtained are shown in Fig. 2.

- For the case of relatively small pump ($I_L < 10^8$ W/cm^2, average energy conversion $\eta < 1\%$) the Stokes pulse energy distributions are close to the negative exponential form (Fig. 2a).

- For the case where the laser depletion is not negligible ($\eta > 1\%$) a second maximum appears in the distribution $P(W)$ at a nonzero value $W = W_2^{max}$ (Fig. 2b). There is a laser field bistability region for which the two maxima coexist.

- At a larger laser field ($\eta \approx 5 \div 6\%$) the first maximum at $W_1^{max} = 0$ disappears. The second peak has a sharp bound at the low-W side of $P(W)$ at the point $W = W_b$ (Fig. 2c). This threshold phenomenon is realized only for the broadband laser excitation (in our experiment we observed a sharp bound at $\Delta\omega_L \approx 0.2$ cm^{-1}, whereas at $\Delta\omega_L \approx 0.06$ cm^{-1} it was absent. It should be noted that W_b corresponds to the constant conversion efficiency ($\eta = \eta_b$) (in our experiment $\eta_b = 5.6$).

- For a highly nonlinear regime ($\eta \approx 40\%$) the data show stabilization of the Stokes pulse energy at about W_2^{max}. Fig. 2d shows stabilization of about 8%.

The theoretical explanation (which is now in progress) of the observed peculiarities of the Stokes pulse energy distribution is based on the consideration of SRS in the nonlinear regime under broadband laser excitation. The 3-dimensional problem [22] must be taken into account also for the explanation.

References

1. P. A. Apanasevich and S. Ya. Kilin, Zh. Prikladn. Spektrosk. 24, 738 (1976); ibid, 29, 252 (1978); Izv. Akad. Nauk SSSR, ser. fiz. 43, 1533 (1979).
2. F. Y. Wu, R. E. Grove, S. Ezekiel, Phys. Rev. Lett., 35, 1426 (1975).
3. H. J. Carmichael, D. F. Walls, J. Phys., B9, 1199 (1976).

4. H. J. Kimble, M. Dagenais and L. Mandel, Phys. Rev., $\underline{A18}$, 201 (1978).
5. P. A. Apanasevich and S. Ya. Kilin, J. Phys. $\underline{B12}$, L83 (1979).
6. A. Aspect, G. Roger, S. Reynaud, J. Dalibard and C. Cohen-Tannoudji, Phys. Rev. Lett., $\underline{45}$, 617 (1980).
7. P. A. Apanasevich and S. Ya. Kilin, Phys. Lett., $\underline{A62}$, 83 (1977).
8. D. F. Walls, P. Zoller, Phys. Rev. Lett., $\underline{47}$, 709 (1981).
9. L. Mandel, Optics Lett., $\underline{4}$, 205 (1979).
10. R. Short and L. Mandel, Phys. Rev. Lett., $\underline{51}$, 384 (1983).
11. S. Ya. Kilin (to be published).
12. P. Kumar and J. H. Shapiro, Phys. Rev., $\underline{A30}$, 1568 (1984).
13. S. Ya. Kilin, Optics Comm., $\underline{53}$, 409 (1985).
14. M. G. Raymer, K. Rząźewski and J. Mostowski, Optics Lett., $\underline{7}$, 71 (1982).
15. I. A. Walmsley and M. G. Raymer, Phys. Rev. Lett., $\underline{50}$, 962 (1983).
16. N. Fabricius, K. Nattermann and D. von der Linde, Phys. Rev. Lett., $\underline{52}$, 113 (1984).
17. F. Haake, Phys. Lett., $\underline{A90}$, 127 (1982).
18. M. Lewenstein, Z. Phys., $\underline{B56}$, 69 (1984).
19. M. Trippenbach, K. Rząźewski, Phys. Rev., $\underline{A31}$, 1932 (1985).
20. I. A. Walmsley, M. G. Raymer, T. Sizer II, I. N. Duling III, J. D. Kafka, Optics Comm., $\underline{53}$, 137 (1985).
21. First reported by A. S. Grabchikov, S. Ya. Kilin, V. P. Kozich, N. N. Jodo at the XII All-Union conf. on Coher. and Nonlinear Optics, Moscow, August 1985.
22. M. G. Raymer, I. A. Walmsley, J. Mostowski, B. Sobolewska, Phys. Rev., $\underline{A32}$, 332 (1985).

Fritz HAAKE

Universität-GHS Essen
Fachbereich Physik
4300 Essen, Deutschland

QUANTUM PENCILS AND A KICKED SPINNING TOP: TWO DIFFERENT TYPES OF LARGE FLUCTUATIONS

First Lecture: Quantum Pencils [1]

Contrary to classical fiction, marginally stable states in macroscopic systems cannot be infinitely long-lived. External perturbations which would induce decay can, at least in principle, be excluded but quantum fluctuations are, however small, always present. The uncertainty principle forbids, e.g., the notion of simultaneous immobility (sharp momentum) and marginality (sharp location) for a pendulum.

I propose to show that the fluctuations (quantum or classical) which initiate the decay of a marginal equilibrium get amplified as the decay goes on. A suitably defined decay time t_d, e.g., has, in an ensemble of decay trajectories, a mean and a r.m.s. of comparable magnitude. The probability density of t_d can be calculated in closed form for various types of systems which are of experimental interest.

The simplest prototype of the processes in question involves a single coordinate S obeying the classical equation of motion

$$\dot{S} = S + \xi(t). \qquad (1)$$

We encounter a linear systematic force and a random $\xi(t)$ competing to destroy the initial "state"

$$S(0) = 0. \qquad (2)$$

The random force can be taken to represent Gaussian white noise of tiny strength according to

$$\langle \xi(t) \rangle = 0, \quad \langle \xi(t)\xi(t') \rangle = \frac{2}{N}\delta(t-t'), \quad N \gg 1. \tag{3}$$

The solution of (1, 2),

$$S(t) = e^t \int_0^t dt' \, e^{-t'} \xi(t'), \tag{4}$$

allows for an asymptotic simplification if we restrict our interest to values of $S(t)$ above the noise level, $S(t) \gg 1/\sqrt{N}$,

$$S(t) \approx e^t \int_0^\infty dt' \, e^{-t'} \xi(t') \equiv e^t \sigma. \tag{5}$$

In this limit the variable $S(t)$ follows a deterministic trajectory (e^t solves $\dot{S} = S$) originating from an effective random initial value σ. Since σ is linearly related to the noise force $\xi(t)$ we find its probability density from (3) to be the Gaussian

$$W(\sigma) \sim e^{-N\sigma^2/2}. \tag{6}$$

Typical values of σ are thus of the order $1/\sqrt{N}$, i.e. tiny on the scale of interest for $S(t)$.

We may now characterize the decay process by the time at which $S(t)$ first reaches some reference value M of order unity $|M| \gg 1/\sqrt{N}$. This decay time,

$$t_d = \ln|M/\sigma|, \tag{7}$$

is a random quantity with the probability density

$$P(t_d) = \int d\sigma \, \delta(t_d - \ln|M/\sigma|) \, W(\sigma)$$
$$= \left\{ NM^2 e^{-2t_d}/2\pi \right\}^{1/2} \exp(-NM^2 e^{-2t_d}/2). \tag{8}$$

It is now a matter of doing simple integrals to evaluate the ratio of the r.m.s. and the mean of t_d,

$$\left\{ \langle (\Delta t_d)^2 \rangle / \langle t_d \rangle^2 \right\}^{1/2} = \sqrt{\psi'(1/2)} / \left\{ \ln(NM^2/2) - \psi(1/2) \right\}, \tag{9}$$

where ψ and ψ' denote the digamma and the trigamma function, respectively. <u>The relative variance of the decay time thus scales</u> with the noise strength as $1/\ln N$. Fluctuations of such large a relative size are easily detectable experimentally.

The $1/\ln N$ scaling of the relative variance of t_d is a rather universal behavior, to be met for beyond the innocent example for which I have derived it here. It prevails, for instance, when (i) an inertial term $\sim \ddot{S}$ is included in (1), (ii) the single variable S is elevated to an n-component vector $\vec{S}$ or (iii) even to a vector field $\vec{S}(\vec{x})$ capable of diffusion in a d dimensional space. We may even replace the linear systematic force in (1) by a nonlinear one like $S - S^3$ or $\sin \theta$, provided only that the reference value M is kept well away (on the scale $1/\sqrt{N}$) from both the marginal equilibrium and any stable equilibrium the variable S might be attracted to by the nonlinear force. For all the generalizations mentioned the probability density of t_d preserves the simple structure (8) up to minor modifications.

I would now like to show explicitely that the above arguments go through unaltered if S represents a set of quantum mechanical operators. With the eventual goal of the comparing theoretical and experimental results I shall present the discussion for the phenomenon of superfluorescence. In that case we have to consider a collection of identical atoms, each excited to the upper one of a pair of relevant energy levels. At the same time, the electromagnetic field must be prepared in the vacuum state. Were it not for the effect of spontaneous emission of photons, i.e. for quantum fluctuations, the coupled system would remain in its initial state forever. The first few spontaneously generated photons will induce further emission and eventually the atoms may deliver all of their excitation energy into an intense outburst of radiation. Such radiation processes are called superfluorescence if their peak intensity is proportional to the square of the number of initially excited atoms.

An interesting situation to investigate arises when N atoms fill an active volume of the shape of a long thin cylinder with Fresnel number close to unity. The radiation will then predominantly go into the diffraction solid angle around the cylinder axis and have sufficiently little transverse structure to allow for a one dimensional description. If, moreover, each transverse section of the cylinder one wavelength in length contains a large number of excited atoms the atomic medium can be regarded as a continuum. Quantum electrodynamics then yields the following Heisenberg equations of motion for the operators representing the positive frequency parts of the electric field and the electric polarization density

$$\left(\frac{\partial}{\partial x} + \frac{1}{c}\frac{\partial}{\partial t}\right) E(x, t) = P(x, t)$$
$$\frac{\partial}{\partial t} P(x, t) = E(x, t). \tag{10}$$

Units have been chosen such that all quantities occurring in (10) are dimensionless and such that an expectation value $\langle E^\dagger E \rangle$ of order unity corresponds to a radiation intensity of order N^2. Actually, the Heisenberg equations retain the simple linear form (10) only during the early stages of the emission process while the depletion of atomic excitation is not yet appreciable.

In the initial state of full excitation the initial mean polarization vanishes

$$\langle P(x, 0) \rangle = 0 \tag{11}$$

while the independence of the fully excited atoms is reflected in the initial second moment

$$\langle P(x, 0) P^\dagger(x', 0) \rangle = \frac{2}{N} \delta(x - x'). \tag{12}$$

Due to the assumed high-density limit the central limit theorem secures Gaussian factorization properties for all higher moments of the initial polarization. Normally ordered products of $E^\dagger(x, 0)$ and $E(x, 0)$, on the other hand, have vanishing expectation values since the initial state of the field is the vacuum.

Up to additive pieces which do not contribute to means of normally ordered products of $E^\dagger(x, t)$ and $E(x, t)$ the solution of (1)

$$E(x, t) = \int_0^x dx' \, I_0(2\sqrt{x'(t - x'/c)}) \, P(x - x', 0), \tag{13}$$

expresses the electric field operator at time t as a linear functional of the initial polarization operator, the kernel involving the modified Bessel function of order zero. The mean intensity $I(t) \equiv \langle E^\dagger(1, t) E(1, t) \rangle$ at the end of the cylinder, $x = 1$, then follows as

$$I(t) = \frac{4}{N} I_0(2\sqrt{t})^2 - I_1(2\sqrt{t})^2 \tag{14}$$

while the mean field itself vanishes at all times. Higher order moments of $E^\dagger(x,t)$ and $E(x,t)$ have, like the initial polarization, Gaussian factorization properties.

Due to the linearity of the early-stage radiation process we can reinterpret the quantum mechanical results in terms of a classical stochastic process. In fact, the quantum means of normally ordered products of $E^\dagger$, E, P, and $P^\dagger$ at arbitrary times are correctly reproduced if these fields are taken as classical random fields obeying deterministic equations of motion of the form (10), E and $E^\dagger$ being zero initially and the initial polarization densities P and $P^\dagger$ having a Gaussian distribution defined by the elementary means (11) and (12). Quantum averages like (14) then appear as classical functional averages over initial polarization configurations.

The reinterpretation of the quantum ensemble by a formally equivalent classical one becomes especially fruitful at times at which the mean intensity $I(t)$ has risen well above the noise level i.e. the level of the initial polarization uncertainty, $I(t) \gg 1/N$. For fields that intense we know from the correspondence principle that an experimentally observed trajectory is indistinguishable from a classical one. The ensemble of classical trajectories can then be identified with an ensemble of experimentally recorded trajectories. Fluctuations in the latter ensemble thus appear as macroscopic blow ups of the tiny quantum uncertainties from which the radiation pulses originate.

We now proceed to introducing a decay time. In analogy with (7) we set some reference intensity $I_{ref} \gg 1/N$ and define t_d by

$$|E(1, t_d)|^2 = I_{ref}, \qquad (15)$$

where E is taken as a complex quantity given by (13). Even though we cannot explicitely solve (15) for the decay time it is possible to construct its probability density from

$$P(t_d) = \frac{d}{dt_d} \langle \Theta(|E(1, t_d)|^2 - I_{ref}) \rangle, \qquad (16)$$

where Θ denotes the unit step function. The average in (16) is a functional one, to be carried out over initial polarization configurations with the Gaussian ensemble specified by the moments (11, 12). There is no difficulty in carrying out the Gaussian functional integral and to obtain the simple result

$$P(t_d) = \frac{d}{dt_d} e^{-I_{ref}/I(t)}, \qquad (17)$$

where $I(t)$ is the mean intensity (14). The smooth curve in the plot below shows the distribution (14) for $N = 1.5 \times 10^8$, $1/c = 0.3$, and a choice for I_{ref} to be specified presently. It is obvious from this plot that the mean and the r.m.s. of t_d are comparable in magnitude. Once again, we have found a behavior typical of the decay of marginal equilibria: while each decay trajectory is effectively deterministic in nature, an ensemble of such trajectories displays large fluctuations.

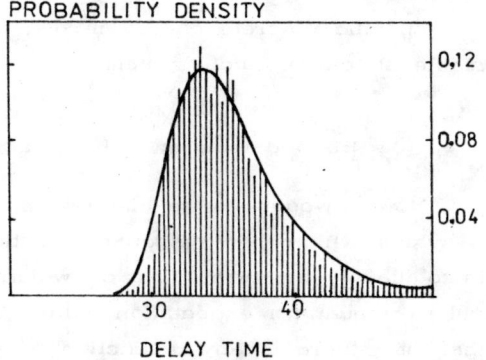

The linearized Heisenberg equations (10) loose their validity when a noticeable fraction of the initially excited atoms have returned to their ground states, i.e. when $I(t)$ has grown to order unity. This happens well after the rise of $I(t)$ above the quantum noise level. In the nonlinear regime during which the radiation intensity is near its maximum value $I_{max} = O(1)$ the classical description of each pulse continues to be accurate. The nonlinear generalizations of (10) can thus be taken as deterministic equations (the socalled Maxwell Bloch equations) and be solved numerically. To obtain an useful ensemble of solutions we must employ a Gaussian random number generator to generate many initial polarization configurations. Each trajectory so initiated reaches its maximum intensity at a particular time which now serves as a natural definition of the decay time. The histogram in the above plot is based on some 1500 such trajectories. The excellent agreement with the result of the smooth curve (linear theory) has been reached by adjusting I_{ref}. Similarly fine is the agreement with the experimental decay time statistics obtained by Vrehen et al. [2].

In concluding this first lecture I should note that the decay of marginal equilibria has been observed and quantitatively explained in terms of the ideas presented here in numerous experiments on superfluorescence [2, 3], the switch-on of lasers [4, 5], and the onset of Stokes radiation in Raman scattering [6, 7, 8, 9]. These experiments invariably exploit the discreteness of atomic energy levels all of which, except for the ground level, are marginally stable from a macroscopic point of view. There are as yet no experiments on fluctuations in the decay of marginal equilibria in hydrodynamical or mechanical systems nor in spinodal decomposition; we have no practical means available to populate specific discrete energy levels, i.e. to precisely prepare marginally stable states for such systems.

Second lecture: A Kicked Spinning Top [8]

Now, I would like to discuss the quantum mechanical behavior of a system which displays chaotic motion in the classical limit. The goal is to identify the time scale on which classically chaotic behavior dies out from quantum expectation values and to find out whether, beyond that time, there is any distinction between regular and irregular behavior in quantum means, i.e. any possibility to define something like quantum chaos.

As a convenient system to carry out such an investigation for we have chosen a spinning top subjected to a periodic sequence of nonlinear impulsive kicks. The appropriate dynamical variables are the components J_i of an angular momentum vector obeying the commutation relations $[J_i, J_j] = i \varepsilon_{ijk} J_k$. They allow the Hamiltonian to be expressed as

$$H(t) = \hbar p J_y + (\hbar k/2j) J_z^2 \sum_{n=-\infty}^{+\infty} \delta(t - n). \qquad (18)$$

The first term in (18) induces a precession of $\vec{J}$ around the y axis with angular velocity p. Each of the kicks described by the second term may be interpreted as giving rise to a precession, too. However, the rotation axis points in the z direction and, most importantly, the angular velocity is proportional to J_z rather than simply a constant; we here have the nonlinearity without which we could not expect chaos even classically. The kick strength is characterized by the parameter k/j.

The process in question obviously has the squared length $\vec{J}^2$ of the angular momentum as a conserved quantity, $[H, \vec{J}^2] = 0$. We may therefore restrict our quantum mechanical investigation to a $2j + 1$ dimensional Hilbert space in which

$$\vec{J}^2 = j(j + 1). \qquad (19)$$

Nearly classical behavior must be expected for $j \gg 1$. We might think of $\hbar j$ as having some fixed macroscopic value; the classical limit then formally corresponds to $j \to \infty$ with $\hbar \to 0$ such that $\hbar j$ = const.

It is appropriate and convenient to describe the dynamics of our top in a stroboscopic fashion and to introduce the unitary evolution operator

$$U = e^{-ip J_y} e^{-i(k/2j) J_z^2}, \qquad (20)$$

which transports the wave function from kick to kick. Actually, this operator could also have resulted from a Hamiltonian which has the nonlinear precession as the time independent piece and the linear precession as the externally driven one. Such a reinterpretation suggests an experimental realization of our process with the help of a compass needle bound to the $z = 0$ plane by a potential $\sim J_z^2$ [9].

In order to find the classical behavior of our top we may start from the Heisenberg picture operators $\vec{J}(n) = U^{\dagger n} \vec{J} U^n$ and write down the map relating $\vec{J}(n+1)$ to $\vec{J}(n)$. As j goes to infinity the rescaled operators

$$\vec{X}_n = \vec{J}(n)/j \qquad (21)$$

become c-number vectors of unit length. Their discrete time evolution is governed by the quantum map $\vec{J}(n) \to \vec{J}(n+1)$ with all commutators taken to vanish. We thus find the map

$$\begin{aligned} X_{n+1} &= Z_n \cos k X_n + Y_n \sin k X_n \\ Y_{n+1} &= Y_n \cos k X_n - Z_n \sin k X_n \\ Z_{n+1} &= -X_n, \end{aligned} \qquad (22)$$

which is two dimensional because of $\vec{X}_n^2 = 1$.

I shall not go into a detailed discussion of the properties of the classical map (22) here but rather state what we need for the following, specializing to $p = -\pi/2$ and $k = 3$. There are two hyperbolic fixed points at the poles defined by the precession axis, $y = \pm 1$. Moreover, the hemisphere $y \geq 0$ accomodates two elliptic fixed points related to one another by a rotation around the precession axis by π. Upon rotating each of these elliptic points by π around the kick axis we obtain

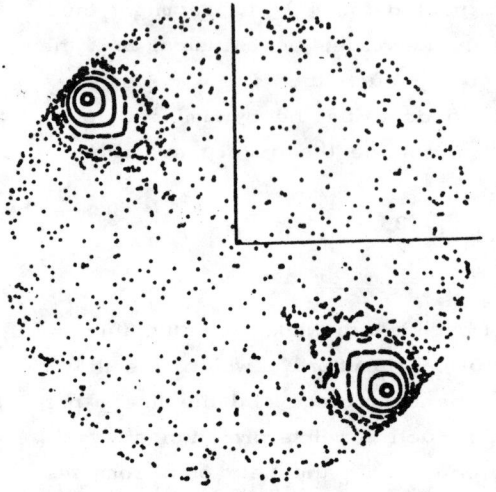

two new points in the hemisphere $y < 0$ which are part of an elliptic period 2 solution. Around each of the four stable points there is a region, about 1/2 radians in size, within which all trajectories are regular. These regions are surrounded by narrow bands containing both chaotic and regular trajectories. Further out, and in fact on the rest of the sphere, much smaller islands of stability excepted, we find global chaos with a Lyapounov exponent $\lambda \approx 0.33$. The following plot displays typical trajectories projected onto the plane $y = 0$.

Let us now turn to the quantum case. We have done calculations for $j = 100$. In order to resolve as much structure as possible we have chosen coherent states [10] as initial states. Such states assign to the expectation value $\langle \vec{J} \rangle$ a direction with minimum uncertainty, the relative variance of $\vec{J}$ being $1/\sqrt{j}$. The finest details of the classical may we can expect to resolve are thus about 1/100 radians in angular aperture. For various such coherent initial states we have numerically calculated the expectation value $\langle J_y \rangle_n / j$ for times n up to 1000. Each of these means was compared to a classical average of y over a bundle 1000 classical trajectories, the bundle originating from an initial cloud of points on the unit sphere corresponding in location and size to the quantum initial state. The following results were obtained:

(i) For initial states well within a region of fully regular classical motion the classical and quantum means agree well for times up to $n \lesssim j = 100$. For larger times the classical means become stationary while the quantum ones go through rather periodic alternations of collapse and revival [11]. A revival typically brings $\langle J_y \rangle$ back close to its initial value and the period is of the order $j = 100$. Such revivals are typical quantum effects, this period just being the time scale on which the spacing of neighboring quasi energy levels (roughly $2\pi/(2j+1)$) can be resolved.

(ii) For initial states well within the region of global classical chaos the agreement between classical and quantum means is lost already at times of the order ln j. This can be understood as a manifestation of chaos: tiny perturbations such as quantum fluctuations, order-of-magnitude-wise equivalent to an initial separation $1/\sqrt{j}$ of two classical trajectories, entail separations growing exponentially in time. On the time scale j which reveals the quasiperiodicity of the quantum motion we now find somewhat of a surprise. There are still recurrences of $\langle J_y \rangle_n$ to the neighborhood of its initial value; however, these recurrences now appear to take place completely irregularly rather than nearly periodically; there is no trace of regular sequences of collapse and revival [11].

(iii) As the initial state wanders from the region of classically regular motion into that of global chaos the regular modulations of $\langle J_y \rangle_n$ get lost gradually. The transition is smeared out in correspondence to the "size" of a coherent state and the width of the classical transition region.

(iv) A spectral analysis corroborates the above findings. While a coherent state located at a classical elliptic fixed point is exhausted in its normalization to within 1% by the eight most important eigenvectors of the time evolution operator, at least about 60 eigenvectors are need to represent a coherent state well within the classically chaotic region. Again, the transition is smeared out similarly as the one seen in the time dependent expectation values.

It may be appropriate to conclude by emphasizing that the results discussed above yield a qualitative criterion for quantum chaos in our model: both the periodic alternations of collapse and revival (which indicate regular motion) and the erratic recurrences (which signal "chaos") are typical quantum effects.

References

1. The material reported in the first lecture has been extracted from
    a) R. Glauber and F. Haake, Physics Lett., 68A, 29 (1978).
    b) F. Haake, Phys. Rev. Lett., 41, 1685 (1978).
    c) F. Haake, H. King, G. Schröder, J. Haus, R. Glauber, F. Hopf, Phys. Rev. Lett., 42, 1740 (1979).
    d) F. Haake, H. King, G. Schröder, J. Haus, R. Glauber, Phys. Rev., 20A, 2047 (1979).
    e) F. Haake, J. Haus, H. King, G. Schröder, R. Glauber, Phys. Rev. Lett., 45, 558 (1980) and Phys. Rev., A23, 1322 (1981).
    f) F. Haake, J. Haus, R. Glauber, Phys. Rev., A23, 3255 (1981).

2. Q. H. F. Vrehen in Laser Spectroscopy IV, edited by H. Walther and K. W. Rothe (Springer, Berlin, 1979).
3. M. Gross and S. Haroche, Physics Reports, $\underline{93}$, 301 (1982).
4. F. T. Arecchi and V. Degiorgio, Phys. Rev., $\underline{A3}$, 1108 (1971).
5. D. Meltzer and L. Mandel, Phys. Rev., $\underline{A3}$, 1763 (1971).
6. I. A. Walmsley and M. G. Raymer, Phys. Rev. Lett., $\underline{50}$, 962 (1983).
7. N. Fabricius, K. Nattermann, and D. von der Linde, Phys. Rev. Lett., $\underline{52}$, 113 (1984).
8. The material presented in the second lecture was generated in collaboration with M. Kuś, J. Mostowski, and R. Scharf. It will be published elsewhere in more detail.
9. H. Frahm and H. J. Mikeska, Z. Physik, $\underline{60B}$, 117 (1985).
10. a) F. T. Arecchi, E. Courtens, R. Gilmore, and H. Thomas, Phys. Rev., $\underline{A6}$, 2211 (1972).
    b) R. Glauber and F. Haake, Phys. Rev., A13, 357 (1976).
11. J. H. Eberly, N. B. Narozhny, J. J. Sanchez-Mondragon, Phys. Rev. Lett., $\underline{44}$, 1323 (1980).
12. R. Graham and M. Höhnerbach also find, in an analysis of the Jaynes Cummings model, revivals to disappear for sufficiently strong couplings. See R. Graham in "Proceedings of the Topical Meeting on Instabilities and Dynamics of Lasers", Rochester, 1985, to be published.

A. V. MASALOV

Lebedev Physical Institute,
Moscow 117924, USSR

SPECTRAL AND TEMPORAL FLUCTUATIONS OF BROAD-BAND LASER RADIATION

In the field of nonlinear optics, multiphoton atomic and molecular spectroscopy, etc., an ideal laser source is a laser generating a single longitudinal mode of lowest transversal index. The intensity distribution in time and space of a single-mode laser pulse is smooth (bell-shaped) function of coordinates and can be measured with a sufficient accuracy. A single-mode laser radiation has the highest degree of monochromaticity and in some light-matter interactions realizes the dream of theoretician about a monochromatic light field. However, for numerous high power pulse lasers this operation regime is hardly achieved, while a broad-band regime is common. A laser generating many longitudinal modes is a well-known example of broad-band operation. A specific feature of broad-band laser radiation is that its temporal and spectral pictures have a random character: intensity changes randomly in time during laser pulse and the spectrum exhibits a random structure (Fig. 1). In the processes of nonlinear interaction with matter the broad-band radiation is essentially nonmonochromatic. Broad-band radiation requires a more complicated statistical description. In spite of this broad-band lasers are widely used for the study of nonlinear light-matter interaction because the broad-band operation regime is realized in lasers of a rather simple construction and in its energetic characteristics it is better than other regimes, in particular, than a single-longitudinal-mode operation. In this lecture the statistical properties of the temporal and spectral pictures of broad-band laser radiation are de-

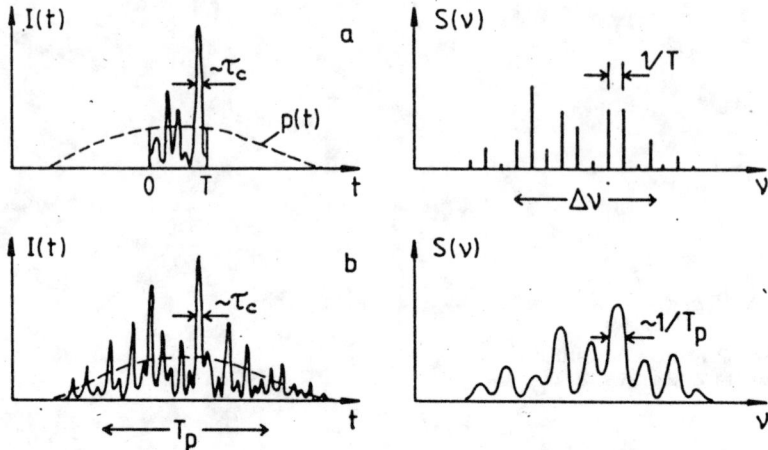

Fig. 1. Temporal $I(t)$ and spectral $S(\nu)$ pictures of broad-band laser radiation with well-pronounced (a) and without (b) longitudinal modes

scribed. The consideration is made within a laser light model, which is aimed first of all at describing the properties of radiation of Q-switched solid-state lasers (neodymium-glass, ruby, YAG:Nd) and dye lasers pumped by nanosecond pulses. The main attention is given to the radiation properties which are essential for nonlinear interaction, namely, to the probability distribution of intensity $P(I)$ and high-order spectra $S_K(\nu)$ [1].

The intensity distribution $P(I)$ is required in the description of inertialess interaction with matter, where the matter response time is less than the duration of intensity fluctuations i.e. the coherence time. Examples of such interaction are: multiphoton ionization of atoms, tunnel ionization of atoms, harmonic generation, etc. In these processes the broad-band radiation interacts with matter by the same manner as a monochromatic field with a slowly-varying intensity. The yield of a nonlinear process in this case is equal to that in a monochromatic field averaged over intensity fluctuations of broad-band radiation by means of probability distribution.

The K-order spectrum of radiation $S_K(\nu)$ is required in the description of K-photon interaction in the case where the radiation coherence time is smaller than the matter response time. An example of such an interaction is the multiphoton excitation of atoms under condition that the radiation spectrum width exceeds the atomic transition width.

The K-order spectrum is the Fourier-transform of the K-order coherence function $\langle [E^*(t)E(t+\tau)]^K \rangle$. The expression for the yield of K-photon excitation in a frequency language has a form of integral over K-order spectrum multiplied by atomic absorption line.

Statistical Model of Broad-Band Laser Radiation

In pulse lasers the output radiation originates from the spontaneous noise of an active medium provided that this noise is affected by a number of factors: amplification, spatial and frequency filtration, saturation of amplification in an active medium, nonlinear interaction with resonator elements. The statistical model of laser radiation, which includes the effects of all these factors does not exist. Here in formulating the statistical model of broad-band radiation only those lasers are considered, in which a nonlinear interaction of radiation with resonator elements is negligibly small, and the role of active medium saturation is reduced to a stabilization of the total pulse energy. Besides, the spatial filtration of the initial spontaneous noise is assumed to be sufficiently effective to select the lowest transverse mode, and the frequency filtration - to be not so effective and to select in the noise spectrum a broad interval $\Delta\nu$ exceeding the inverse pulse duration T_p: $\Delta\nu > 1/T_p$. According to this model the broad-band laser radiation is a noise inscribed into a pulse envelope and differed from the initial spontaneous noise by the determinate value of energy.

Mathematically the model of broad-band laser radiation is described by energy-normalized Gaussian noise. In the spectral language the broad-band laser field is a sum over the fields of longitudinal modes with random amplitudes and phases; modes are statistically independent, but the sum over mode energies is constant. Although the spontaneous noise is associated with the Gaussian statistics, the output radiation statistics is not Gaussian because the energy stabilization is a nonlinear transform of initial noise. The model under discussion was analysed in a number of papers (for references see [2]).

Intensity Distribution

The intensity distribution of radiation in the energy-normalized Gaussian noise model is given in [3]:

$$P_N(I) = \frac{N-1}{N\langle I \rangle}\left(1 - \frac{I}{N\langle I \rangle}\right)^{N-2}, \quad 0 \leqslant I \leqslant N\langle I \rangle,$$

where N is the number of generated longitudinal modes. The average intensity $\langle I \rangle$ is practically unchanged during the axial period of radiation, but varies from period to period like the pulse envelope. The main property of the distribution $P_N(I)$ is the dependence on the number of generated modes N.

Besides, the distribution $P_N(I)$ differs from exponential one, which corresponds to Gaussian noise. While in the exponential distribution the intensity is unlimited, in this distribution the range of possible intensity values is limited from above by the value N times larger than the average intensity. The maximum intensity realizes if the radiation energy of one axial period is concentrated in a single overshoot. Obviously the difference is caused by the condition that the energy of laser radiation is constant. In the Gaussian noise the energy of N overshoots fluctuates in a scale of $1/\sqrt{N}$.

At $N = 1$ the distribution $P_1(I)$ tends to the delta-function; this is the case of a single-mode generation.

At $N = 2$ the distribution $P_2(I)$ is uniform in the interval from 0 to $2\langle I \rangle$. Such a distribution is realized when the amplitudes of both generated modes are random and their mean energies are equal.

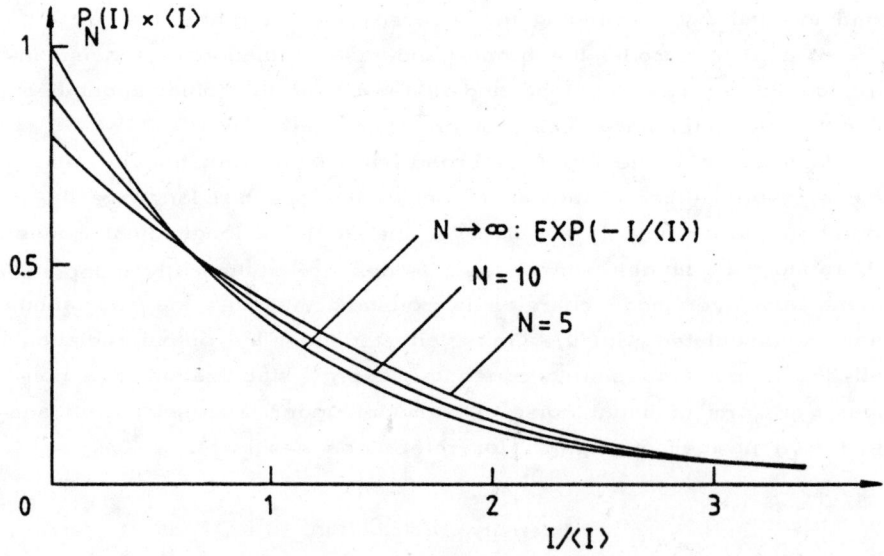

Fig. 2. The intensity distribution $P_N(I)$ of broad-band laser radiation for various numbers of modes N

At $N = 3$ in the distribution $P_3(I)$ there arise features which make it resemble the exponential distribution: the most probable intensities are zero and as the intensity increases the probability density decreases.

Some other examples of the distribution $P_N(I)$ are shown in Fig. 2. When the number of modes increases the distribution tends to the exponential one. This circumstance is the basis for the formulation of a validity criterion of Gaussian statistics for the broad-band laser radiation (this will be done later).

Thus, the distribution $P_N(I)$ describes in a universal form the laser radiation properties in quite polar operation regimes: the single-mode operation regime, where intensity fluctuations are absent, and a multimode regime, where the intensity fluctuations are practically unlimited.

Statistical Factors of Radiation

When the nonlinear light-matter interaction has the form of K-photon process, the yield of the process is proportional to I^K and it is sufficient to characterize broad-band radiation by the Kth moment of intensity distribution. The relative value of moment is called a statistical factor of radiation in K-photon process: $g_K = \langle I^K \rangle / \langle I \rangle^K$. The statistical factor g_K shows how much the K-photon yield under the broad-band laser radiation exceeds the yield under the single-mode radiation, the mean intensities being equal.

The moments of intensity distribution $P_N(I)$ and corresponding statistical factors are known: $g_K = K! N^K (N-1)! / (N+K-1)!$. Like the intensity distribution the statistical factors of broad-band radiation depend on the number of generated modes N. As N increases from 1 to infinity the statistical factor also increases from 1 to its asymptotic value K!. Both limiting cases are obvious: radiation with $N = 1$ corresponds to a single-mode laser, and at large N the radiation obeys the Gaussian statistics. Figure 3 illustrates the dependence of the statistical factors on the number of modes at several K. The higher the nonlinearity K the larger the range of N-dependent variation of the statistical factor. For example, for 5-photon process the statistical factor changes within two orders of magnitude, and for 11-photon process - within more than seven orders.

Asymptotic behavior of statistical factors at large N makes it possible to formulate in terms of mode number the validity criterion of Gaussian

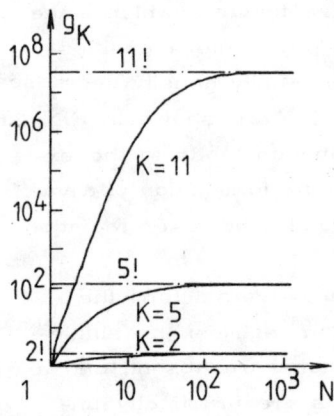

Fig. 3. Dependence of the statistical factor g_K on the number of modes N at K = 2, 5 and 11

statistics for broad-band laser radiation in K-photon process. In this process the laser statistics cannot be distinguished from the Gaussian one if the relative deviation of statistical factor from K! does not exceed the relative accuracy of yield measurements $\delta Y/Y$. Using the asymptotic behavior of statistical factor one can conclude that if the number of modes N exceeds the value $N^* = K(K-1)/2(\delta Y/Y)$ then in the K-photon process the laser radiation is equivalent to that with Gaussian statistics. The radiation spectrum of Q-switched neodymium-glass laser or dye laser reaches $\sim 10^4$ modes, and that of ruby and neodymium-YAG laser $\sim 10^2$ modes. The radiation of these lasers can be equivalent to that with Gaussian statistics in K-photon processes within a wide range of K.

A direct experimental measurement of the intensity distribution of broad-band laser radiation is rather difficult, it requires the apparatus capable of tracing radiation intensity with temporal resolution in a sub--picosecond scale. For this reason the way of verifying the applicability of the distribution $P_N(I)$ to real broad-band lasers consists of measuring the statistical factors of laser radiation.

Experimental measurements of statistical factors answer different questions depending on the number of laser modes. Measurements at $N < N^*$ solve the question about applicability of the distribution $P_N(I)$ to a real broad-band laser and verify the dependence of statistical factors on the number of modes. Measurements of factors at $N > N^*$ answer the question on the possibility to construct a laser emitting a radiation with Gaussian statistics.

Measurements of statistical factors of broad-band laser radiation consists of comparison of the yield of K-photon inertialess process under a laser field with that under the field of a single-mode laser of the same intensity. Such a comparison is practically made in the study of yield dependence on the average radiation intensity in a double logarithmic scale (Fig. 4).

Statistical factors of laser radiation at $N > N^*$ were measured in [4, 5]. The factor g_2 of neodymium-glass laser radiation with 300 modes has been measured in the process of 2-photon absorption in Rodamine 6G solution [4]. The value $g_2 = 2,1 \pm 0,3$ coincided with 2!. The factor g_5 of neodymium-glass laser radiation with N = 4000 has been measured in the process of 5-photon ionization of sodium atoms [5] (Fig. 4). The obtained value $g_5 = 10^{2,04 \pm 0,25}$ coincided with $5! = 10^{2,08}$. Results of these measurements show that broad-band laser radiation can be equivalent to that with Gaussian statistics in nonlinear processes at least for $K \leq 5$.

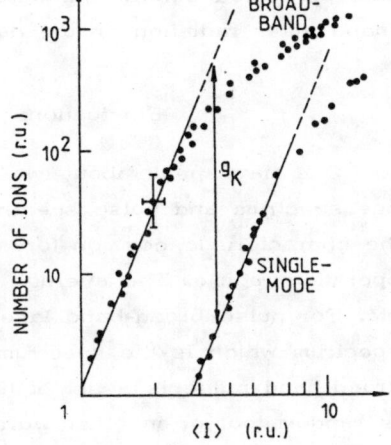

Fig. 4. Number of ions produced in the process of 5-photon ionization of sodium atoms as a function of average intensity of broad-band (upper curve) and single-mode (lover curve) laser radiation according to [5]. The deviation from linear dependence is due to the depletion of sodium atoms

The most complete data on statistical factors of laser radiation at $N < N^*$ are presented in [6]. The factors g_{11} of neodymium-glass laser radiation with various numbers of modes were measured in the process of 11-photon ionization of xenon atoms. The data are summarized in Fig. 5. Experimental values are compared with the calculations where the averaging has been made over logarithm of ion signals, but not

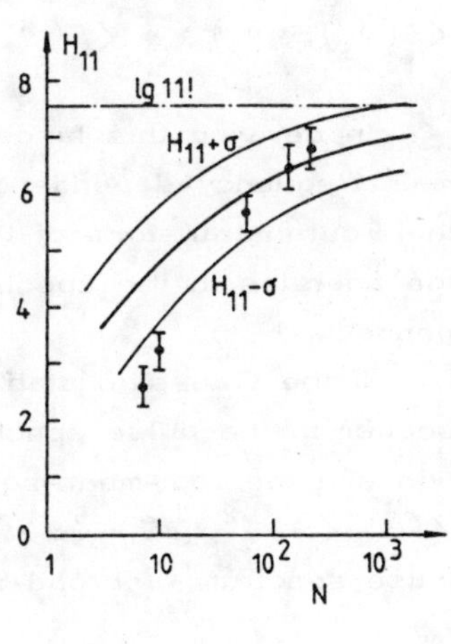

Fig. 5. Dependence of "pseudostatistical factor" H_{11} of broad-band laser radiation on the number of generated modes N. Solid lines: calculations of H_{11} and root-mean--square deviation from [7]; points: experimental data [6]

over ion signals directly. This procedure of averaging to the actual experiments and is caused by the high nonlinearity of the process. Thus, Fig. 5 shows not the statistical factors, but the factors corrected for the procedure of averaging. The experimental values are seen to agree satisfactorily with the calculations. Some discrepancy at small mode numbers may be due to a roughness of spectral width estimation. The data presented confirm the dependence of statistical properties of broad--band laser radiation on the number of generated modes.

Fluctuations in Ordinary Spectrum

One may speak about two spectra of pulse laser radiation: average spectrum and pulse spectrum. Average spectrum is understood as the characteristic common for all pulses emitted by a laser in a certain operation regime. The average spectrum is described by shape, width, etc. For pulse broad-band lasers one can also speak about the pulse spectrum which is the spectrum of one separate laser pulse. Since a broad-band pulse is a set of fluctuation overshoots, the pulse spectrum is random too, or in other words, has a random structure. The principal characteristics of the pulse spectrum are: the frequency scale of structure, fluctuation scatter, etc.

Qualitatively the statistical properties of a pulse spectrum are known: the frequency scale of random structure is determined by the inverse pulse duration, and the scatter is close to the value of average. Quantitatively the fluctuations of a pulse spectrum are described by the frequency correlation function [8]:

$$\langle S(\nu_1) S(\nu_2) \rangle = \langle S(\nu_1) \rangle \langle S(\nu_2) \rangle \left[1 + \frac{\left| \int p(t) e^{-2\pi i (\nu_1 - \nu_2) t} dt \right|^2}{\left(\int p(t) dt \right)^2} \right].$$

In deriving this function the Gaussian statistics were used. The main frequency dependence of the correlation function is described by the Fourier-transform of the pulse envelope $p(t)$. When $|\nu_1 - \nu_2| > 1/T_p$, the energies in the spectrum $S(\nu_1)$ and $S(\nu_2)$ are statistically independent.

If the Gaussian statistics is assumed for the radiation the energy scatter in the pulse spectrum corresponds to the exponential distribution and the root-mean-square deviation equals to the mean value: $\langle (S - \langle S \rangle)^2 \rangle^{1/2} = \langle S \rangle$. The probability distribution of energy in the pulse spectrum of broad-band radiation was studied in [9]. The energy distribution of a dye laser pulse was measured after passing through a narrow-band frequency filter - a confocal Fabry-Perot interferometer. The filter bandwidth was several times less than the inverse pulse duration. The obtained distribution turned out to be close to exponent with a good accuracy. It should be noted that one can ignore the effect of energy stabilization in these measurements because the width of the spectrum was large enough. Besides, the dye laser used did not exhibit the structure of longitudinal modes.

For the laser with a pronounced mode structure the given relation means that the amplitudes of individual modes are independent of each other and the energy distribution is exponential. These both facts were verified in the experiment with the radiation of neodymium-glass laser [8]. According to the results of this experiment the pulse spectrum of broad-band laser radiation with a pronounced mode structure is a set of modes of chaotic energies, among which zero-energy modes are encountered more often than others.

Spectra of High Orders

Like in the case of an ordinary spectrum, one should distinguish between pulse and average high-order spectra: the pulse spectrum characterizes the radiation of a single pulse, while the average spectrum - a set of pulses. Mathematical definitions of pulse and average spectra are equivalent to those of ordinary spectra with a simple replacement of the field amplitude $E(t)$ by $E^K(t)$. From the mathematical point of view the spectra of high orders are equivalent to the particular correlation functions of laser noise. However in a K-order process of light-matter interaction a K-order spectrum has some advantages. In the process of K-photon absorption a K-order spectrum plays the same role as an ordinary spectrum in one-photon absorption. This make it possible to extend the usual language of one-photon absorption to a multiphoton case. Besides, high-order spectra are connected directly with experimentally measurable ordinary spectrum, whereas correlation functions are related to the ordinary spectrum through a Fourier-transformation.

On a frequency scale the K-order spectrum is located near the frequency K times larger than the central frequency of radiation (Fig. 6). The K-order spectrum has a simple interpretation: it coincides with the radiation spectrum of K-order harmonic produced by an ideal harmonic generator.

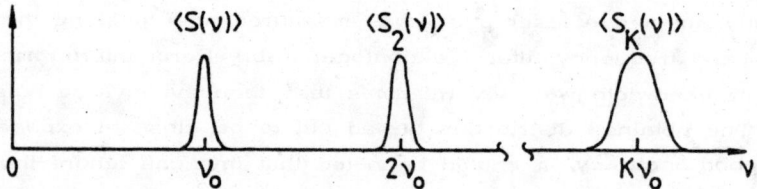

Fig. 6. High-order laser radiation spectra $\langle S_K \rangle$

In the case of Gaussian statistics the average K-order spectrum is connected with $\langle S \rangle$ by the $(K-1)$-fold convolution: $\langle S_K \rangle \sim \langle S \rangle \otimes \langle S \rangle \otimes \ldots \otimes \langle S \rangle$. Hence the K-order spectrum is broader than the ordinary one. Only one example of experimental measurement of average high-order spectrum is known [10]. The authors have measured $\langle S_2(\nu) \rangle$ of a broad-band dye laser in the process of two-photon sodium atom excitation. The central frequency of laser radiation was tuned over two-photon resonance, the atomic line being much smaller than the radiation spectral width. The registered dependence corresponded to the convolution of two ordinary average spectra.

The high-order pulse spectra of broad-band radiation exhibit a random structure. Quantitative characteristics of this structure are described by the frequency correlation function [2]:

$$\langle S_K(\nu_1) S_K(\nu_2) \rangle =$$
$$= \langle S_K(\nu_1) \rangle \langle S_K(\nu_2) \rangle \left[1 + \frac{A_K}{\Delta \nu T_K} + \frac{|\int p^K(t) e^{-2\pi i (\nu_1 - \nu_2) t} dt|^2}{(\int p^K(t) dt)^2} \right].$$

Again the Fourier-transform of the pulse envelope $p^K(t)$ defines the frequency scale of the structure; T_K is the width of $p^K(t)$. In contrast to the ordinary spectrum there is some correlation between $S_K(\nu_1)$ and $S_K(\nu_2)$ even for the frequency difference larger than $1/T_K$. This correlation is attributed to the smooth "base" in the spectrum which fluctuates from pulse to pulse. In the ordinary spectrum such a base is absent. Values A_K responsible for this correlation can be calculated when the Gaussian statistics are applicable to the laser radiation [11]: $A_2 = 4$, $A_3 = 15,6$, $A_4 = 51,6$,... The scatter of energy in the K-order spectrum exceeds the value of average. This shows that the probability distribution of S_K differs from exponential. Examples of experimental measurements of random structure in the K-order spectra are not known.

The K-order spectrum is a very convenient laser radiation characteristic for the analysis of yield fluctuations in K-photon process. The analysis is based on the expression for the yield, which is the overlap integral of K-order spectrum and atomic line $\mathcal{A}(\nu)$: $Y \sim \int S_K(\nu) \mathcal{A}(\nu) d\nu$. This relation describes, for example, the yield of K-photon excitation of atoms in the low intensity regime, where atomic levels are not shifted or broadened by a laser field [1]. In the analysis of yield fluctuations three particular cases should be considered: atomic line width (i) is less than the random structure of S_K, (ii) covers several components of the structure and (iii) is broader than the spectral width (Fig. 7).

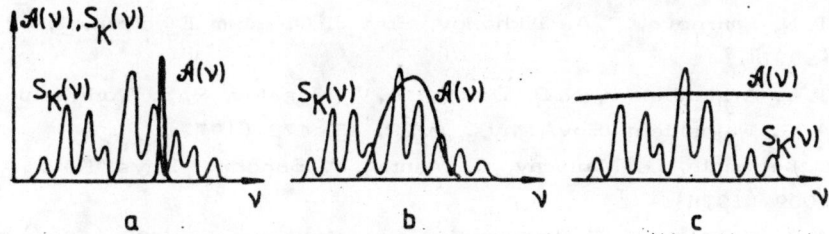

Fig. 7. The overlap of atomic line $\mathcal{A}(\nu)$ and K-order spectrum

The maximum fluctuations are expected in the first case, and the minimum - in the last one.

In the intermediate case, where for the atomic line width $\Delta\nu_a$ an inequality holds: $1/T_K < \Delta\nu_a < \Delta\nu$, the yield fluctuations depend on the atomic line width. This circumstance is the fundamental principle of fluctuation spectroscopy (sub-laser-line spectroscopy): using the data on the signal scatter in the K-photon excitation process, one can measure the atomic line width $\Delta\nu_a$ when it is smaller than the laser radiation spectrum width [11].

Concluding Remarks

Two characteristics of broad-band laser radiation have been analyzed: the intensity distribution $P_N(I)$ and the spectra of different orders $S_K(\nu)$. The distribution $P_N(I)$ is necessary to calculate the statistical factor of radiation in the measurements of multiphoton cross-sections using broad-band laser. High order spectra are convenient in analyzing the frequency dependence in the process of multiphoton excitation of atoms (molecules) and, in particular, in estimating yield fluctuations of these processes. The concept of high-order spectrum may prove useful for describing the radiation properties not only of broad-band lasers, but also of other sources.

References

1. N. B. Delone, V. A. Kovarski, A. V. Masalov, N. F. Perelman, Sov. Phys.-Usp., 23, 472 (1980).
2. A.V. Masalov, in: Progress in Optics XXII, ed. E. Wolf (North-Holland, Amsterdam) 1985, p. 145.
3. A. V. Masalov, Sov. J. Quantum Electron, 6, 902 (1976).
4. T. N. Smirnova, E. A. Tikhonov, Sov. J. Quantum Electron. 7, 621 (1977).
5. T. U. Arslanbekov, N. B. Delone, A. V. Masalov, S. S. Todirashku, A. G. Fainshtein, Sov. Phys. JETP, 45, 473 (1977).
6. C. Lecompte, G. Mainfray, C. Manus, F. Sanchez, Phys. Rev., A11, 1009 (1975).
7. A. V. Masalov, S. S. Todirashku, Opt. Commun., 32, 497 (1980).
8. V. I. Malyshev, A. V. Masalov, A. I. Milanich, Sov. J. Quantum Electron. 5, 1066 (1975).
9. S. M. Curry, R. Cubeddu, T. W. Hänsch, Appl. Phys., 1, 153 (1973).
10. S. J. Smith, P. B. Hogan; in: Proc. IV Int. Conf. on Laser Spectroscopy, 1979 (Springer, Berlin) p. 360.
11. A. V. Masalov, L. Allen, J. Phys., B15, 2375 (1982).

Peter ZOLLER

Institute for Theoretical Physics,
University of Innsbruck,
6020 Innsbruck, Austria

RYDBERG THRESHOLDS IN LASER FIELDS: APPLYING MULTICHANNEL QUANTUM DEFECT THEORY TO QUANTUM OPTICS PROBLEMS

1. Introduction

A good deal of our understanding of the resonant interaction of laser radiation with atoms is based on the "two-level atom". In this model the coupling between the two (near) resonantly coupled atomic states is treated exactly (within the rotating wave approximation) while excitation of all other nonresonant atomic levels is either ignored or calculated in perturbation theory. A necessary condition for the validity of the two-level approximation is that the Rabi frequency of the resonant atomic transition is much smaller than the separation to the neighboring (nonresonant) states. Obviously, when this condition is violated, we are led to consider "three-level systems" etc. But it is clear that this philosophy breaks down for excitation near a Rydberg threshold where the mixing on an infinite number of bound states converging to the threshold and the adjacent continuum has to be considered (for a summary of various aspects of Rydberg states see Ref. 1 and 2).

Here we suggest a scattering type theory to deal with this problem of near threshold excitation of Rydberg series by laser radiation. Briefly speaking, the idea is to solve the problem of interaction of the Rydberg electron with the laser light in the continuum <u>above</u> threshold and then to <u>extrapolate</u> by analytic continuation this scattering information to the <u>bound</u> part of the Rydberg series. The essential feature of this theory is that it considers the interaction with a Rydberg series as a whole, characterized by a few parameters (as opposed to treating the interaction with each Rydberg level separately). Thus from the outset we include the laser mixing between all bound and continuum states of one series (n-mixing) and relate processes across the Rydberg threshold.

The first question to be answered is how this extrapolation of scattering information across the Rydberg threshold can be achieved. As will be shown in Sec. 3 (and in a special case in Sec. 2), the interaction of a Rydberg electron with optical laser radiation can be formulated as a scattering problem where Coulomb type dissociation channels are coupled by a laser induced <u>short range</u> potential (in this context see also Ref. 3 and 4). This allows the well-developed formalism of scattering theory with Coulomb functions plus short range potentials, known to atomic physicists as multichannel quantum defect theory (MQDT) [2], to be applied (which so far has only been used in the context of parametrizing configuration mixing of many electron atoms).

The present theory assumes that -as a first step - we are able to solve a scattering problem in the laser field. As our second question we may, therefore, ask to what extent we are able to do better in solving a scattering problem than a bound state problem directly. It turns out, however, that even simple perturbation theory for the laser field in the continuum leads below threshold to nontrivial results which are certainly better than standard perturbation theory in the bound region.

Finally, the above discussion can be extended to develop a unified theory of autoionizing Rydberg series in the presence of laser radiation. This may be achieved by treating electron correlations and interaction with the light on the same footing in an MQDT-scattering formalism, where part of the short range interaction is due to configuration mixing, coupling channels with different ionic excitations; and the laser induced short range coupling leads to transitions between channels differing by photon numbers.

2. Strong Field Excitation of an Autoionizing Rydberg Series

Instead of introducing a general formalism at the very beginning, we start with a specific example. We consider excitation of an autoionizing Rydberg series in a double optical resonance configuration (Fig. 1), a problem which for an isolated resonance has been discussed in detail in the literature [5]: a weak laser of frequency excites atoms from the ground state g with energy E_g to one of the excited states e (energy E_e) which is coupled by a strong laser of frequency to an autoionizing Rydberg series. For weak excitation by the first laser the ionization rate is

$$R = -\frac{1}{2}\Omega^2_{eg} \operatorname{Im}\left\{\left[E_g + \omega_1 - E_e - \sum\right]^{-1}\right\} \qquad (1)$$

with

$$\sum = \langle e|D\left[E_g + \omega_1 + \omega_2 - H_A + i\epsilon\right]^{-1}D|e\rangle \qquad (2)$$

the resonant part of the complex self energy of $|e\rangle$. H_A is the atomic Hamiltonian, D the dipole operator describing interaction with the strong light field. For $E_I < E_e + \omega_2$, i.e., in the unstructured continuum (see Fig. 1), $\sum$ is a slowly varying function of energy; in the autoionizing region $0 < E_e + \omega_2$ contains the infinite series of Rydberg resonances. To extract this rapid energy dependence we relate the "above" and "below threshold" self energies in the following way: first we write $\sum = \langle e|D|F^\lambda\rangle$ where $|F^\lambda\rangle$ is a solution of the inhomogeneous equation

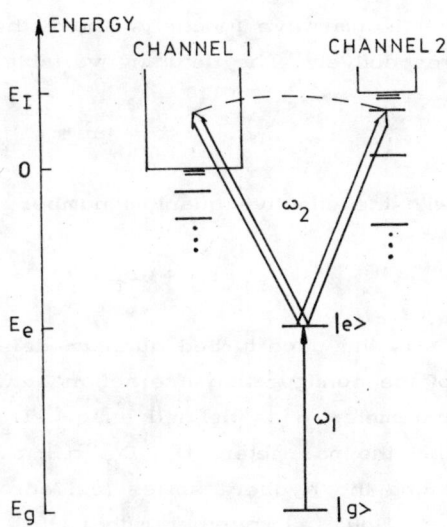

Fig. 1. Atomic configuration

$$(E_g + \omega_1 + \omega_2 - H_A + i\epsilon)|F^\lambda\rangle = D|e\rangle. \qquad (3)$$

In the flat continuum $|F^\lambda\rangle$ has outgoing waves in both channels 1 and 2; in the autoionizing region the physical solution of (3), we call it $|G^\lambda\rangle$, has outgoing waves only in the open channel 1 and is exponentially decaying in the closed channel 2. If we analytically continue the physical solution $|F^\lambda\rangle$ above the second threshold to the autoionizing region (using the properties of the Coulomb functions, see Appendix) it will contain unphysical exponentially growing contributions.

We can relate $|F^\lambda\rangle$ to the physical solution $|G^\lambda\rangle$ (and, therefore, Σ above and below threshold), however, by adding to $|F^\lambda\rangle$ regular solutions of the homogeneous equation to enforce the physical boundary conditions in the autoionizing region. Some of the details of this calculation are collected in the Appendix. As a result we obtain

$$\Sigma = \begin{cases} \delta\omega_{bg} - i\pi(|D_1|^2 + |D_2|^2) & \text{(flat continuum)} \\ \delta\omega_{bg} - i\pi|D_1|^2 \left\{1 + (q-i)^2/(x+i)\right\} & \text{(autoionizing region).} \end{cases} \quad (4)$$

Here $\delta\omega_{bg}$ is the smooth background shift in the flat continuum; D_1 and D_2 are dipole matrix elements from $|e\rangle$ to <u>energy normalized</u> continuum wave functions above the threshold 2 to channel 1 and 2 respectively. The detuning variable is defined by

$$x = \tan\pi(\nu_2 + \alpha)/\tanh\pi\tau, \quad (5)$$

with the effective quantum number ν_2 given by

$$E_g + \omega_1 + \omega_2 = E_I - Ry/\nu_2^2 \quad (<E_I) \quad (6)$$

α is the unperturbed quantum defect of series 2 and τ is a measure of the configuration interaction between the two channels. The Fano parameter q is defined as $q = -D_2/D_1\sqrt{\tau}$. It is of central importance that the parameters D_1, D_2, α, τ and $\delta\omega_{bg}$ are essentially constant along the Rydberg series and across the threshold.

Equ. (1) together with (4) is the central result of this section. We emphasize the following features:
- Σ contains the n-mixing by the laser for the infinite number of Rydberg states in terms of a small set of parameters which vary slowly with energy,
- The two-level result is obtained when we expand the detuning variable x around one of the Rydberg states with main quantum number n,

$$x \to (E_g + \omega_1 + \omega_2 - E_I + Ry/\nu_2(n)^2)/\tfrac{1}{2}\Gamma, \quad (7)$$

where $\Gamma \sim \tau/\nu_2(n)^2$ is the autoionization width. This is valid when Γ, the ionization width and the Rabi frequency for the transition $e \to n$ (which scale proportional to ν_2^{-3} and $\nu_2^{-3/2}$ respectively) are much smaller than the energy separation between neighboring Rydberg states. It is remarkable that the generalization from two-level theory to infinite Rydberg series can be obtained by the simple substitution rule (7).

- When averaged over the resonances, $\sum$ is continuous across the threshold 2 as function of energy.

The AC-Stark splitting of the autoionizing Rydberg series described by Equ. (1) is plotted in Fig. 2. In Fig. 3 we show results of a similar

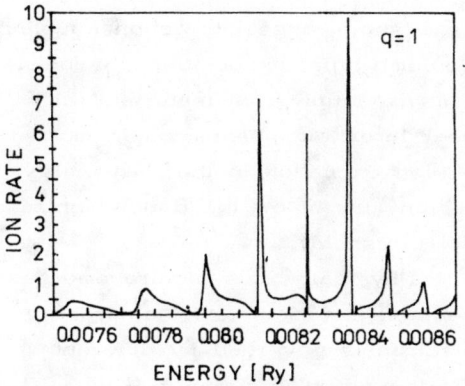
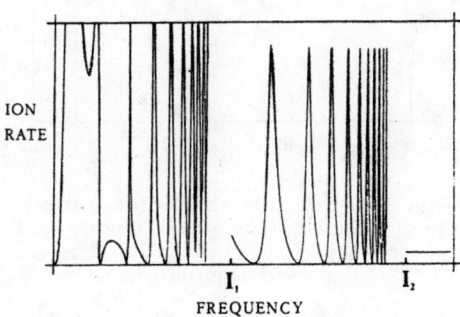

Fig. 2. Ionization rate as a function of the detuning of the first laser: AC-Stark splitting

Fig. 3. Nonresonant two-photon ionization rate in a two-channel MQDT analysis for the intermediate state

calculation for a nonresonant two-photon excitation amplitude with configuration mixed intermediate resonances, obtained by analytic continuation from the flat continuum to the autoionizing and bound Rydberg series.

Obviously many of the two-level calculations performed in the past can be generalized in the same simple way to infinite Rydberg series.

3. Rydberg Electrons in an Optical Laser Field: a Short Range Interaction Problem

It is of central importance for the analytic continuation procedure outlined in Sec 2 that the source term of the inhomogeneous equation decays sufficiently fast in the asymptotic region (see Appendix). This is guaranteed for the problem in Sec. 2 because $|e\rangle$ is a bound state. At first sight it seems questionable if similar arguments can be invented to perform the extrapolation from the continuum to the bound region when $|e\rangle$ is a continuum (or high lying) Rydberg state.

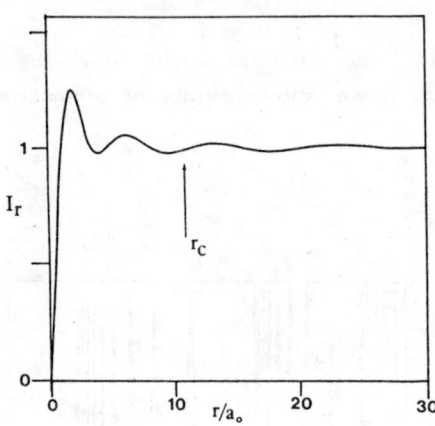

Fig. 4. Accumulated radial transition integral from the Rydberg state n = 10 s to the E = 0.3 Ry p continuum

Here we show that these ideas can be considerably generalized: the key point is that the interaction of laser radiation with Rydberg states can be formulated as a multichannel scattering problem, where the effect of the laser is identified as a short range interaction coupling Coulomb type dissociation channels. This is a universal feature of all near threshold phenomena in multiphoton excitation in the low energy region (far below the double ionization threshold).

Physically, this picture emerges because the motion of the Rydberg electron is governed by different forces close to the atomic core and in the asymptotic region. Close to the core the atomic forces dominate; in the asymptotic domain the electron vibrates rapidly in the time dependent optical laser field while the weak atomic force (the asymptotic Coulomb potential) is responsible for a slow mean motion of the electron. Transitions from Rydberg states to different bound or free orbits by absorption or emission of laser photons occur in the transition region between the two different types of motion. In other words, optical photons are absorbed by Rydberg electrons in a finite volume of space.

Let us consider an electron in the presence of a time dependent harmonic electric laser field $\vec{E}(\vec{x} = 0, t) = \mathcal{E}_0 \vec{\epsilon} \exp(-i\omega t) + \text{c.c.}$, bound in a potential $V(\vec{x})$ which for large $r = |\vec{x}|$ behaves like a Coulomb potential. Far away from the atomic core, where the atomic forces are weak, we expect the electron to oscillate with amplitude $\vec{\alpha}(t)$ according to the solution of Newton's equation $m\ddot{\vec{\alpha}}(t) = e \vec{E}(0, t)$. Accordingly, as a first step, we remove these asymptotic oscillations of the atomic charge distribution by transforming the usual minimal coupling Schrödinger equation to an oscillating frame by the time dependent transformation

$\Psi(x, t) \rightarrow \exp(-i\vec{\alpha}(t)\hat{\vec{p}})\psi(x, t)$ with $\hat{\vec{p}}$ the momentum operator. This leaves us with the space translated Schrödinger equation [3]

$$i\frac{\partial}{\partial t}\psi(\vec{x}, t) = \left\{\hat{\vec{p}}^2/2m + V(\vec{x} + \vec{\alpha}(t)) + \frac{1}{2}m\dot{\vec{\alpha}}^2\right\}\psi(\vec{x}, t), \quad (8)$$

where $\frac{1}{2}m\dot{\vec{\alpha}}(t)^2$ is the kinetic oscillation energy. The periodically oscillating atomic potential has in this frame the multipole expansion

$$V(\vec{x} + \vec{\alpha}(t)) \sim 1/r + \vec{\alpha}(t)\vec{x}/r^3 + \ldots . \quad (9)$$

The asymptotically dominating term is a Coulomb potential. This defines the threshold of our system modulo $\hbar\omega$. Note that these thresholds are shifted by the average oscillation energy. Equ. (8) nad (9) can be interpreted as describing the motion of an electron moving in a Coulomb potential and being scattered off from a nonstatic potential which is responsible for transitions between the Coulomb type dissociation channels.

The second and essential step in our arguments is now to identify the dipole and higher order terms in Equ. (9) as a <u>short range</u> interaction (in the sense of scattering theory). It is obvious that a Fourier- (Floquet-) expansion and an angular decomposition of the electron wave function $\psi(\vec{x}, t)$ in the oscillating frame gives a system of time independent close coupling equations for the radial wave functions (Equ. (3) is an example of a truncated set of equations of this sort). The dipole term in Equ. (9) (which only contains terms oscillating with frequency ω) now only couples dissociation channels differing by one photon. Correspondingly, for a given energy the associated radial wave functions oscillate for large r with different wave numbers. The effective range of the $1/r^2$ term, which in principle is a long range potential, is thus cut-off by destructive interference. This is in complete analogy to the situation in low energy electron ion scattering (with the exception of $e^- - He^+$, where the ion has degenerate energy levels).

This argument of a cut-off by destructive interference and an estimate of the cut-off radius as a function of the laser frequency is best discussed and illustrated in the weak field limit. For simplicity we confine ourselves to a hydrogenic system. In first Born approximation the coupling matrix element responsible for transitions between two channels differing by one photon accompanied by the change of angular momentum $1 \rightarrow 1 \pm 1$ is proportional a radial integral

$$\int_0^\infty dr\, f(\epsilon, 1, r)\, 1/r^2\, f(\epsilon + \omega, 1 \pm 1) \qquad (10)$$

between <u>energy normalized</u> Coulomb functions (ϵ is measured in Rydbergs relative to the shifted thresholds). Equ. (10) is readily identified as a dipole matrix element in the acceleration form. For energies close to threshold and low values of angular momenta the effective range of contribution to the integral (10) is confined to values up to a cut-off radius r_c,

$$\Delta k\, r_c \simeq \omega r_c / v_{orb}(r_c), \qquad (11)$$

with Δk the difference of the electron wave numbers in the two channels and v_{orb} the local orbital velocity of the Rydberg electron. Note that the cut-off occurs at the r-value where the frequency of the driving field equals the instantaneous angular frequency of the electron. To the extent this cut-off radius is small the matrix element (10) is a slowly varying function of energy across the Rydberg threshold. For l = 0, for example, Equ. (11) leads to the estimate ("frequency cut-off") [6]

$$r_c \simeq a_0 (2 Ry / \hbar \omega)^{2/3}. \qquad (12)$$

For sufficiently high frequency (i.e., optical frequencies) this cut-off is a few Bohr radii. This is illustrated in Fig. 3 by plotting the accumulated radial integral as a function of the upper integration limit for a p $\rightarrow$ s transition. With increasing l the main contribution to the integral (10) comes from increasingly larger distances; at the same time, however, the magnitude of the integral is strongly reduced. Terms decaying faster than the dipole in Equ. (9) a short range anyway and need no further discussion. Finally, with increasing intensity the interaction volume expands as is apparent from Equ. (8); we can speak of a short range interaction therefore only for intensities such that the oscillation amplitude is not much larger than the Bohr radius ("intensity cut-off") which for typical optical frequencies corresponds to intensities of the order of 10^{13} W/cm^2.

3.1. Ionization width and Starkshift of Rydberg states as complex quantum defects

We illustrate the use of MQDT by two simple examples. We first consider a Rydberg series which couples to a single continuum by the absorption of one photon. Although this assumption is an over-simplification, it illustrates the idea. A simple perturbation calculation yields for the complex energies of the Rydberg states

$$E = \frac{1}{2} e^2 \mathcal{E}_0^2 / m \omega^2 - Ry/(n - \sigma - \mu)$$

with

$$\mu = \sum \int d\epsilon' |\langle \epsilon' l' m' | \vec{\epsilon} \cdot \vec{\alpha}_0 / r^2 | \epsilon lm \rangle|^2 / (\epsilon + \hbar\omega/Ry - \epsilon' + i0), \quad (13)$$

where $|\epsilon lm\rangle$ denote energy normalized Coulomb functions. The real part corresponds to a continuum phase shift above threshold and describes a Stark shift of the Rydberg state in addition to the overall shift of the threshold. The imaginary part of corresponds above threshold to an inverse Bremsstrahlung cross section and is associated with an ionization width in the bound region. The familiar expressions for the second order Stark shift and ionization width is obtained by expanding the denominator in Equ. (13); this gives us the familiar n^3 scaling laws for the lifetime of the Rydberg states.

3.2. AC-Stark splitting of Rydberg series

As our second example we consider AC-Stark splitting. Usually AC-Stark splitting is formulated within a two-level language as the resonant coupling of two atomic states by strong laser radiation.

This leads to a splitting of the energy levels proportional to the electric field amplitude. The quantum defect approach allows us to replace both the lower und upper state of two-level theory by Rydberg series. Defining two effective quantum numbers by

$$E = \frac{1}{2} e^2 \mathcal{E}_0^2 / m \omega^2 - Ry / \nu_1^2 = \frac{1}{2} e^2 \mathcal{E}_0^2 / m \omega^2 + \hbar\omega - Ry / y_2^2 \quad (14)$$

for a given energy, a first order perturbation calculation above threshold leads to

$$\tan \pi (\nu_1 + \sigma_1) \tan \pi (\nu_2 + \sigma_2) + |\langle \epsilon l'm' | \vec{\epsilon} \cdot \vec{\alpha}_0 / r^2 | \epsilon + \omega\, lm \rangle|^2 \quad (15)$$

Eqs. (14) and (15) are a system of equations for the dressed eigenenergies of the system (including the n-mixing of the two Rydberg series). In spectroscopic work a graphical representation of $\nu_1 \bmod(1)$

versus ν_2 mod (1) according to Equ. (15) is usually referred to as Lu-Fano plot 2.

Appendix

We expand the many electron wave function in the form

$$\Psi_j(E) = \sum_{i=1}^{N} \phi_i \frac{F_{ij}(r)}{r}, \tag{A1}$$

where $E = I_1 + \epsilon_1 = I_2 + \epsilon_2 = \ldots$ denotes the energy the I's ionization limits, ϕ_i is the wave function of the ion core including the angular momentum part of the Rydberg electron, F_{ij} are radial wave functions of the Rydberg electron and N is the number of channels included (N = 2 in Sec. 2). Equ. (3) can thus be reduced to a system of coupled radial equations

$$\sum_j (h_{ij} - \epsilon_i \delta_{ij}) F_j^\lambda = Q_i \tag{A2}$$

with

$$h_{ij} = \left\{ -\frac{1}{2}\frac{d^2}{dr^2} + \frac{1}{2}\frac{l_i(l_i+1)}{r^2} \right\} \delta_{ij} + U_{ij}(r). \tag{A3}$$

The diagonal part of the potential U_{ij} goes asymptotically like a Coulomb potential; its nondiagonal part describes the short range configuration coupling. The inhomogenous term has been denoted by Q. It is zero for distances larger than the size of the orbital e. Above both thresholds (see Fig. 1) $F_i^\lambda(r)$ behaves asymptotically like

$$F_i^\lambda(r) \longrightarrow -2\pi i\, \varphi_i^+(r) D_i^{(-)}(\chi), \tag{A4}$$

where $D_i^{(-)}(\chi)$ (i = 1, 2) are the slowly varying photoionization amplitudes from $|e\rangle$ to the unstructured continuum and $\varphi^\pm$ are Coulomb functions corresponding to outgoing (incoming) waves. In the autoionizing region, on the other hand, we have

$$G_1^\lambda(r) \longrightarrow -2\pi i\, \varphi_1^+(r) D_1^{(-)}(s)$$
$$G_2^\lambda(r) \longrightarrow 0, \tag{A5}$$

where $D_1^{(-)}(s)$ is the photoionization amplitude to the open channel containing the autoionizing Rydberg structure. The outgoing (incoming) wave Coulomb functions are related to the (below threshold) exponentially decreasing Coulomb function (J) and increasing function (P) by

$$\varphi^{\pm} = e^{\pm i\pi\nu} \left[J \mp i \left(\frac{\nu^3}{2}\right)^{1/2} P \right], \qquad (A6)$$

where $\epsilon = -1/\nu^2 < 0$. Accordingly, below the second threshold F_i^λ will contain unphysical exponentially growing contributions; we can relate it to the physical G^λ, however, by adding regular solutions $F_{ij}^{(+)}(r)$ (i, j = 1, 2) of the homogeneous equation (A2) above both thresholds with scattering boundary conditions,

$$G_i^\lambda = F_i^\lambda + \sum_{i=1}^{2} F_{ij}^{(+)} L_j$$

and determine the coefficients L_j by the requirement (A5).

References

1. C. Fabre and S. Haroche, in "Rydberg states of atoms and molecules", ed. R. F. Stebbings and F. B. Dunning (Cambridge University Press, 1983); G. Leuchs in "Multiphoton Processes", ed. P. Lambropoulos and S. J. Smith (Springer-Verlag, Berlin, Heidelberg, New York, Tokyo 1984), p. 48; R. R. Freeman, L. A. Bloomfield, W. E. Cooke, J. Bokor and R. M. Jopson, ibid., p. 42.
2. M. J. Seaton, Rep. Progr. Phys., $\underline{46}$, 167 (1983)
3. M. Gavrila and J. Z. Kamiński, Phys. Rev. Lett., $\underline{52}$, 613 (1984).
4. P. Avan, C. Cohen-Tannoudji, J. Dupont-Roc and C. Fabre, J. Phys., $\underline{9}$, 993 (1976).
5. V. S. Lisitsa and S. I. Yakovlenko, Sov. Phys. JETP, $\underline{39}$, 975 (1974); K. Rzążewski and J. H. Eberly, Phys. Rev. Lett., $\underline{47}$, 408 (1981) and Phys. Rev., $\underline{A27}$, 2026 (1983); G. S. Agarwal, S. L. Haan, K. Burnett and J. Cooper, Phys. Rev. Lett., $\underline{48}$, 1164 (1982); A. I. Andryushin, A. E. Kazakov and M. V. Fedorov, Sov. Phys. JETP, $\underline{55}$, 53 (1982); Z. Białynicka-Birula, Phys. Rev., $\underline{A28}$, 836 (1983); M. Crance and L. Armstrong jr., J. Phys. $\underline{B15}$, 4637 (1982); P. E. Coleman and P. L. Knight, J. Phys., $\underline{B15}$, L235 (1982); G. Alber and P. Zoller, Phys. Rev., $\underline{A29}$, 2290 (1984); for a review see P. Lambropoulos and P. Zoller in "Multiphoton Ionization of Atoms", ed. S.L. Chin and P. Lambropoulos (Academic Press, 1984); see also contribution in this volume.
6. I am indebted to U. Fano for a discussion of this point.

Bruce W. SHORE

Lawrence Livermore
National Laboratory
Livermore, CA 94550

STOCHASTIC PROCESS IN QUANTUM OPTICS

Introduction

This lecture will summarize some aspects of the theory of stochastic processes in quantum optics that I have examined during the last few years in collaboration with colleaques J. H. Eberly of the University of Rochester and K. Wódkiewicz of Warsaw University [1-4]. I want to discuss a particular aspect of quantum optics: the description of atomic excitation (and fluorescence) by an intense radiation field which I shall refer to as a laser. The term "quantum optics" in this talk therefore refers to quantum mechanical behavior of a system excited by optical radiation; I shall not discuss the broader issues of light propagation. Let us consider an atom (more specifically a two-state atom) immersed in nearly monochromatic, nearly resonant, radiation. The radiation is not truly monochromatic, however: it has a finite bandwidth. Nor is the atom truly isolated: it undergoes collision with other atoms as well as with photons. Our task is to predict the observable behavior of the atom under these conditions. These predictions are, of course, limited by the fundamental constraints imposed by quantum theory. In the present example they are also limited because the finite laser bandwidth and the collisions introduce an element of uncontrollable randomness in the atomic environment - this is the stochastic part of the problem.

I shall begin by presenting the basic time evolution equations of quantum excitation. I shall then suggest ways in which the coefficients of these equations fluctuate randomly in time, thereby providing examples of stochastic processes. I shall comment upon some of the approaches that are used to treat such stochastic differential equations, and shall discuss one of these approaches in more detail: the model of a Poisson jump process. This process, an idealization of the stochastic processes that occur in nature, allows application of straightforward methods of linear algebra, through formulation as the Burshtein stochastic equation. I shall discuss this equation and some examples of its use.

The Equations of Motion

Consider an atom subjected to near resonant radiation as mentioned above. The regime of interest here occurs when the excitation is strong, so that traditional time dependent perturbation theory does not provide an adequate description of the excitation dynamics. Instead, all properties must be deduced from the appropriate time dependent Schrödinger equation,

$$\hbar \frac{\partial}{\partial t} \Psi(t) = -i H(t) \Psi(t). \tag{1}$$

An equivalent description, often more convenient, is provided by the Heisenberg equations

$$\hbar \frac{d}{dt} \sigma(t) = -i \left[H(t), \sigma(t) \right] \tag{2}$$

for the elementary Heisenberg operators $\sigma(t)$ that evolve from initialization $\sigma_{jk}(0) = |j\rangle \langle k|$. Often one extends these equations to allow relaxation processes (an expression of stochastic fluctuations) by employing the quantum Liouville equation

$$\hbar \frac{\partial}{\partial t} \rho(t) = -i \left[H(t), \rho(t) \right] - \hbar \Gamma \rho(t) \tag{3}$$

for the (reduced) density matrix $\rho(t)$.

In all of these equations the influence of the environment enters through the Hamiltonian $H(t)$. I shall consider near resonant excitation, in which one can apply a rotating wave approximation (RWA) to separate time variations at the nominal laser frequencies. The RWA Hamiltonian appropriate to an N-state atom has, as off-diagonal elements, Rabi frequencies. As diagonal elements it has detunings of the laser frequencies from the Bohr transition frequencies. These two types of frequencies – Rabi frequencies measuring the strength of the atom-field interaction, and detuning frequencies dependent upon stationary energies – provide the basic parameters of the problem.

Let us exclude, for simplicity, the realm of pulsed excitation. Then, in the absence of disturbances of either the atom or the ideally monochromatic laser, the RWA Hamiltonian consists of a matrix of constants. As an example, the two-state Bloch equations, obtained from the Heisenberg equations in the RWA, can be written.

$$\frac{d}{dt}\begin{bmatrix} \langle \sigma_{11} \rangle \\ \langle \sigma_{22} \rangle \\ \langle \sigma_{12'} \rangle \\ \langle \sigma_{21'} \rangle \end{bmatrix} = -\frac{i}{2}\begin{bmatrix} 0 & i2A & \Omega^* & -\Omega \\ 0 & -i2A & -\Omega^* & \Omega \\ \Omega & -\Omega & 2\Delta - iA & 0 \\ -\Omega^* & \Omega^* & 0 & -2\Delta - iA \end{bmatrix}\begin{bmatrix} \langle \sigma_{11} \rangle \\ \langle \sigma_{22} \rangle \\ \langle \sigma_{12'} \rangle \\ \langle \sigma_{21'} \rangle \end{bmatrix} \quad (4)$$

where $\Delta = E_2 - E_1 - \hbar\omega$ is the detuning, Ω is the Rabi frequency (permitted here to be complex valued), and $\langle \sigma_{12}'(t) \rangle = \langle \sigma_{12}(t) \rangle \exp(i\omega t)$. For simlicity I have omitted all time arguments in this equation. In general we shall wish to consider situations in which both the Rabi frequency $\Omega(t)$ and the detuning $\Delta(t)$ are (fluctuating) functions of time. I have included spontaneous emission, as expressed by the Einstein A coefficient, but I have not included other relaxation processes: their effects follow from fluctuations in the parameters.

We can see that the equations of interest fit the general pattern of coupled first order equations, say

$$\frac{d}{dt} Y(t) = -i W(t) Y(t), \quad (5)$$

where $Y(t)$ is a vector and $W(t)$ is a matrix. In our present discussion the time dependence of $W(t)$ occurs because the Hamiltonian varies with time, due to uncontrollable fluctuations in the environment or in the laser.

Quantities of Interest

The basic quantities of interest, to be predicted by appropriate equations of motion and compared with experiment, are the populations (diagonal elements of the reduced density matrix)

$$P_n(t) = \langle \sigma_{nn}(t) \rangle = \rho_{nn}(t), \quad (6)$$

the correlation function

$$c(\tau, t) = \langle \sigma_{21}(\tau + t) \sigma_{12}(t) \rangle, \qquad (7)$$

which reduces to the excited state population $P_2(t) = c(0, t)$ at $\tau = 0$, and the spectral profile of the steady-state (Wiener-Khinchine) spectrum:

$$g(\omega - \omega_0) = \frac{1}{\pi} \operatorname{Re} \int_0^\infty d\tau \exp(i\omega\tau) c(\tau, \infty). \qquad (8)$$

Each of these quantities appears as a quantum mechanical expectation value, denoted by $\langle \ldots \rangle$. When fluctuations occur in the Hamiltonian, when we must augment the customary quantum mechanical expectation value with an average over those fluctuations, a procedure that I shall denote by $\{\ldots\}$. Thus the correlation function, in the presence of fluctuations, is to be evaluated as

$$c(\tau, t) = \{\langle \sigma_{21}(\tau + t) \sigma_{12}(t) \rangle\}. \qquad (9)$$

I have emphasized the distinction between two types of averages $\langle \ldots \rangle$ and $\{\ldots\}$, because the two require different treatments in the theory, although both represent the results of repeated observations on samples drawn from an ensemble. It is important to realize that one cannot immediately obtain the stochastic average of some variable $Y(t)$ by averaging the equation for $d/dt \, Y(t)$. This is because that procedure would require the evaluation of the average of the product $W(t)Y(t)$, and this is not equal to the product of the average of $W(t)$ times the average of $Y(t)$. Such averages can sometimes be evaluated more readily if the time evolution equation is rewritten in integral form, involving a single variable rather than couplings with a set of variables. For example, the equation for the population inversion $w(t) = \langle \sigma_{22}(t) \rangle - \langle \sigma_{11}(t) \rangle$ reads

$$\frac{d}{dt} w(t) = -A - Aw(t) - \int_0^t dt' \exp\left[-\frac{A}{2}(t - t')\right] \operatorname{Re}\left[\Omega^*(t)\Omega(t')\right] w(t'). \qquad (10)$$

We have found this equation a useful starting point [2].

Stochastic Processes [5]

Collisions between the atom of interest and surrounding particles cause momentary shifts of the Bohr frequencies as the passing particle perturbs the stationary energy states of the atom. As a result of these collisions, the accumulated phase, being the time integral of the instantaneous Bohr frequency, is altered. These collisions occur at random; the only control exercised by the experimenter is upon such gross average properties as the density, temperature, and chemical composition of the perturbers. Within those broad constraints, the perturbers essentially produce at the atom a randomly fluctuating microfield. It is these fields that are responsible for pressure broadening of spectral lines.

The laser too is subject to random fluctuations - these are the source of the laser bandwidth. These fluctuations affect the phase of the laser radiation, and they may also shift the frequency, as in mode hopping. The laser field enters the expression for the Rabi frequency, so that a fluctuation in laser phase is equivalent to a fluctuation in the phase of the complex-valued Rabi frequency.

All of these examples fit the pattern of a stochastic process [5]: a randomly fluctuating function of time, $\tilde{x}(t)$, defined by the possible values that the variable may take at any instant (the sample space), and by the set of probabilities that these values occur. If the possible values form a discrete set (say N discrete frequencies) we may take the values of the stochastic variable $\tilde{x}$ to be $x_1, x_2, \ldots x_j, \ldots x_N$. For continuous values we let the value be x. The relevant probabilities include the probability $P(x, t)$ of observing value x at time t; the joint probability $P(x, t; x't')$ of observing the value x at time t and the value x' at time t'; and further multiple-sequence probabilities.

Although probabilities serve to define a stochastic process, greater practical interest attaches to various stochastic averages. These include the mean value of the variable (at time t)

$$\{\tilde{x}(t)\} = \int dx \, x \, P(x, t) \qquad (11)$$

and the autocorrelation function

$$\{\tilde{x}(t) \, \tilde{x}(t')\} = \int dx \int dx' \, x \, x' \, P(x, t; x't'). \qquad (12)$$

If we have analytic expressions for the probabilities, then these stochastic averages are readily evaluated. More often, however, the probabilities themselves may be unavailable: we may, for example, possess information only about the stochastic average.

In the application to atomic excitation we can pose the following theoretical problem. Given a stochastic evolution equation of the form

$$\frac{d}{dt} \tilde{Y}(t) = - i \tilde{W}\left[\tilde{x}(t)\right] \tilde{Y}(t), \tag{13}$$

where $W(x)$ is a function of the variable x, and given stochastic properties of the process $\tilde{x}(t)$, to find properties of the stochastic process $\tilde{Y}(t)$, such as the average $\{\tilde{Y}(t)\}$ and $\{\tilde{Y}(t)\tilde{Y}(t')\}$. These averages represent such observables as the population inversion, subject to noisy excitation.

Stochastic Models

Several theoretical approaches have been developed to incorporate stochastic properties into the quantum time-evolution equations. The simplest technique is to introduce phenomenological damping into the Bloch equation by means of relaxation times T_1 and T_2. The first of these parameters, T_1, expresses the inverse of the rate coefficient for shifting population via inelastic collisions. The second parameter, T_2, expresses the inverse of the rate coefficient for elastic collisions that destroy phase memory. These parameters can, of course, be related to more fundamental properties such as atomic cross sections and the density and temperature of the perturbers. Nevertheless, the use of a single parameter, T_2, to incorporate all elastic stochastic properties, places undesirable constraints on the stochastic processes.

A second approach proves valuable when the fluctuations are the result of many small, independent, and frequent changes of a variable. Brownian motion offers a prototype of such processes. Such Gaussian processes have the property that all multiple-time correlation functions are expressible in terms of the one- and two-time correlation functions. Most commonly, the _Gaussian_ process takes the form of a zero-mean Gaussian white noise process, for which the autocorrelation function is idealized as a Dirac delta:

$$\{\tilde{x}(t)\tilde{x}(t')\} = \Gamma\delta(t-t'). \tag{14}$$

This simple form makes possible a number of analytic manipulations, including justification of the use of relaxation times in the Bloch equation, and so it has underlain the major portion of theoretical modelings of noisy excitation. For example, it is the basis of the phase diffusion model of laser noise.

I want to discuss a third approach, less commonly found in quantum optics, but for which considerable literature exists. This is the <u>Poisson</u> process, in which changes of the variable occur only at discrete, random times.

The Poisson Jump Process [6-11]

In the jump process that I shall discuss, the Poisson distribution $p_n(\tau)$ describes the probability of observing n jumps during the time interval τ if the means time between jumps is T:

$$p_n(\tau) = \frac{1}{n!}\left(\frac{\tau}{T}\right)^n \exp\left(-\frac{\tau}{T}\right). \tag{15}$$

That is, we consider the variable $\tilde{x}(t)$ to maintain a constant value, say x_0, until, at the random time t_1, it suddenly takes the new value x_1. To define the process we must specify several things. First, we must specify the mean time between jumps, T (also termed the dwell time). Next we must specify the values that the variable $\tilde{x}(t)$ may take, say the discrete values $x_1,\ldots x_N$ or a range of continuous values x. We must specify the probability distribution of these values, $p(x_k)$ or $p(x)$, assumed to be independent of time. And finally, we must specify the transition probability $w(x \leftarrow x')$ or $w(x_k \leftarrow x_j)$ that, when a jump occurs, the variable $\tilde{x}$ makes the specified change. This probability too we take to be independent of time.

The simplest example of such a process is the bivalued random telegraph signal, in which the variable $\tilde{x}(t)$ takes only two values, say +a and -a. This process has a number of useful properties, including the expressions [11]

$$\left\{\tilde{x}(t+\tau)\tilde{x}(t)\right\} = a^2 \exp\left(-\frac{2\tau}{T}\right), \tag{16}$$

$$\left\{\exp\left[i\tilde{x}(t+\tau) - i\tilde{x}(t)\right]\right\} = \cos^2 a + \sin^2 a \exp\left(-\frac{2\tau}{T}\right). \tag{17}$$

These properties allow one to derive an equation for the population inversion (a third-order ordinary differential equation) when the laser phase or the laser frequency has a component represented by a binary random telegraph. In either case the population inversion exhibits damped oscillations, as can be described by an effective Rabi frequency Ω and a damping coefficient Γ. Both models permit evaluation of Ω and Γ for the two extremes of long T (rate interruptions, for which Γ varies inversely with T) and short T (frequent interruptions, for which varies directly with T).

It is a straightforward matter to generalize the jump process from a single bivalued signal to the sum of several independent bivalued signals. Each additional signal makes possible a larger range of values for $\tilde{x}(t)$, amongst evenly spaced values, but jumps always occur only between adjacent values. In the limit of a large number of signals, each jumping frequently between two slightly different values, the process becomes a Gaussian process (specifically, it becomes the Ornstein--Uhlenbeck process).

The Burshtein Equation

The Poisson process lends itself quite generally to both analytical and numerical modeling. Suppose that the values for $\tilde{x}(t)$ are discrete, $x_1, \ldots x_k, \ldots x_N$. We can define a set of discrete conditional probabilities $g_k(t) = P(x_k, t; x_0 t_0)/P(x_0, t_0)$. Because this is a Markov process, these conditional probabilities satisfy a Chapman-Kolmogorov-Smoluchowski equation or, in differential form, a master equation. For the Poisson process, in particular, the master equation is

$$\frac{\partial}{\partial t} g_k(t) = -\frac{1}{T} + \frac{1}{T} \sum_j w(k \leftarrow j) g_j(t). \qquad (18)$$

We might imagine solving this equation for the conditional probabilities, and in turn evaluating the stochastic averages of $\tilde{x}(t)$. However, our interest lies not with variable $\tilde{x}(t)$, which may represent a phase or frequency, but with some observable $\tilde{Y}(t)$, such as a population inversion. Thus the master equation does not appear to directly assist us. Nevertheless, the structure of this equation is central to the definition of the stochastic process, and it must therefore appear in some form in any stochastic description of the excitation.

Let us write the stochastic average of the variable $\tilde{Y}(t)$ as the sum

$$\{\tilde{Y}(t)\} = \sum_k \{\tilde{Y}(t)\}_k \qquad (19)$$

in which the marginal average $\{\tilde{Y}(t)\}_k$ is a stochastic average under the constrain that at time t the stochastic variable $\tilde{x}(t)$ has the particular value x_k. Let us suppose that $\tilde{Y}(t)$ satisfies the equation

$$\frac{\partial}{\partial t}\tilde{Y}(t) = -iM[\tilde{x}(t)]\tilde{Y}(t). \tag{20}$$

Then, as shown by Burshstein [6, 7] and by Zoller and Ehlotzky [10], the marginal average obeys the equation

$$\frac{\partial}{\partial t}\{\tilde{Y}(t)\}_k = -iM(x_k)\{\tilde{Y}(t)\}_k - \frac{1}{T}\{\tilde{Y}(t)\}_k + \frac{1}{T}\sum_j w(k \leftarrow j)\{\tilde{Y}(t)\}_j. \tag{21}$$

The first line of this equation expresses the change that occurs in the marginal average in the absence of interruptions, due to the Hamiltonian with the fixed parameter value x_k. The second line represents the change caused by the abrupt interruptions; it is just the change embodied in the master equation. This equation embodies all of the stochastic properties of the fluctuation phenomena, and it represents a useful basis for both analytical and numerical study. This equation was first applied to the equations of strong excitation (the Bloch equation) by Burshtein, and so we have referred to it as the Burshtein equation, although Anderson [8] and Kubo [9] dealt with the same process in context of weak excitation.

The Burshtein equation can be applied both to discrete-valued variables $\tilde{x}(t)$ and to continuous-valued variables. When the values are discrete, as we assume above, the various marginal average can be arranged into a vector $Y(t)$, with components $\{\tilde{Y}(t)\}_1,\ldots,\{\tilde{Y}(t)\}_N$. The Burshtein equation then takes the form of a set of coupled first order ordinary differential equations with constant coefficients,

$$\frac{d}{dt}Y(t) = -iWY(t). \tag{22}$$

The matrix W has elements

$$-iW_{jk} = -iM(x_j)\delta_{jk} - \frac{1}{T}[\delta_{jk} - w(j \leftarrow k)]. \tag{23}$$

Thus standard techniques of linear algebra and ordinary differential equation theory may be used to solve the equations for the marginal averages. A final summation provides the desired stochastic average.

I shall discuss a few examples of the effects of phase, amplitude, and frequency fluctuations on populations and upon fluorescence spectra. These are taken from more extensive discussions in the literature [1-4].

Acknowledgments

Research reported here was performed, in part, under the auspices of the U. S. Department of Energy by Lawrence Livermore National Laboratory under contract No. W-7405-Eng-48. My participation in this school was made possible, in part, by a travel grant from the National Science Foundation.

References

1. B. W. Shore, JOSA $\underline{B1}$, 176 (1984).
2. K. Wódkiewicz, B. W. Shore, and J. H. Eberly, Phys. Rev., $\underline{A30}$, 2390 (1984).
3. K. Wódkiewicz, B. W. Shore, and J. H. Eberly, JOSA $\underline{B1}$, 398 (1984).
4. J. H. Eberly, K. Wódkiewicz, and B. W. Shore, Phys. Rev., $\underline{A30}$, 2381 (1984).
5. N. G. van Kampen, Stochastic Processes in Physics and Chemistry (North-Holland, New York, NY, 1981).
6. A. Burshtein, Sov. Phys. JETP, $\underline{22}$, 939 (1966).
7. A. I. Burshtein and Yu. S. Oseldchik, Sov. Phys. JETP, $\underline{24}$, 716 (1967).
8. P. W. Anderson, J. Phys. Soc. Jap., $\underline{9}$, 316 (1954).
9. R. Kubo, J. Phys. Soc. Jap., $\underline{9}$, 935 (1954).
10. P. Zoller and F. Ehlotzky, J. Phys., $\underline{B10}$, 3023 (1977).
11. G. S. Agarwal, Z. Phys., $\underline{B33}$, 111 (1979).

D. L. SHEPELYANSKY

Institute of Nuclear Physics
Siberian Division
of the USSR Academy
of Sciences
630090 Novosibirsk, USSR

INTRISIC CHAOS IN QUANTUM SYSTEMS

Quantum limitations of the dynamical chaos in various simple models are discussed. The quantum dynamics turns out to be stable and time reversible even if the corresponding classical system is fully chaotic. Consequently, there is always a residual correlation which does not decay in time. The chaos localisation length is determined by the diffusion rate in the classical limit ($1 \approx D_{cl}$). Peculiarities of the quantum localisation near the chaos border are also described.

As a physical example the localisation of the hydrogen atom excitation in a monochromatic electrical field is considered. Results of some numerical experiments are compared with the theory which, particularly, predicts the delocalisation and diffusive excitation above a certain critical field strength. A new phenomenon of a multiphoton resonance pattern is described. Laboratory experiments on the quantum chaos are suggested and discussed.

Joseph W. HAUS

Physics Dept.
Rensselaer Polytechnic Institute
Troy, New York 12180-3590
U. S. A.

QUANTUM BEAT SUPERFLUORESCENCE

I. Introduction

Superfluorescence is a fundamental phenomenon of quantum optics [1]. Initially, N atoms confined to a volume are excited and there is no dipole moment. In the superfluorescent process the atoms decay, first by spontaneous emission of radiation then by amplification through stimulated emission processes. However, the process is also cooperative and the electric field developed during the decay of the excited states is coherent and the intensity is proportional to N^2, rather than N as expected for N incoherent sources (Fig. 1). Of course, since the total energy is fixed, the pulse of radiation is released in a superfluorescent time $\tau_{SF} \sim \tau_{sp}/N$ where τ_{sp} is the decay time for an independent atom to decay.

From a broader perspective, the superfluorescent process is a particular example of the decay of an unstable state. There are many other examples which fall into this field including the laser which is suddenly from below the lasing threshold to above this threshold [2] or stimulated Raman scattering [3]. Outside the field of quantum optics, decay of unstable states are observed in the development of Taylor vortices [4] or the rolls in Rayleigh-Benard [5] convection and in the phenomenon of spinodal decomposition [6], ie. phase separation of materials quenched into the two-phase region. The common feature of all these processes is the amplification of tiny fluctuations initiating the

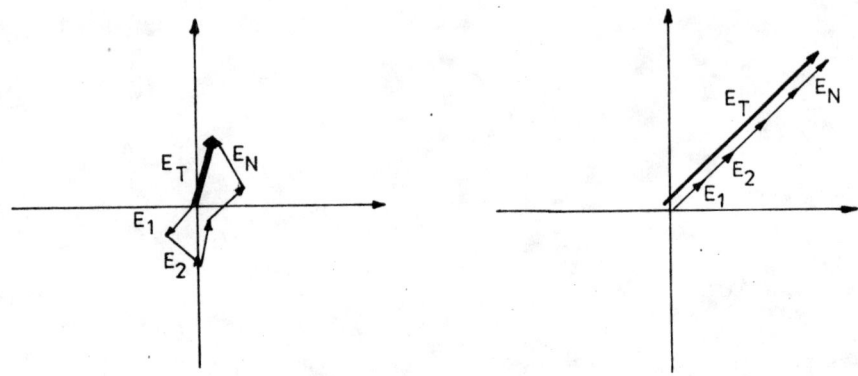

Fig. 1.

(a) Electric field produced by addition of incoherent fields, the electric field undergoes a random walk in the complex plane

(b) Coherent electric field produced when all the atoms emit radiation in phase

decay of the unstable state. If the system is precisely prepared in the unstable state, then the tiny fluctuations are magnified to macroscopic levels and <u>huge</u> fluctuations can be observed.

Quantum beat superfluorescence, the subject of this paper, is a superfluorescent pulse of coherent electromagnetic radiation emitted from a volume containing two species of excited atoms. The two species are distinguished by a slight detuning of their atomic transition frequencies which gives rise to a beating of the electromagnetic field (Fig. 2).

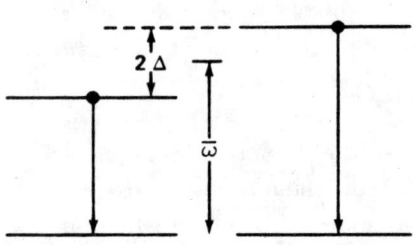

Fig. 2. Schematic of the level spacing for two atomic species. $\bar{\omega}$ is the average transition frequency and 2Δ is the detuning between the levels

Beating is a familiar phenomenon in electromagnetic theory when two monochromatic sources, eg. lasers, are slightly detuned and their radiation is combined. Nevertheless, if the sources are each of arbitrary phase independent of one another, then the time average intensity of their emitted radiation does not exhibit the beats. For individual detuned atoms the beats will be washed out by the randomness in their phase. Beats are observable if the excited atoms are prepared in a manner which preserves the phase relation between their radiated electromagnetic fields.

In quantum beat superfluorescence [7, 8, 9], the radiation is emitted in phase even trough the atoms have no special preparation of their excited states. This occurs because the radiation emitted by one atomic species induces the other to oscillate at that frequency in addition to their own transition frequency. Once the first few photons have been emitted, the atoms are coherently forced by the common radiation field and beats develop in the coherently radiated intensity.

In the following we use two models to investigate the statistical properties of the coherent superfluorescent pulses emitted from a collection of two species of excited atoms. The first model consists of two inverted harmonic oscillators which are detuned from one another. This is a prototypical model for a nonsaturated amplifier [10] and it does not include the propagation effects which are so important in other experimental situations [11]. Nevertheless, this model is completely soluble and gives insight into the beating and amplification process.

The second model includes propagation and saturation effects. The model can be used as a basic for deriving quantitative statistical properties of the emitted electromagnetic field.

II. Inverted Oscillators

Inverted harmonic oscillators models have already been discussed in Glauber's lecture [12]. Now the Hamiltonian is generalized to include two species of inverted oscillators:

$$H = -\sum_{j=1}^{2} \hbar \omega_j a_j^+ a_j + \sum_k \hbar \omega_k b_k^+ b_k + \hbar \sum_{j=1}^{2} \sum_k (\lambda_{jk} a_j^+ b_k^+ + h.c.). \qquad (1)$$

The operators a_j^+, a_j are the de-excitation and excitation operators for the jth oscillator; ω_j is the transition frequency of the oscillator. The electromagnetic field modes are expressed in terms of Bose creation and annihilation operators b_k^+ and b_k and each mode k has an energy $\hbar\omega_k$. The final term retains only the resonantly coupled operators; the coupling constants between the electromagnetic field modes and the jth oscillator are λ_{jk}.

At the initial time the system is assumed to contain only the vacuum photon states $|\{0\}\rangle_{EM}$ and all the inverted oscillators are in the highest energy state $|0\rangle_1 |0\rangle_2$:

$$|0\rangle_T = |0\rangle_1 |0\rangle_2 |\{0\}\rangle_{EM}. \qquad (2)$$

The Hamiltonian in Eq. (1) also has an invariance under global rotations in the complex plane and the corresponding conserved operator is (see Glauber's lecture):

$$\sum_j a_j^+ a_j - \sum_k b_k^+ b_k. \qquad (3)$$

For the initial state given in Eq. (2) this can be used to determine the average number of photons:

$$\eta(t) = {}_T\langle 0 | \sum_k b_k^+(t) b_k(t) | 0 \rangle_T,$$
$$= {}_T\langle 0 | \sum_{j=1}^2 a_j^+(t) a_j(t) | 0 \rangle_T. \qquad (4)$$

The time dependence of the operators are determined by solving the Heisenberg equations of motion for a set of linear equations [13]. The average transition frequency is $\bar{\omega} = \frac{1}{2}(\omega_1 + \omega_2)$ and the separation of the levels is $2\Delta = (\omega_1 - \omega_2)$. In the Weisskopf-Wigner approximation the amplification factor is given by:

$$\kappa = \pi \sum_{\vec{k}} |\lambda_k|^2 \delta(\bar{\omega} - \omega_k), \qquad (5)$$

where $\lambda_{jk} = \lambda_k$ has been used as a simplifying assumption. There is a dimensionless quantity which is used to distinguish the beat region from the non-beating region.

$$\varepsilon = \left(\frac{\Delta}{\kappa}\right)^2. \qquad (6)$$

The distinction between the region is seen in the behavior of the average photon number ($\tau = \kappa t$):

$$n(\tau) = 2(e^{2\tau} - 1) + 4e^{2\tau} \begin{cases} \dfrac{\sinh^2 \sqrt{1-\varepsilon}\,\tau}{1-\varepsilon} & \varepsilon < 1, \\ \tau^2 & \varepsilon = 1, \\ \dfrac{\sin^2 \sqrt{\varepsilon-1}\,\tau}{\varepsilon - 1} & \varepsilon > 1. \end{cases} \qquad (7)$$

This first term is the amplification from two decoupled inverted oscillators. There is no cooperation between the amplifiers. The second part is the contribution stemming from the interaction between the two oscillators. For $\varepsilon < 1$ no oscillations occur because the two oscillators have a beat frequency which is within the gain bandwidth. This is shown in Fig. 3 for $\varepsilon = .5$. Also shown in Fig. 3 is the case $\varepsilon = 10$ and beating is apparent in the short dashed curve.

The solid curve represents the contribution of only the first term in Eq. (7), which corresponds to two decoupled amplifiers.

It is apparent in Eq. (7) that the amplitudes of the beats is strong as ε approaches unity. As ε approaches a very large value, the oscillators have transition frequencies that are so widely separated that they only weakly couple to one another. In this limit the amplifiers are nearly independent.

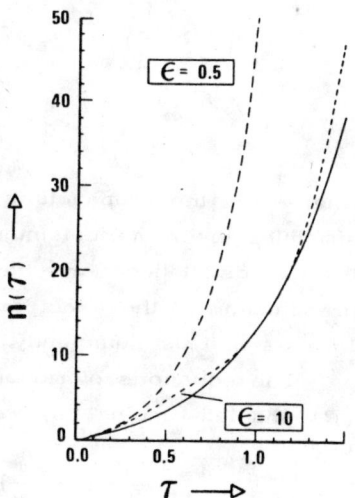

Fig. 3. For $\varepsilon < 1$ no undulations are observed. The beating is apparent for $\varepsilon = 10$. The solid line is the amplification of two decoupled inverted oscillators

III. Propagation Model

A more complete description of quantum beat superfluorescence includes the propagation effects in the Heisenberg equations of motion [14]. The model is simplified by introducing several approximations. Plane-wave propagation can be approximated by choosing the ratio of the geometric angle d/L, where d is the beam waiste and L is the length of the active medium, to the diffraction angle λ/d, where λ is the emission wavelength, to be about unity:

$$F = \frac{d^2}{\lambda L} \simeq 1. \qquad (8)$$

F is called the Fresnel number. This assumption corresponds to approximately one spatial mode being amplified in the medium. This is, of course, not a rigorous argument, but it has been supported by numerical integration of the equations of motion in superfluorescence including transverse effects [15].

The second important approximation introduced in the analysis is the slowly-varying envelope approximation in both time and space.

$$\bar{\omega}|E| \gg \left|\frac{\partial E}{\partial t}\right|,$$
$$\frac{\bar{\omega}}{c}|E| \gg \left|\frac{\partial E}{\partial x}\right|,$$
(9)

where x is the propagation direction. This assumption considerably simplifies the numerical integration of the nonlinear equations, since the fast oscillating parts have been removed and it is also a good representation of the exact solutions in the linear regime, at least as long as λ/L is sufficiently small [16].

The equations of motion for the electric fields using right-running (R) and left-running (L) waves is:

$$\left(\frac{\partial}{\partial x} + \frac{1}{v}\frac{\partial}{\partial t}\right) E_R^{\pm} = \sum_{j=1}^{2} W_j R_j^{\mp},$$
$$\left(\frac{\partial}{\partial x} - \frac{1}{v}\frac{\partial}{\partial t}\right) E_L^{\pm} = \sum_{j=1}^{2} W_j L_j^{\mp},$$
(10)

where the superscripts denote the positive (+) or negative (-) frequency parts of the electromagnetic fields; $R_j^{\pm}$ is the positive and negative frequency components for the right-running polarization of the jth atomic species and W_j are weights coupling the polarizations to the electric fields. If each species has the same number of excited atoms, then $W_1 = W_2 = \frac{1}{2}$. This assumption will be made in the results below.

The Heisenberg equations of motion for the polarization operators is:

$$\frac{\partial R_j^{\pm}}{\partial t} = \pm i \Delta R_j^{\pm} + E_R^{\mp} + E_R^{\mp} Z_j,$$ (11a)

$$\frac{\partial L_j^{\pm}}{\partial t} = \pm i \Delta L_j^{\pm} + E_L^{\mp} Z_j,$$ (11b)

where Z_j is the population inversion of the jth species which satisfies the Heisenberg equation of motion

$$\frac{\partial Z_j}{\partial t} = \frac{1}{2}(E_R^{+} R_j^{+} + E_L^{+} L_j^{+} + \text{h.c.}).$$ (11c)

The initial state of the system has the electromagnetic field in the vacuum state and all two-level atoms in their excited states:

$$|0\rangle_T = |\{\uparrow\}\rangle_1 |\{\uparrow\}\rangle_2 |\{0\}\rangle_{EM}. \tag{12}$$

The inversion operator for this state has the quantum-mechanical average corresponding to complete inversion:

$$_T\langle 0|Z_j(x, 0)|0\rangle_T = 1. \tag{13}$$

The electric field averages and polarization average are zero. This is a state of unstable equilibrium, which fluctuations cause to decay. During the decay process the tiny quantum-mechanical fluctuations are amplified to macroscopic levels. In the following, normal ordering of the operators is chosen so that the electric field second moment is:

$$_T\langle 0|E^+(x, 0)E^-(x', 0)|0\rangle_T = 0, \tag{14}$$

and the polarization second moments, for example, are:

$$_T\langle 0|R_j^+(x, 0)R_j^-(x', 0)|0\rangle_T = \frac{4}{N_j}\delta_{jj'}\delta(x-x'), \tag{15}$$

where N_j is the number of excited atoms of the jth species within the active volume higher order moments are only approximately Gaussian corrections being of the order $\frac{1}{N_j}$ [14]. The equations of motion in the linear regime can be interpreted classically. They are entirely deterministic equations of motion; however, the initial conditions are taken from an ensemble which has the above specific quantum-mechanical averages, Eqs. (13)-(15), for the first and second moments. Since these are field equations, the required probability distribution for the initial conditions is a Gaussian functional of these fields:

$$P\left[\{E_{R,L}^\pm, R_j^\pm, L_j^\pm, Z_j\}\right] \propto \prod_x \{\delta^2(E_R^+)\delta^2(E_L^+)\}$$

$$\prod_{j=1}^{2}\delta(Z_j - 1)\exp\left\{-\frac{\Delta x}{4}(N_1[|R_1^+|^2 + |L_1^+|^2] + N_2[|R_2^+|^2 + |L_2^+|^2])\right\}. \tag{16}$$

In the nonlinear regime, the quantum-mechanical fluctuations are no longer important and the equations of motion may be interpreted classically.

The linearized equations obtained by setting $Z_j = 1$ are analytically soluble. In this case, the right- and left-running waves are independent.

The solution of the linearized equations for zero initial electric field values is $(W_1 = W_2 = W)$:

$$E_R^+(x,t) = \int_0^x dx' \left\{ G_1(x',t) R_1^-(x-x',0) W + G_2(x',t) W R_2^-(x-x',0) \right\}, \quad (17)$$

where

$$G_1(x',t) = \Theta\left(t - \frac{x'}{v}\right) \left\{ I_0\left(2\sqrt{Wx'\left(t-\frac{x'}{v}\right)}\right) e^{i\Delta\left(t-\frac{x'}{v}\right)} + \int_0^{t-\frac{x'}{v}} dt' \, e^{i\Delta\left(t-\frac{x'}{v}-t'\right)} I_0\left(2\sqrt{wx'\left(t-\frac{x'}{v}-t'\right)}\right) \right. \quad (18)$$

$$\left. * e^{-i\Delta t'} \sqrt{\frac{wx'}{t'}} \, I_1\left(2\sqrt{wx' t'}\right) \right\}$$

and $G_2(x',t)$ is the complex conjugate of $G_1(x',t)$. The average intensity at the end face $x = 1$ (in the scaled units used here) is:

$$I(t) = \frac{4}{N} \int_0^1 dx' \left[|G_1(x',t)|^2 + |G_2(x',t)|^2 \right] . \quad (19)$$

where $W = 1$ and $N = N_1 = N_2$ were used.

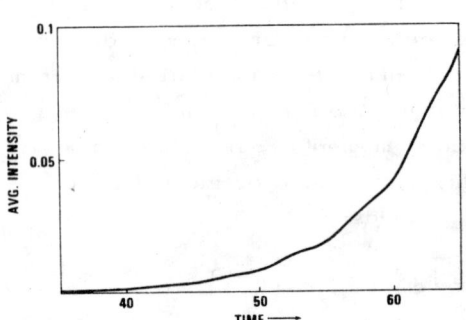

Fig. 4. The linearized average intensity for $N = 1.5 \cdot 10^8$. The detuning is $\Delta = 1$

This is plotted in Fig. 4 and the undulations are clearly visible. These beats are present at any detuning value for this model. The reason for this qualitative difference between the propagation model and the inverted oscillators discussed in the previous section is that the gain has the asymptotic time dependence $e^{\sqrt{t}}$.

The nonlinear equations must be solved numerically, the intensity from one trajectory is shown in Fig. 5. The beats are very deep especially after saturation has been achieved. The intensity averaged over 200 trajectories is exhibited in Fig. 6. The beats are still clearly visible. The maximum of the intensity exhibits huge fluctuations. At any point in time the fluctuations are from 50-150% of its average value.

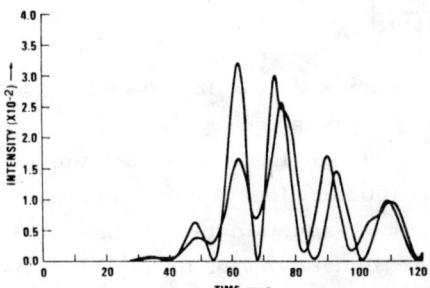

Fig. 5. A typical nonlinear trajectory. Both right- and left-running waves are shown

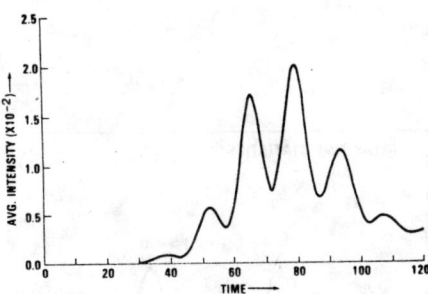

Fig. 6. The average intensity of 200 trajectories

A statistical quantity which can be analytically, calculated is called the delay-time distribution [14, 17]. This is related to the passage-time statistics in random-walk theory [18]. The delay-time statistics are calculated by setting the intensity to a preset reference value I_{ref}. Each time a trajectory first achieved this value (Fig. 7),

$$I_{ref} = |E_R^+(1, t_d)|^2 \qquad (20)$$

the delay-time t_d is noted. After a large number of trajectories have been counted, the distribution of delay-times, $W(t_d)$, can be calculated.

In Fig. 7, a linear trajectory is sketched next to the nonlinear trajectory. The two trajectories are nearly equal except very close to the maximum. The delay-time distribution can be analytically calculated using the linear solutions for the electric field, Eq. (17), in Eq. (20). The delay-time distribution is defined as:

$$W(t) = \langle \delta(t - t_d) \rangle, \qquad (21)$$

where t_d is obtained from Eq. (20) and the average is taken

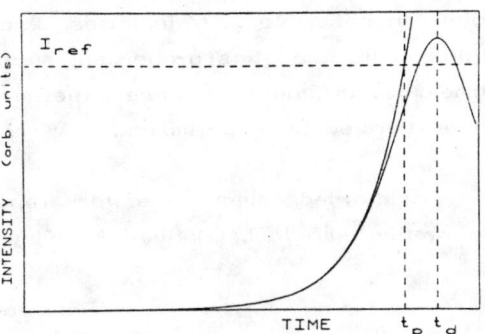

Fig. 7. A nonlinear and linearized trajectory cutting a preset intensity threshold, I_{ref}. t_d is the delay time for the intensity to reach a maximum. The time t_p is the passage time of the linearized trajectory through the threshold

using the weight function in Eq. (16). The calculations can be found in ref. 18, and a good approximation for this quantity is:

$$W(t) = \frac{d}{dt}\left[\left(\exp\left(-\frac{I_{ref}}{I(t)}\right)\right)\right], \quad (22)$$

where $I(t)$ is the average intensity given by Eq. (19). For the quantum-beat model this function has maxima and minima separated by the beat frequency, Fig. 8. This is evidence that the phases of the atomic polarizations are synchronized at an early time in the amplification process. The figure also demonstrates the huge fluctuations of the delay times that are expected.

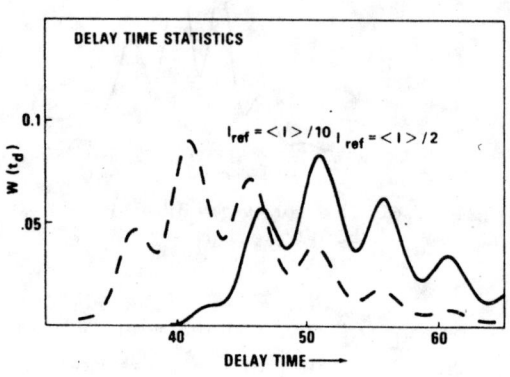

Fig. 8. The delay-time statistics for quantum beat superfluorescence. The parameters used are cited in Fig. 4. $\langle I \rangle$ refers to the average intensity maximum in Fig. 6

IV. Conclusions

Two models have been used to study quantum-beat superfluorescence. Both show that the beating is not smeared out by averaging over an ensemble of trajectories. They both have macroscopic fluctuations in the statistics properties, such as the intensity and the delay times. So far, none of these experimentally accessible features have been verified by experiments.

Acknowledgement: This work reported here has been done in collaboration with R. J. Glauber, H. King and W. Wöger.

References

1. See, Q. H. F. Vrehen and H. M. Gibbs in <u>Dissipative Systems in Quantum Optics</u>, eds. R. Bonifacio, Topics in Current Physics <u>27</u>, (Springer, Berlin, 1982); M. Gross and S. Haroche, Phys. Rep., <u>93</u>, 301 (1982).

2. F. T. Arecchi and V. Degiorgio, Phys. Rev., $\underline{A3}$, 1108 (1971).
3. I. A. Walmsley and M. G. Raymer, Phys. Rev. Lett., $\underline{50}$, 13, 962 (1983).
4. A. C. Newell and J. A. Whitehead, J. Fluid Mech., $\underline{38}$, 279 (1969).
5. J. Swift and P. C. Hohenberg, Phys. Rev., $\underline{A15}$, 319 (1977).
6. H. E. Cook, Acta Metall., $\underline{18}$, 297 (1970); J. W. Haus and H. King, Phys. Rev., $\underline{B25}$, 3298 (1982).
7. G. S. Agarwal, F. Haake and G. Schroder, Opt. Comm., $\underline{34}$, 283 (1980).
8. E. Abraham and R. K. Bullough, Opt. Comm., $\underline{34}$, 345 (1980).
9. C. Leonardi, J. S. Peng and A. Vaglica, J. Phys., $\underline{B15}$, 4017 (1981).
10. R. J. Glauber, in Group Theoretical Methods in Physics, Proceedings of the International Seminar, Zvenigorod, 1982 (Nauka, Moscow, 1983), vol. 2, p. 165.
11. Q. H. F. Vrehens, H. M. J. Hickspoors and H. M. Gibbs, Phys. Rev. Lett., $\underline{38}$, 764 (1977).
12. See lecture of Glauber in this volume.
13. W. Wöger, H. King, R. J. Glauber and J. W. Haus, unpublished (1985).
14. F. Haake, J. W. Haus, H. King, G. Schröder and R. Glauber, Phys. Rev., $\underline{A23}$, 1322 (1981) and references therein.
15. F. P. Mattar, et al. Phys. Rev. Lett., $\underline{46}$, 1123 (1981).
16. M. Lewenstein and K. Rzążewski, Phys. Rev., $\underline{A26}$, 1510 (1982).
17. F. Haake, J. W. Haus and R. Glauber, Phys. Rev., $\underline{A23}$, 3255 (1981).
18. A. J. F. Siegert, Phys. Rev., 81, 617 (1951); D. A. Darling and A. J. F. Siegert, Ann. Math. Stat., $\underline{24}$, 624 (1953).

I. J. BERSONS

Latvian SSR Academy of Sciences
Institute of Physics Riga
Salaspils, USSR

SEMICLASSICAL THEORY OF MULTIPHOTON PROCESSES IN RYDBERG ATOMS

The principal equations of semiclassical theory are derived and the validity of semiclassical theory for description of different multiphoton processes in atoms is discussed. The iterative procedure is used to solve the semiclassical equations and a simple expression for multiphoton ionization cross sections of Rydberg states of atoms is obtained. The comparison with quantum mechanical calculations shows that semiclassical cross sections are sufficiently precise in resonant domain as well as in the case of above threshold ionization.

The general solution of semiclassical equations is obtained and the problem of boundary conditions is reduced to solution of the system of algebraic equations or the integral equation. The solution of these equations is found in the approximation of equidistant atomic levels. The solution is used to calculate the probabilities of radiative transitions between Rydberg states of atoms in the presence of strong low frequency field and to estimate the ionization rate of such states in a microwave field.

P. R. BERMAN

Physics Department
New York University
4 Washington Place
New York, New York 10003

R. G. BREWER

IBM Research Laboratory
5600 Cottle Road, K01/281
San Jose, California 95193, USA

MODIFIED BLOCH EQUATIONS FOR SOLIDS

In a recent experiment of DeVoe and Brewer [1], it was concluded that the optical Bloch equations are incapable of describing the saturation phenomena observed. Optical free induction decay (FID) measurements of the impurity ion crystal Pr^{3+}:LaF_3 were conducted where the Pr^{3+} ions are coherently prepared by a laser field under steady-state conditions and then freely process when the driving field is removed. At low optical fields, the observed Pr^{3+} optical linewidth is dominated by magnetic fluctuations arising from pairs of fluorine nuclear flip-flops. At high optical fields, this nuclear broadening mechanism is quenched and the Bloch equations are seriously violated. On physical grounds, this failure is due to a time-averaging of the magnetic interaction as the optical nutation frequency increases [2]. The phenomenological dipole dephasing time T_2 of the Bloch equations is therefore not a true constant but lengthens with increasing field strength.

Several theories [2]-[8] have been proposed to explain the DeVoe-Brewer results. Most of these theories are an extension of a method developed by Redfield [9] for modifying the Bloch equations to include the effects of local, magnetically-induced fluctuations of the transition frequency of the radiating spins.

In this report, we present theoretical results for the optical FID decay rate obtained by assuming specific models for the local field fluctuations imposed on the ions' optical resonance frequency, and compare the results with "diffusion-type" theories [2]-[5] of frequency fluctuations. Whereas all the theories explain the overall qualitative structure of the experimental results, especially in strong fields, the predictions in the low field regime differ sufficiently to allow for the possibility of an experimental test of one theory over another. The low field data of DeVoe and Brewer are not yet precise enough to provide a critical test of any of the theories, but future experiments may be more conclusive.

Without going into the details of the calculation, a good qualitative understanding of the FID decay rate can still be obtained using simple physical considerations. The FID signal results when an ensemble of ions such as Pr^{3+} is initially prepared by an external optical field and then freely radiates when the field is suddenly removed. A large-scale inhomogeneous width Δ^* characterizes the sample, leading to a first-order FID signal that decays rapidly in a time $(\Delta^*)^{-1}$ - an effect that we do not consider here. The contribution to the FID signal which does concern us arises from a nonlinear interaction with the external field. As in the case of an inhomogeneously broadened vapor, the external field "burns" a hole in the ground state population and creates a bump in the excited state population, corresponding to that subgroup of ions with transition frequencies that resonantly interact with the external field. The FID decay rate is determined by the power broadened linewidth of the hole resulting from the preparative phase plus the free-precession decay when the field is turned off.

In the absence of frequency fluctuations for a two-state system, the FID decay rate given by the Bloch equations is

$$\gamma_F = \left[(1/T_2^0)^2 + \chi^2(T_1^0/T_2^0) \right]^{1/2} + 1/T_2^0, \quad (1)$$

where T_1^0 and T_2^0 are relaxation times in the absence of any frequency fluctuations and χ is the Rabi frequency of the external field. The bracketed term in (1) is the power-broadened hole width and the second term is the contribution of the free-decay period.

Frequency fluctuations modify the FID signal by affecting both the width of the hole and the free-decay process. We adopt a simple model, often applicable to solids, in which local field fluctuations produce frequency shifts $\delta\varepsilon(t)$ in the ion resonance frequency. Only Markoffian processes are considered; i.e., the frequency following a fluctuation depends at most only on the frequency before the fluctuation. As such, the frequency jump processes can be characterized by the three parameters:

Γ = rate at which fluctuations occur ("jump rate"),
$\delta\varepsilon$ = rms frequency shift per fluctuation,
$\varepsilon_0 = \sqrt{2}$ frequency associated with the frequency fluctuations distribution.

The quantity ε_0, in effect, is a measure of the maximum frequency displacement produced by the local field fluctuations. It is assumed that Γ, $\delta\varepsilon$ and ε_0 are much smaller than the large inhomogeneous width Δ^*.

In this work, we restrict the discussion to three types of Markoffian processes: (1) a Gaussian-Markoffian or diffusion-type model; (2) a "Difference" model in which the probability of finding a frequency ε following a fluctuation when the frequency was ε' before the fluctuation is a function of $(\varepsilon - \varepsilon')$ only; and (3) a "Strong" model in which the frequency distribution following a fluctuation is the equilibrium distribution characterized by width ε_0. In the diffusion model, Γ, $\delta\varepsilon$ and ε_0 appear only in the combinations [10]:

$\beta = \Gamma\delta\varepsilon/\varepsilon_0$ = effective jump rate,
$q = \beta\varepsilon_0^2/2$ = diffusion coefficient.

We now proceed to give the FID decay rates calculated for each model. Only the weak field regime is considered since, for strong fields, all models predict the same decay rate $\gamma_F \sim \chi (T_1^0/T_2^0)^{1/2}$ owing to the fact that, for large χ, frequency fluctuations become unimportant on the coherence time scale associated with the Rabi oscillation[1]. As has been noted [1], this result disagrees with the conventional Bloch prediction $\gamma_F \sim \chi(T_1/T_2)^{1/2}$, where the low field values T_1 and T_2 include the effects of frequency fluctuations.

Diffusion Model

Two limits, $\varepsilon_0 \ll \beta$ and $\varepsilon_0 \gg \beta$, can be distinguished. If $\varepsilon_0 \ll \beta$, one finds [2]-[5] an FID decay rate

$$\gamma_F \sim 2/T_2; \quad (1/T_2) = (1/T_2^0) + \varepsilon_0^2/2\beta, \tag{2}$$

---

[1]The zero field intercept extrapolated from the strong field regime differs for the various models and could be used in a consistency check of the theories. Moreover, values of γ_F in the intermediate field regime can also be used to distinguish the theories.

which implies that the T_2 associated with the weak-field regime can be much smaller than T_2^0 if $\varepsilon_0^2/2\beta \gg 1/T_2^0$. When $\varepsilon_0 \gg \beta$, the problem is best solved by the "Difference" model (discussion below) and, for $\varepsilon_0(\beta T_1^0)^{1/2} \gg 1/T_2^0$, we find a decay rate of order

$$\gamma_F \sim 2/T_2; \quad (1/T_2) = \tfrac{1}{2}(\beta T_1^0/2)^{1/2}\varepsilon_0. \tag{3}$$

Difference Model

The difference model [10] does not obey detailed balancing and cannot be used in the limit $\varepsilon_0 \ll \Gamma$ since frequency fluctuations <u>do</u> restore equilibrium in this limit. Thus, use of the difference model, which is characterized by a "kernel" $W(\varepsilon' \to \varepsilon)$ [$W(\varepsilon' \to \varepsilon)$ is the probability density per unit time for fluctuations to change the frequency from ε' to ε] given by

$$W(\varepsilon' \to \varepsilon) = \Gamma(2\pi\delta\varepsilon^2)^{-1/2}\exp\left[-(\varepsilon - \varepsilon')^2/2\delta\varepsilon^2\right], \tag{4}$$

is limited to the case $\varepsilon_0 \gg \Gamma$. Within this restriction, one can distinguish two subcases, $\delta\varepsilon < (1/T_2^0) + \Gamma$, $\delta\varepsilon > (1/T_2^0) + \Gamma$.

If $\delta\varepsilon < (1/T_2^0) + \Gamma$, one recovers the diffusion limit. The broadening of the hole is of order $\sqrt{n}\delta\varepsilon$, where $n = \Gamma T_1^0$ is approximately the number of frequency jumps occurring in time T_1^0. Considering only the interesting case when this width is considerably larger than the natural width $1/T_2^0$, i.e., when

$$(\Gamma T_1^0)^{1/2}\delta\varepsilon T_2^0 = (\beta T_1^0)^{1/2}\varepsilon_0 T_2^0 \gg 1, \tag{5}$$

we find an FID signal which decays as

$$\text{FID signal} \propto \frac{\pi}{2} - \tan^{-1}\left[(\beta T_1^0/2)^{1/2}\varepsilon_0 t\right]. \tag{6}$$

The decay rate (3) agrees qualitatively with the numerical work of Javanainen [5] using a diffusion model.

If $\delta\varepsilon > (1/T_2^0) + \Gamma$, fluctuations take the ion frequency outside the hole burned by the field and simply increase the decay rate, one finds

$$\gamma_F \sim 2/T_2; \quad 1/T_2 = 1/T_2^0 + \Gamma, \tag{7}$$

Strong Model

In the strong model, the kernel is given by

$$W(\varepsilon' \to \varepsilon) = \Gamma \, (\pi \varepsilon_0^2)^{-1/2} \exp(-\varepsilon^2/\varepsilon_0^2), \qquad (8)$$

and is independent of the initial frequency ε'. If $\varepsilon_0 \ll \Gamma$, we find [8] a decay rate

$$\gamma_F \sim 2/T_2; \quad 1/T_2 = 1/T_2^0 + \varepsilon_0^2/2\gamma, \qquad (9)$$

which is similar to (2). For large fluctuation rates, it is unimportant whether a few strong or many weak fluctuations redistribute the frequency. If $\varepsilon_0 \gg \Gamma$, $1/T_2^0$, χ, fluctuations shift the frequency out of the hole burned by the field and simply increase the decay rate as

$$\gamma_f \sim 2/T_2; \quad 1/T_2 = 1/T_2^0 + \Gamma. \qquad (10)$$

The physical mechanism is the same as the leading to (7).

All the weak field results for the FID decay rate give values of $\gamma_F \equiv 2/T_2 \neq 2/T_2^0$. Since the form of T_2 is model dependent, the different models lead to different predictions for the weak-field regime. However, with the exception of (6) which indicates a nonexponential decay, one cannot distinguish the various theories in the weak-field regime unless one an indepent means for experimentally varying the parameters $\delta\varepsilon$, Γ and ε_0. This appears feasible in molecular vapors by varying the perturber-to--active-atom mass ratio or the perturber pressure, but may prove difficult in solids. If an independent variation of the parameters is not possible, one must compare the various theories over the range from weak to strong external fields to obtain a best fit to the data, a procedure that may prove useful when new experimental data become available.

This work is partially support by the U. S. Office of Naval Research.

References

1. R. G. DeVoe and R. G. Brewer: Phys. Rev. Lett., 50, 1269 (1983); 52, 1354 (1984).
2. A. Schenzle, M. Mitsunaga, R. G. DeVoe and R. G. Brewer: Phys. Rev. A30, 325 (1984); R. G. Brewer and R. G. DeVoe: Phys. Rev. Lett., 52, 1354 (1984).

3. M. Yamanoi and J. H. Eberly: Phys. Rev. Lett., $\underline{52}$, 1353 (1984); J. Opt. Soc. Am., $\underline{B1}$, 751 (1984).
4. E. Hanamura: J. Phys. Soc. Jpn., $\underline{52}$, 2258 (1983); $\underline{52}$, 3265 (1983); $\underline{52}$, 3678 (1983).
5. J. Javanainen: Opt. Comm., $\underline{50}$, 26 (1984).
6. K. Wódkiewicz, B. W. Shore and J. H. Eberly: Phys. Rev., $\underline{A30}$, 2390 (1984); K. Wódkiewicz and J. H. Eberly: Phys. Rev., $\underline{A32}$ (in press).
7. P. A. Apanasevich, S. Ya. Kilin, A. P. Nizovtsev and N. S. Onishchenko: Opt. Comm., $\underline{52}$, 279 (1984).
8. P. R. Berman and R. G. Brewer: Phys. Rev. A (in press).
9. A. G. Redfield: Phys. Rev., $\underline{98}$, 1787 (1955).
10. P. R. Berman: Phys. Rev., $\underline{A9}$, 2170 (1974); P. R. Berman, J. M. Levy and R. G. Brewer: Phys. Rev., $\underline{A11}$, 1668 (1975).

R. G. DE FOE
C. FABRE*
R. G. BREWER

IBM Research Laboratory
5600 Cottle Road, K01/281,
San Jose, California 95193, USA

LASER FREQUENCY DIVISION AND STABILIZATION

The current method for measuring an optical frequency relative to the primary time standard, the cesium beam standard at ~9.2 GHz, utilizes a complex frequency synthesis chain involving harmonics of laser and klystron sources. The method has been extended recently to the visible region [1], to the 633 nm He-Ne laser locked to a molecular iodine line, with an impressive accuracy of 1.6 parts in 10^{10}. With the new definition of the meter, the distance traversed by light in vacuum during the fraction 1/299 792 458 of a second, the speed of light is now fixed and both time and length measurements can be realized with the same accuracy as an optical frequency measurement. In view of the complexity of optical frequency synthesis, these developments set the stage for originating complementary techniques for stabilizing and measuring laser frequencies which are more convenient.

This paper reports a sensitive optical interferometric technique <u>dual frequency modulation</u> (DFM) for measuring and stabilizing a laser frequency by comparison, in a single step, to a radio frequency (rf) standard. Conversely, a low-noise rf source can be stabilized by a laser frequency reference. A prototype [2] has demonstrated a resolution of 2 parts in 10^{10}, but devices currently under development should have a resolution of 10^{-12} and an absolute accuracy of ~10^{-10}. The meth-

---

*On leave from the Laboratoire de Spectroscopie Hertzienne de l'ENS, Paris.

od may be competitive with the optical frequency synthesis chain in accuracy and its simplicity suggests its convenient use in metrology, high-precision optical spectroscopy, and gravity wave detection.

The principle of the technique rests on phase-locking the mode spacing c/2L of an optical cavity to a radio frequency standard and simultaneously phase-locking a laser to the n-th order of the same cavity. When these two conditions are satisfied, the optical frequency ω_0 and the radio frequency ω_1 are simply related

$$\omega_0 = n\omega_1, \qquad (1)$$

neglecting for the moment diffraction and mirror phase-shift corrections. The idea of locking a laser to a cavity is of course a well-established subject [3, 4], but the concept of phase-locking an optical cavity to a radio frequency source is new. Interferometric rf-optical frequency comparisons of lower sensitivity have previously been performed by Bay et al. [5] using a related idea based on amplitude modulation (AM) rather than frequency modulation.

To introduce the DFM technique, first consider a single frequency modulation scheme. An electrooptic phase modulator driven at ω_1 generates a comb of optical frequencies $\omega_0 \pm m\omega_1$ which are compared to cavity modes of frequency $n\sigma$, where σ is the cavity free spectral range, $m = 0, 1, 2,...$ and n is a large integer $\sim 10^6$. The cavity response perturbs the balanced phase relationships between the sidebands and transforms frequency modulation into intensity modulation at ω_1. A photodetector, viewing the cavity either in reflection or transmission, then generates an error signal at the heterodyne beat frequency ω_1. This signal yields a null when the laser frequency ω_0 equals $n\sigma$ and the radio frequency ω_1 matches the mode spacing σ. In this circumstance, the comb of optical frequencies all resonate with their corresponding cavity modes. Although simple in principle, this single FM technique generates a complex error signal which depends not only on the rf detuning $\delta = \omega_1 - \sigma$ but also on the optical detuning $\Delta = \omega_0 - n\sigma$ and is therefore unsuitable for locking.

The DFM technique overcomes this problem by using two phase modulators, driven at frequencies ω_1 and ω_2 respectively, where $\omega_2/2\pi$ is nonresonant with c/2L. Dual frequency modulation creates, in lowest order, sidebands at $\omega_0 \pm \omega_1$, $\omega_0 \pm \omega_2$, and $\omega_0 \pm \omega_1 \pm \omega_2$. A photodetector views the cavity in reflection and two error signals are derived, one at ω_2 and the other at the <u>intermodulation</u> frequency $\omega_1 \pm \omega_2$. The first signal at ω_2 allows locking the laser to the reference cavity as described elsewhere [3, 4] and is independent of ω_1 tuning. The second signal at $\omega_1 \pm \omega_2$ allows locking the cavity to the rf reference. This signal varies directly with the rf detuning $\delta = \omega_1 - \sigma$ and provides the desired null at the rf resonance condition $\omega_1 = \sigma$, while being independent of laser detuning Δ.

As has been shown elsewhere [2], this intermodulated DFM signal is given by

$$I(\omega_1 \pm \omega_2) = I_0 \cdot C \cdot \sin\omega_2 t \cdot \{ \text{Im}[g(\Delta + \delta) - g(\Delta - \delta)]\cos\omega_1 t \},$$

where I is the light intensity on the photodiode, g is the complex cavity lineshape function, and $C = 4J_0(\beta_1)J_0(\beta_2)J_1(\beta_1)J_1(\beta_2)$, where β_1 and β_2 are the modulation indices of the phase modulators. Assuming a Lorentzian lineshape function g

$$g(\Delta) = \frac{\Delta(\Delta - i\Gamma)}{(\Delta^2 + \Gamma^2)},$$

the error signal becomes, for $\delta < \Gamma$

$$I(\omega_1 \pm \omega_2) = I_0 \cdot C \cdot \frac{\delta}{\Gamma} \cdot [\sin(\omega_1 + \omega_2)t + \sin(\omega_1 - \omega_2)t]$$

By detecting the beat either at $\omega_1 + \omega_2$ or $\omega_1 - \omega_2$ in a double balanced mixer, an error signal proportional to the rf detuning δ can be derived for locking an optical cavity to an rf standard, or conversely an rf source to a cavity. Second, the error signal is independent of laser detuning Δ and thus optical frequency jitter. Third, there is no background signal. Fourth, the DFM signal has excellent signal-to--noise ratio since $C = 4J_0(\beta_1)J_1(\beta_1)J_0(\beta_2)J_1(\beta_2) = .45$ at $\beta_1 = \beta_2 = 1$.

An examination of the above equations shows that the intermodulated DFM technique realizes an FM <u>Differential Interferometer</u> which produces a locking signal similar to conventional FM laser locking, but in which the optical tuning parameter $\Delta = \omega_0 - 2\pi \cdot n \cdot c/2L$ is replaced by the radio frequency (differential) tuning parameter $\delta = \omega_1 - 2\pi \cdot c/2L$. The DFM sideband structure creates an optical subtraction in the photodiode of the $(n+1)$-th cavity resonance curve from the $(n-1)$-th curve and thus permits accurate, low noise measurements of the cavity mode spacing.

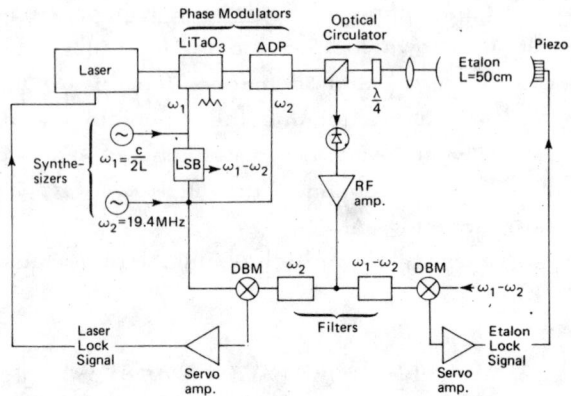

Fig. 1. Block diagram of an optical frequency divider showing two servo loops where the laser is locked to a reference cavity and the cavity to a radio frequency standard. The $LiTaO_3$ modulator is driven at ω_1 and the ADP modulator at ω_2. LSB denotes a mixer and filter which generates the difference frequency $\omega_1 - \omega_2$ for the double balanced mixer (DBM) in the cavity-rf servo.

A prototype DFM standard (Fig. 1) [2] was constructed to verify the principle and study resolution and systematic errors. A home-made cw dye ring laser contains an intracavity ADP crystal for laser phase-locking to an external cavity at 6000Å. DFM sidebands are applied by two phase modulators driven at $\omega_1/2\pi = c/2L \sim 300$ MHz and $\omega_2/2\pi = 19.4$ MHz. The laser beam is then mode-matched onto the reference cavity through an optical circulator which both isolates the laser from the cavity and directs the reflected light, the signal, to a high-speed photodiode. The detector photocurrent contains rf beats at ω_2 and $\omega_1 \pm \omega_2$ which are amplified and then separately filtered. The error signal at ω_2 is sent to an FM sideband servo which locks the dye laser to the n-th cavity fringe so that $\omega_0 = n\sigma$ with a short-term error $\Delta/2\pi \sim 300$ Hz RMS or less [4]. The DFM signal at $\omega_1 - \omega_2$ is coherently detected in a double-balanced mixer. This signal can control the cavity length L via a piezo so that the resonance condition $\sigma = \omega_1$ is satisfied. In initial tests the piezo control loop was opened and noise levels of 60 millihertz RMS, corresponding to a fractional frequency deviation σ_y (Allan variance) of 2×10^{-10} in 0.1 sec. integration time, were measured. The measurement of divider noise of 60 millihertz in the presence of 300 Hz of optical jitter verifies the differential behavior to the DFM standard and indicates that laser jitter is cancelled at the 10^{-3} to 10^{-4} level.

Four improvements have been made to the prototype to increase the resolution to 10^{-12} and to make possible absolute frequency comparisons at the 10^{-10} level. First, the interferometer itself has been placed in an acoustically isolated vacuum chamber and constructed from a 5 cm diameter by 47.5 cm long Zero-Dur spacer. Super-polished ring laser gyro quality mirrors have been used to achieve a finesse of > 20.000 or a linewidth $\Gamma/2\pi < 10$ kHz. Secondly, $\omega_1/2\pi$ has been raised to 4095 MHz to resonate with two cavity modes separated by 13 times the free spectral range of 315 MHz. Although this does not change the locking error δ, it reduces the _fractional_ error and noise δ/ω_1 by a factor of 13. A simple but highly efficient resonant cavity microwave phase modulator has also been developed which will be described elsewhere. Third, two I_2-stabilized He-Ne lasers of an NBS design have been constructed to serve as optical frequency standards. Fourth, a low noise microwave source has been built whose frequency is known to $\sim 10^{-11}$ by radio comparison to WWVB and Loran C.

These improvements described above have increased the effective Q of the system by a factor of > 200, and thus resolution of 10^{-12} is expected. The theoretical shot-noise limited Allan variance of the new system is $\Gamma/(\omega_1 \sqrt{N\tau})$ where N is the number of photoelectrons/sec and τ is the integration time. For $N = 10^{15}$ sec^{-1} and $\tau = 0.1$ sec $\sigma_y = 2 \times 10^{-13}$.

New techniques have also been developed to measure systematic errors common to precision interferometry. Equation (1) must be modified by corrections due to phase shifts in the cavity arising from (a) the multilayer dielectric coatings on the mirrors and (b) diffraction effects. Previous workers [6] have corrected for (a) by using two different cavity lengths, which require disassembly and realignment of the interferometer. The DFM standard, on the other hand, is designed to reach 10^{-10} accuracy with mirrors which can be quickly moved in vacuum, without disturbing alignment, by only 4 cm. In addition, the diffraction phase shift (b) can also be determined by measuring the higher order transverse mode spacing.

The DFM technique, therefore, should provide an independent and accurate measurement of an optical frequency that can be compared to the results of the optical frequency synthesis chain [1].

We are indebted to H. P. Layer for providing us with a detailed design of the NBS He-Ne iodine stabilized laser, to J. L. Hall for informing us of his design of an acoustically isolated interferometer chamber, to K. L. Foster for assistance in all phases of their fabrication, and to F. Walla for help with ultra-stable crystal oscillators. This work is supported in part by the U. S. Office of Naval Research.

References

1. D. A. Jennings, C. R. Pollack, F. R. Peterson, R. E. Drullinger, K. M. Evenson, and J. S. Wells: Optics Letters $\underline{8}$, 136 (1983).
2. R. G. DeVoe and R. G. Brewer: Phys. Rev. $\underline{A30}$, 2827 (1984).
3. R. W. P. Drever, J. L. Hall, E. V. Kowalski, H. Hough, G. M. Ford, A. J. Munley and H. Ward: Appl. Phys. $\underline{B31}$, 97 (1983).
4. R. G. DeVoe and R. G. Brewer: Phys. Rev. Lett. $\underline{50}$, 1269 (1983).
5. Z. Bay, G. G. Luther and J. A. White: Phys. Rev. Lett. $\underline{29}$, 189 (1972).
6. H. P. Layer, R. D. DesLattes and W. G. Schweitzer: Appl. Optics $\underline{15}$, 734 (1976).

Stephen J. SMITH[*]

Joint Institute
for Laboratory Astrophysics
University of Colorado
and National Bureau
of Standards Boulder,
Colorado 80309

EXPERIMENTAL INVESTIGATIONS OF THE ROLE OF LASER FIELD FLUCTUATIONS IN NON-LINEAR OPTICAL ABSORPTION PROCESSES

Introduction

In the experimental program I will describe, we deliberately broaden a well-stabilized single mode laser beam by introducing fluctuations to the laser frequency, in order to synthesize laser power spectra for which the fluctuations are well-characterized to all orders in a statistical sense [1, 2]. With this technique we are able to produce single mode laser fields which have nearly Lorentzian power spectra at one limit, essentially Gaussian power spectra at the other limit, and which may be varied continuously between these two limits.

Our objective in this program is to make it possible to carry out fully quantitative investigations of certain non-linear atomic absorption processes. The role of fluctuations in non-linear absorption processes has been discussed in a large number of theoretical papers [3]. Relevant experimental evidence has been rather limited and mostly rather qualitative, from experiments [4-6] with muti-mode pulsed lasers with correspondingly broadened power spectra.

---

[*] Staff Member, Quantum Physics Division, National Bureau of Standards.

I will outline results of some calculations concerning the power spectrum of a carrier frequency modulated by Gaussian noise. Next I will show how one may generate Gaussian radio frequency noise, and I will describe a system which we have developed for transferring this radio frequency noise to an initially nearly monochromatic laser beam in order to control laser power spectra out to ~ 1 GHz from line center. Finally, I will show results of several measurements made using this system.

Power Spectrum of Frequency Modulation Using Gaussian Noise

We describe the electric field of a continuous wave laser as

$$E(t) = E_0 e^{-i\left[\omega_0 t + \phi(t)\right]} . \tag{1}$$

We assume throughout this paper that the amplitude E_0 is a constant in time t. The central frequency is ω_0, also constant. The instantaneous frequency is $\omega(t) = \omega_0 + \dot{\phi}(t)$. We assume that $\dot{\phi}(t)$ fluctuates randomly with respect to its most probable value $\dot{\phi}(t) = 0$. We prescribe that the frequency fluctuations derive from some statistically stationary normal random process – that the noise follows a Gaussian distribution. This is a most important point, since it implies that all higher order even correlation functions of the field may be constructed from permutations of the lowest-order correlation function and that odd order correlation functions are zero [7]. The Wiener-Khintchine theorem states the equivalence of the lowest-order correlation function and the power spectrum. It follows that, for Gaussian noise, the laser power spectrum, which can be measured directly, provides a full statistical characterization of the frequency noise on the laser.

We want to describe the laser power spectrum which results from frequency modulation using Gaussian noise. This problem was first worked out by Middleton [8-9]. The autocorrelation function

$$R_E(\tau) = \tfrac{1}{2} \mathrm{Re} \langle E(t) E^*(t+\tau) \rangle \tag{2}$$

lead to a form for the laser power spectrum

$$P_E(\omega) = \frac{E_0^2}{4} \int_{-\infty}^{\infty} d\tau \, \exp\left\{-i(\omega - \omega_0)\tau\right\} \times$$

$$\times \exp\left\{-\frac{1}{2} \int_{-\tau}^{\tau} \langle \omega(0)\ \omega(t)\rangle\ (\tau - |t|)\ dt\right\} \tag{3}$$

in terms of the correlation function for the frequency.

The frequency correlation function for the so-called "phase diffusion" model

$$R_F(\tau) = \langle \omega(t)\ \omega(t+\tau)\rangle = 2b\ \delta(\tau) \tag{4}$$

correspond to a frequency power spectrum which is perfectly flat with spectral density 2b for all frequencies (the "white noise" case). Application of this correlation function leads directly to a pure Lorentzian laser power spectrum. A more realistic and applicable frequency correlation function [7] is

$$R_F(\tau) = \langle \omega(t)\ \omega(t+\tau)\rangle = b\ \beta\ e^{-\beta|\tau|}, \tag{5}$$

which corresponds to a power spectrum of frequency fluctuations

$$P_F(\omega) = 2b \left[1 + \left(\frac{\omega}{\beta}\right)^2\right]^{-1}. \tag{6}$$

Here β specifies the bandwidth of the frequency fluctuation spectrum. Together, b and β describe the Gaussian frequency fluctuations completely.

The laser power spectrum, that of a carrier frequency ω_0 on which are imposed the frequency fluctuations corresponding to Eq. (5), is found to be

$$P_E(\omega) = \frac{E_0^2}{4} \int_{-\infty}^{\infty} d\tau\ \exp\left\{-i(\omega - \omega_0)\tau\right\} \times$$

$$\times \exp\left\{-b\left[|\tau|\ (e^{-\beta|\tau|} - 1)/\beta\right]\right\}. \tag{7}$$

Two physically interesting cases correspond to the small and large limits of the ratio β/b.

If $\beta \gg b$, the frequency noise power bandwidth is very large compared to the spectral density and we obtain a laser power spectrum of the form

$$P_E(\omega) \to \frac{E_0^2}{2} \frac{b}{(\omega - \omega_0)^2 + b^2}, \tag{8}$$

which is a Lorentzian profile with full width at half maximum $\Delta_{FWHM} = 2b$.
If $\beta \ll b$,

$$P_E(\omega) \to \frac{E_0^2}{2} \left(\frac{2\pi}{b\beta}\right)^{1/2} \exp\left[-(\omega - \omega_0)^2/2b\beta\right], \qquad (9)$$

with $\Delta_{FWHM} = (8b\beta \ln 2)^{1/2}$. In this limit the laser power spectrum is Gaussian.

We generate the Gaussian frequency fluctuations with controllable values of b and β from a Gaussian noise voltage source using standard voltage-to-frequency conversion techniques. A typical voltage-to-frequency transfer function is shown in Fig. 1. We can impose the form of Eq. (6) on the voltage noise power spectrum by using a simple RC filter on "white" voltage noise if $RC = \beta^{-1}$. This form will then be realized in the frequency noise power spectrum with the appropriate parameters.

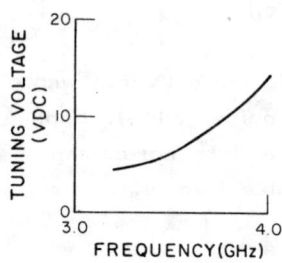

Fig. 1. Typical voltage controlled oscillator tuning curve. For frequency modulation ≤ 10 MHz the transfer from voltage to frequency modulation is approximately linear

Experimental Realization of the Phase Diffusion Model

In our system the source of random voltage fluctuations is based on use of a reverse-biased avalanche diode, in which the noise voltage corresponds to the traversal of electrons across the p-n junction. Since the electron avalanches are randomly distributed in time the noise voltage follows a Poisson distribution which, for large electron currents, approximates a Gaussian distribution very closely.

Figure 2 schematically represents a system in which the noise voltage is converted to frequency fluctuations, after the spectral density b is controlled by linear broadband amplifiers (or attenuators) and after the cut-off frequency $\beta/2\pi$ is imposed with an RC filter. All devices are linear and have dynamic ranges at least 10 times the rms noise voltage to avoid clipping. A voltage controlled oscillator (VCO) with a tuning curve of the form shown in Fig. 1 converts the voltage fluctuations to frequency fluctuations with center frequency ~ 3.5 GHz. The frequency modulated signal is heterodyned down from 3.5 GHz to ~ 200 MHz, the optimal center frequency of the acousto-optic modulator

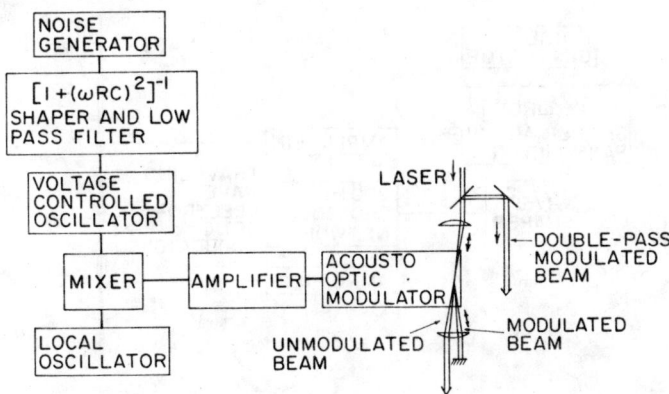

Fig. 2. Schematic diagram of the system used to drive the acousto-optic modulator

(AOM), and amplified to an rf power level of about 1 Watt for efficient operation of the AOM. A double pass arrangement eliminates the frequency-dependent deflection and doubles the frequency deviations.

The major deficiency of this system lies in the limited effective bandwidth of the AOM. While the nominal bandwidth of the device is ~ 100 MHz, restriction to a flat portion of the response curve limits us to a much narrower bandwidth. We limit the AOM to ± 6 MHz and rely on electro-optic modulators (EOM) to achieve control of the remainder of the power spectrum ($\pm \sim 1$ GHz). Such a system is represented in Fig. 3.

It can be shown that a phase modulator (EOM) can be made compatible with frequency modulation if the signal from the noise generator into the EOM is attenuated $\propto \omega^{-2}$. The resulting noise power spectrum produced by successive action of the AOM and EOM is observed by optical heterodyning of the modulated signal with an unmodulated signal. The power spectrum is displayed on an rf analyzer and compared with an ideal calculated power spectrum. System parameters are adjusted to achieve a fit within ± 1 dB.

Fluctuations in Non-Linear Atomic Absorption Processes

I illustrate measurements of the role of fluctuations in non-linear absorption processes with three types of measurements using laser beams broadened in the manner discussed above.

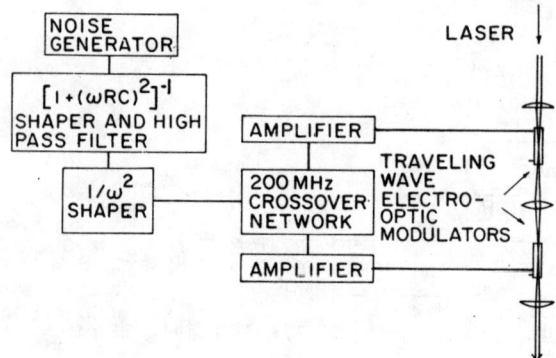

Fig. 3. Schematic diagram of the system used to drive two electro-optic modulators. The power spectrum from 6 MHz to 1 GHz is split at 200 MHz for power economy

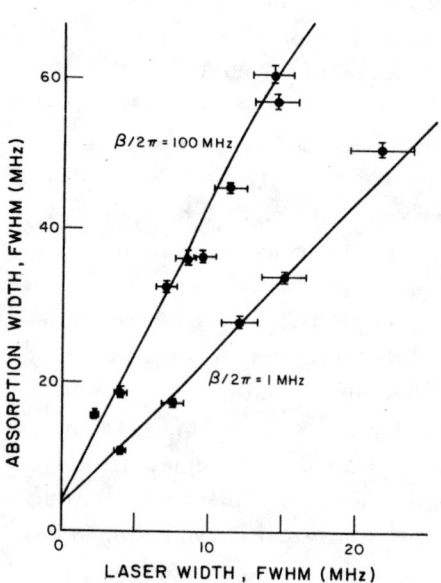

Fig. 4. The development of the absorption width (FWHM) of the 3s--5s two-photon Doppler free transition in atomic sodium is shown as the laser power spectrum is broadened: a) for $\beta/2\pi$ = 100 MHz (nearly Lorentzian case), and b) for $\beta/2\pi$ = 1 MHz (nearly Gaussian case). The finite ordinate at zero laser width corresponds to the natural linewidth plus small Doppler and collision effects

(1) <u>Two-photon Doppler-free unsaturated absorption in the 3s-5s transition in atomic sodium</u> [10]. This is, perhaps, the simplest conceivable non-linear process. This measurement was carried out in a sodium vapor cell with counter-propagating laser beams using retrore-

flection. Absorption profiles were measured as the laser was tuned over the two-photon resonance, using various values of b (noise spectral density) and $\beta/2\pi$ (noise bandwidth). In Fig. 4 we show results for the nearly Lorentzian case $\beta/2\pi$ = 100 MHz, and the nearly Gaussian case $\beta/2\pi$ = 1 MHz. Absorption width in laser frequency units is plotted against laser bandwidth. Of particular interest is the slope of 4 in the case of the Lorentzian power spectrum. The absorption width is four times the laser linewidth. This is in agreement with the theoretical work of Mollow [11], who used a second-order time-dependent perturbation theory. The Gaussian case has a slope of 2, more consistent with intuition.

(2) <u>Double-optical resonance (DOR) in the 3s-3p-4d system in atomic sodium</u> [4, 12]. Here the effect of a strong fluctuating field on the $3s_{1/2} - 3p_{3/2}$ transition was studied using a weak monochromatic laser to couple to the $4D_{5/2}$ state. A series of DOR doublets produced in this way is shown in Fig. 5a. These traces show the evolution of a Rabi split doublet as the strong laser is progressively detuned negatively from resonance. Of interest is the reversal of asymmetry in the peak heights obtained for small detunings. Here the laser power spectrum was approximately Lorentzian.

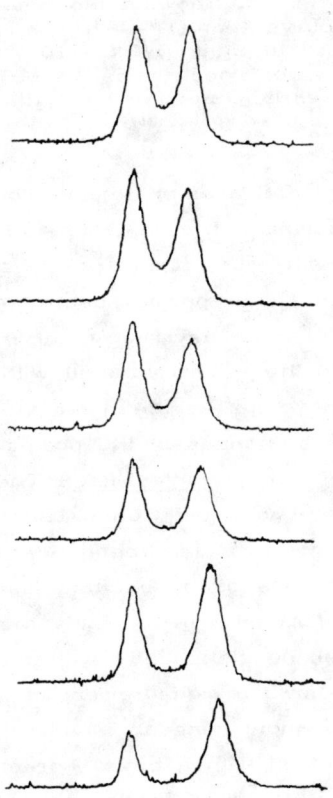

Fig. 5a. Showing the development of the DOR doublet with negative detuning of the saturating laser from exact resonance (top curve) in ~25 MHz steps. The width of the laser power spectrum is ~25 MHz and it is strongly Lorentzian. Rabi splitting is ~100 MHz. The reversal of asymmetry is a characteristic of Lorentzial laser fields with power spectra broader than the natural width of the transition (~12.5 MHz in this case, including Doppler width)

Figure 5b shows the evolution of the peak-height asymmetry parameter

$$\left(A = \frac{I_1 - I_2}{I_1 + I_2}\right),$$

272

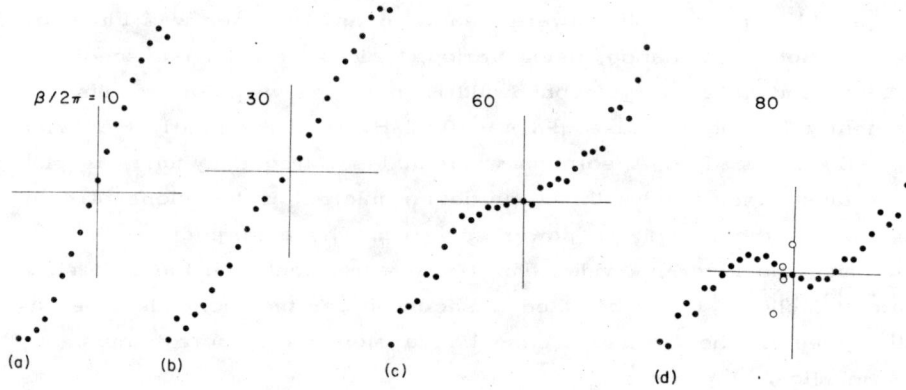

Fig. 5b. Showing the development of the asymmetry versus detuning curve $(A = (I_1 - I_2)/(I_1 + I_2))$ as the noise bandwidth is changed. From left to right $\beta/2\pi$ = 10, 30, 60, and 80 MHz. The width of the laser power spectrum is 14 MHz in very case. For $\beta/2\pi$ = 10, the most nearly Gaussian case, the asymmetry curve is nearly identical to that for a monochromatic laser

plotted against detuning of the strong laser, as the ratio of b to β is changed from a Gaussian power spectrum on the left, in steps, to the nearly Lorentzian power spectrum on the right. The widths (FWHM) of the laser power spectra are held constant.

The asymmetry curve for the Gaussian case is virtually identical to the result obtained with a nearly monochromatic laser. The reversal develops as the noise voltage bandwidth is increased, with correspoding changes in the power spectra toward the Lorentzian case.

(3) <u>Hanle Effect</u>. Theoretical studies by Cohen-Tannoudji [13] and by Avan and Cohen-Tannoudji [14] have shown that Hanle zero-field level-crossing resonances are highly dependent on the statistical properties of the exciting laser field. We have recently started a series of measurements using the method described in this paper for synthesizing statistically well-characterized laser power spectra. Figure 6 shows a measurement in atomic ytterbium, comparing a $^1S_0 \to {}^3P_1$ resonance using an effectively monochromatic laser with a resonance obtained with a broad Lorentzian laser profile.

Further work is planned in all three of these areas. The development of a technique for introducing amplitude fluctuations is under way so that effects of amplitude and frequency fluctuations may be compared.

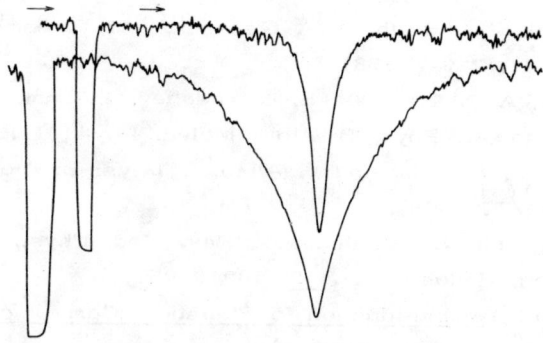

Fig. 6. An example of a Hanle zero-field $J = 0 \rightarrow J = 1$ level crossing resonance in atomic ytterbium. The dc level of light intensity is observed at right angles to the plane of the laser beam and the (perpendicular) applied magnetic field. The lower curve (FWHM = 398 milligauss) is obtained with an effectively monochromatic laser. The upper curve (FWHM = 104 milligauss) is obtained with an 8 MHz wide (FWHM) laser power spectrum of approximately Lorentzian form. At the left end of the traces the laser beam was blocked momentarily to establish the baseline

Acknowledgments: The contributions of a number of colleagues and collaborators is acknowledged. D. Elliott and Rajarshi Roy played leading roles in the development of the modulation techniques. J. Brandenberger, of Lawrence University, Wisconsin, and H. Metcalf of the State University of New York at Stony Brook developed the Hanle effect measurements. M. Hamilton and J. Hall were responsible for the highly--stabilized ring laser. K. Arnett gave invaluable technical assistance and assistance in the generation of the data presented here.

This work was supported by the U.S. Department of Energy, Office of Basic Energy Sciences. In addition, collaboration in the Hanle Effect studies (by J.B. and H.M.) was supported by the U.S. National Science Foundation.

References

1. D. S. Elliott, R. Roy, and S. J. Smith, Phys. Rev., A26, 12 (1982).
2. D. S. Elliott, R. Roy, and S. J. Smith, in Spectral Line Shapes, Vol. 2, edited by K. Burnett (de Gruyter, Berlin, New York, 1983), pp. 989-998.
3. For a current overview, see P. Zoller in Multiphoton Processes, edited by P. Lambropoulos and S. J. Smith (Springer-Verlag, Berlin, Heidelberg, New York, Tokyo 1984), pp. 68-75.

4. P. B. Hogan, S. J. Smith, A. T. Georges, and P. Lambropoulos, Phys. Rev. Lett., 41, 229 (1978).
5. P. Agostini, A. T. Georges, S. E. Wheatley, P. Lambropoulos, and M.D. Levenson, J. Phys. B: Atom. Molec. Phys., 11, 1733 (1978).
6. B. R. Marx, J. Simons, and L. Allen, J. Phys. B: Atom. Molec. Phys., 11, L273-L277 (1978).
7. M. C. Wang and G. E. Uhlenbeck, Rev. Mod. Phys., 17, 323 (1945).
8. D. Middleton, Philos. Mag., 42, 689 (1951).
9. D. Middleton, An Introduction to Statistical Communication Theory (McGraw-Hill, New York, 1960).
10. D. S. Elliot, M. W. Hamilton, K. Arnett, and S. J. Smith, Phys. Rev. Lett., 53, 439 (1984); Phys. Rev., A32, 887 (1985).
11. B. R. Mollow, Phys. Rev., 175, 1555 (1968).
12. M. W. Hamilton, D. S. Elliot, K. Arnett, and S. J. Smith, to be submitted to Phys. Rev. A.
13. C. Cohen-Tannoudji, in Atomic Physics, Vol. 4, edited by G. zu Putlitz, E. W. Weber, and A. Winnacker (Plenum, New York and London, 1975), pp. 589-614.
14. P. Avan and C. Cohen-Tannoudji, J. Phys. B: Atom. Molec. Phys., 10, 171 (1977).

Stanisław KIELICH
and Ryszard TANAŚ

Nonlinear Optics Division
Institute of Physics
A. Mickiewicz University
60-780 Poznań, Poland

SELF-SQUEEZING AS A NOVEL POTENT SOURCE OF QUANTUM FIELD

Introduction

In recent years, an increasing amount of work has been devoted to studies of the quantum and stochastic properties of the electromagnetic field, in particular to the feasibility of producing quantum fields exhibiting photon anticorrelation [1-5] and squeezing [5-7]. The effect of photon antibunching, revealing univocally the quantum nature of light, has hitherto been observed in experiment in processes of atomic resonance fluorescence [1] only. These experiments have directed attention to the possibilities of producing squeezed states of the radiation field [6, 8, 9].

Squeezed states of light are characterized by a decrease in quantum fluctuations in one component of the radiation field at the expense of an increase in fluctuations in the other (non-commuting) component. The crucial role of nonlinearity in photon anticorrelation is a well established fact. Hence, as one would expect, theoretical work on squeezing started from processes involving a nonlinear response of the quantum system to the field signal. Here, several nonlinear optical processes should be mentioned: parametric amplification [8, 11, 12], resonance fluorescence [13-17], four-wave mixing [18, 19], harmonics generation [20, 21], multi-photon absorption [22-25], hyper-Raman scattering [26], and nonlinear propagation of light [27].

The effect of photon anticorrelation is dependent on a number of favourable as well as destructive factors [4, 28]. Squeezing is particularly sensitive to the phase of the field: the fluctuations in phase (or amplitude) strongly reduce squeezing of the signal field [29], whereas interference acts favourably [30]. Other factors as well have an influence on these quantum effects [16, 17, 31]. Despite the differences between photon anticorrelation and squeezing, the two effects have one property in common: they are purely quantum in nature, and fields with such properties have no counterparts in classical optics. The two effects can exist concurrently in certain regions, or can exist independently of each other in different regions [15], whereas certain processes are accompanied by one of the two effects only.

The efficiency of individual nonlinear optical effects in producing photon antibunching and squeezing varies from one case to another. Here, it would appear that a particular role is played by self-induced effects like self-squeezing, which resides in the circumstance that a strong laser beam traversing an atomic medium endows it with optical nonlinearity owing to which the light undergoes self-squeezing of its quantum fluctuations. Herein may well reside a novel method of producing quantum fields with as efficiency of almost 100 per cent of that predicted by quantum mechanics. The process of self-squeezing is universal in nature, because it takes place in all matter even in systems composed of centrosymmetric molecules, or atoms.

Effective interaction Hamiltonian

Consider N noninteracting microsystems (atoms or molecules) in a volume V, acted on by an electromagnetic field with electric vector $\vec{E}$ and magnetic vector $\vec{B}$. The total Hamiltonian of the system as a whole is, in standard form

$$H = H_N + H_F + H_I, \qquad (1)$$

with H_N the Hamiltonian of the system of N atoms (molecules), and H_F that of the free radiation field.

We are interested in the explicit form of the Hamiltonian H_I describing the interaction between the material system and the field of radiation. The interaction is, in general, nonlinear and involves all multipolar electric and magnetic transitions [32, 33]. Since the system is optically isotropic in the absence of external fields, the effective Hamiltonian is a function in even powers of $\vec{E}$ and $\vec{B}$. Hence, the second-order Hamiltonian $H_I^{(2)}$ is the lowest non-zero contribution to H_I; if spatial dispersion is weak, $H_I^{(2)}$ can be written as follows [32] (we omit the purely diamagnetic term, proportional to B^2 [33]):

$$H_I^{(2)} = -\frac{N}{2}\left\{\alpha_{ij}E_iE_j + 2\varrho_{ij}E_iB_j + \frac{2}{3}\eta_{ijk}E_i\nabla_k E_j + \ldots\right\}, \qquad (2)$$

where the Einstein convention regarding ijk is assumed.

In (2), the second-rank tensor α_{ij} defines the linear electric polarizability of the molecule for an electric dipole-dipole quantum transition. The second-rank pseudotensor ϱ_{ij} describes the electro-magnetic polarizability for an electric dipole-magnetic dipole quantum transition. Whereas the third-rank tensor η_{ijk} accounts for the linear electric polarizability for an electric dipole-electric quadrupole quantum transition.

In the same multipole approximation the fourth-order Hamiltonian (which is the first nonlinear contribution) has the following form [32] (we retain terms linear in $\vec{B}$ only):

$$H_I^{(4)} = -\frac{N}{24}\left\{\gamma_{ijkl}E_iE_jE_kE_l + 4\sigma_{ijkl}E_iE_jE_kB_l + \right.$$
$$\left. + \frac{4}{3}\mathcal{H}_{ijkm}E_iE_jE_k\nabla_m E_l + \ldots\right\}, \qquad (3)$$

where the fourth-rank tensor γ_{ijkl} defines the nonlinear electric polarizability for four-fold electric dipole quantum transitions. The fourth-rank pseudotensor σ_{ijkl} determines the nonlinear electro-magnetic polarizability for three-fold electric dipole-magnetic dipole quantum transitions. And the tensor of rank 5, $\mathcal{H}_{ijkm}$ expresses the nonlinear electric polarizability for three-fold electric dipole-electric quadrupole quantum transitions.

In the classical case, one can split the electric vector of the electromagnetic field at the space-time point $(\vec{r}, t)$ into two complex parts [34]:

$$\vec{E}(\vec{r}, t) = \vec{E}^{(+)}(\vec{r}, t) + \vec{E}^{(-)}(\vec{r}, t), \qquad (4)$$

where the component $\vec{E}^{(+)}(\vec{r}, t)$ is related with the time-function $\exp(-i\omega t)$ (ω positive), with $\vec{E}^{(-)}(\vec{r}, t) = \left\{\vec{E}^{(+)}(\vec{r}, t)\right\}^*$.

The transversal electric field can be expressed as a superposition of plane waves:

$$\vec{E}(\vec{r}, t) = \sum_{\vec{k}} \left\{\vec{E}^{(+)}(\vec{k})\, e^{i(\vec{k}\cdot\vec{r} - \omega t)} + \vec{E}^{(-)}(\vec{k})\, e^{-i(\vec{k}\cdot\vec{r} - \omega t)}\right\} \tag{5}$$

The same holds with regard to the magnetic vector $\vec{B}(\vec{r}, t)$, with

$$B_i^{(+)} = \frac{c}{\omega}\, \varepsilon_{ijk}\, k_j\, E_k^{(+)}, \tag{6}$$

where ε_{ijk} is the antisymmetric Levi-Civita tensor.

We assume the configuration in which the light beam propagates along the Z-axis of the Cartesian reference frame XYZ and introduce the circular representation of the field with the unit vectors (on the angular momentum convention):

$$\vec{e}_{\pm} = \frac{1}{\sqrt{2}}(\vec{x} \pm i\vec{y}), \tag{7}$$

where $\vec{e}_+$ and $\vec{e}_-$ refer, respectively, to the right- and left-polarized waves, with $\vec{x}$ and $\vec{y}$ unit vectors in the direction of X and Y.

In quantum electrodynamics, the field vectors (4)-(6) are dealt with as operators in Hilbert space; we have

$$\vec{E}^{(+)}(\vec{k}) = i \sum_{\lambda} c(\omega_k)\vec{e}^{(\lambda)}(\vec{k})\, \hat{a}_{\vec{k}\lambda}. \tag{8}$$

Above, in cgs units, the normalizing factor is $c(\omega_k) = (2\pi\hbar\omega_k/V)^{1/2}$ with V the quantisation volume.

In (8), $\hat{a}_{\vec{k}\lambda}$ is the annihilation operator of a photon with the momentum $\hbar\vec{k}$ and polarisation state λ given by the unit vector $\vec{e}^{(\lambda)}(\vec{k})$. The photon annihilation and creation operators $\hat{a}_{\vec{k}\lambda}$ and $\hat{a}_{\vec{k}\lambda}^+$ fulfill the boson commutation rules

$$\left[\hat{a}_{\vec{k}\lambda}, \hat{a}_{\vec{k}'\lambda'}^+\right] = \delta_{kk'}\, \delta_{\lambda\lambda'}, \quad \left[\hat{a}_{\vec{k}\lambda}, \hat{a}_{\vec{k}\lambda}\right] = \left[\hat{a}_{\vec{k}\lambda}^+, \hat{a}_{\vec{k}\lambda}^+\right] = 0,$$

The unit vectors describing the state of polarisation of the wave are in general complex quantities satysfying the ortho-normalisation conditions

$$e_{k\sigma}^{(\lambda)*}\, e_{k\tau}^{(\lambda')} = \delta_{\sigma\tau}\, \delta_{\lambda\lambda'}, \quad e_{k\sigma}^{(\lambda)} k_\sigma = 0.$$

For a quasi-monochromatic wave with the frequency ω propagating along the Z-axis one may omit in (5) the summation over the index k. Thus, by (8), we get the field component in the form:

$$E_\sigma^{(+)}(z,t) = i\, c(\omega)\exp\{-i(\omega t - kz)\} \sum_{\lambda=1,2} e_\sigma^{(\lambda)} \hat{a}_\lambda, \qquad (9)$$

where $k = \omega/c$.

In fact, the formula (9) provides a two-mode description of a field being the coherent superposition of two modes with mutually orthogonal polarisations.

By analogy with the circular representation of the polarisation of a classical field we may write the transformations for the field operators as:

$$\hat{a}_+ = \tfrac{1}{\sqrt{2}}(\hat{a}_x - i\hat{a}_y), \qquad \hat{a}_- = \tfrac{1}{\sqrt{2}}(\hat{a}_x + i\hat{a}_y). \qquad (10)$$

Both representations are well adapted to the description of the interaction between elliptically polarized light and a material medium. However, as shown previously [27], the circular representation $(\hat{a}_+, \hat{a}_-)$ is radically more convenient than the Cartesian representation $(\hat{a}_x, \hat{a}_y)$ in that the former enables us to obtain the solution of the equation of motion for the field operators in general propagator form.

Since the microsystems are freely oriented in space, we may perform an unweighted averaging of the interaction Hamiltonians (2) and (3) over all their possible orientations. By (4), (9) and (10) and taking the operators in normal order we obtain the effective Hamiltonian in the form [35]

$$H_I^{(2)} = -\tilde{\chi}_R^L(\hat{a}_+^+\hat{a}_+ + \hat{a}_-^+\hat{a}_-) - i\,\tilde{\chi}_A^L(\hat{a}_+^+\hat{a}_+ - \hat{a}_-^+\hat{a}_-), \qquad (11)$$

$$H_I^{(4)} = -\tfrac{1}{2}\tilde{\chi}_R^{NL}(\hat{a}_+^{+2}\hat{a}_+^2 + \hat{a}_-^{+2}\hat{a}_-^2) - \tilde{\varkappa}_R^{NL}\hat{a}_+^+\hat{a}_-^+\hat{a}_-\hat{a}_+ - \tfrac{1}{2}\tilde{\chi}_A^{NL}(\hat{a}_+^{+2}\hat{a}_+^2 - \hat{a}_-^{+2}\hat{a}_-^2), \qquad (12)$$

where we have introduced the coupling constants [36]

$$\tilde{\chi}_R^L = \tfrac{1}{3}c(\omega)^2\, N\, \text{Re}\,\alpha_{\alpha\alpha}, \qquad \tilde{\chi}_A^L = \tfrac{i}{3\omega}c(\omega)^2\, N\, \text{Im}\,\varrho_{\alpha\alpha}, \qquad (13)$$

$$\tilde{\chi}_R^{NL} = c(\omega)^4 \, \frac{N}{15} \, \text{Re}(3\gamma_{\alpha\beta\alpha\beta} - \gamma_{\alpha\alpha\beta\beta}),$$

$$\tilde{\varkappa}_R^{NL} = c(\omega)^4 \, \frac{N}{15} \, \text{Re}(\gamma_{\alpha\beta\alpha\beta} + 3\gamma_{\alpha\alpha\beta\beta}),$$

$$\tilde{\chi}_A^{NL} = -i\, c(\omega)^4 \, \frac{4Nk_z}{15} \left\{ \frac{1}{\omega} \text{Im}(3\sigma_{\alpha\beta\alpha\beta} - \sigma_{\alpha\alpha\beta\beta}) + \frac{1}{3} \text{Re}\, \varkappa_{\alpha\beta\gamma\beta\delta}\, \varepsilon_{\alpha\gamma\delta} \right\}. \quad (14)$$

Thus, the effective Hamiltonians (11) and (12) derived by us are expressed in general by 5 invariants of the field operators (2 linear and 3 nonlinear) and 5 rotational molecular invariants (2 linear (13) and 3 nonlinear). The coupling parameters $\tilde{\chi}_R^{NL}$ and $\tilde{\varkappa}_R^{NL}$ are related with self-induced rotation of the polarisation ellipse whereas the parameters $\tilde{\chi}_A^L$ and $\tilde{\chi}_A^{NL}$ - with optical activity and its nonlinear variation, respectively.

Solution of the Equation of Motion for the Field Operators

The evolution in time of the field operators is described by Heisenberg's equation of motion:

$$\frac{\partial \vec{E}^{(\pm)}(\vec{r}, t)}{\partial t} = \frac{1}{i\hbar} \left[\vec{E}^{(\pm)}(\vec{r}, t), H \right]. \quad (15)$$

In the case of a field, the Hamiltonian H occurring above is nothing but that (H_F) of the free field of radiation. In this case the solution of Eq. (15) is given by the electric field (5) corresponding to "rapidly variable" time-evolution.

Due to interaction between the field of radiation and the material medium the Hamiltonian H of Eq. (15) contains, in addition to H_F, the interaction Hamiltonian

$$H_I = H_I^{(2)} + H_I^{(4)} + \ldots \quad (16)$$

In the present case the solution of (15) is no longer given by the free field (15). Additional time-dependent terms appear, and this additional "slowly variable" time-dependence due solely to the interaction (16) causes the amplitudes $\vec{E}^{(\pm)}(\vec{k})$ given by Eq. (8) to become functions of time.

Generally, considerations bear on the time-evolution of the operators of a field in a cavity of volume V. In our case, however, we deal with the propagation of a wave through an active medium in which it traverses a path of well defined length z i.e. the field emerging from the medium is dependent on z. In the case of plane waves the transition from the problem of a field in a cavity to that of propagation requires but the replacement of the time t by +z/c. Once this is done, the Heisenberg equations of motion (15) become equations describing the z-dependence of the operators of the k-th mode of the field:

$$\frac{\partial E^{(\pm)}(k,z)}{\partial z} = \frac{1}{i\hbar c}\left[\vec{E}^{(\pm)}(k,z), H\right]. \qquad (17)$$

The next step, which is crucial, consists in the choice of the interaction Hamiltonian (16) in a form enabling us to solve Eq. (17) analytically without loss of generality. It has been shown [27, 35] that the general analytical solution is obtained directly if the interaction between the field and the system is described by the effective Hamiltonian (16) in conjuction with the components (11) and (12). On eliminating free evolution, the equation of motion for the field operators reads:

$$\frac{d}{dz}\hat{a}_\pm(z) = \frac{i}{\hbar c}\left\{\tilde{\chi}^L_R \pm i\,\tilde{\chi}^L_A + \tilde{\varkappa}^{NL}_R \hat{a}^+_\mp(z)\hat{a}_\mp(z) + \right.$$
$$\left. + (\tilde{\chi}^{NL}_R \pm i\,\tilde{\chi}^{NL}_A)\hat{a}^+_\pm(z)\hat{a}_\pm(z)\right\}\hat{a}_\pm(z). \qquad (18)$$

Since the photon number operators $\hat{a}^+_+\hat{a}_+$ and $\hat{a}^+_-\hat{a}_-$ are constants of motion (they commute with the Hamiltonian), the above equation possesses a strict formal solution in the form of the translation operator:

$$\hat{a}_\pm(z) = \exp\left\{iz\left[\varphi_\pm + \varepsilon_\pm \hat{a}^+_\pm(0)\hat{a}_\pm(0) + \right.\right.$$
$$\left.\left. + \delta \hat{a}^+_\mp(0)\hat{a}_\mp(0)\right]\right\}\hat{a}_\pm(0), \qquad (19)$$

where we have introduced the following notation:

$$\varphi_\pm = \frac{1}{\hbar c}(\tilde{\chi}^L_R \pm i\,\tilde{\chi}^L_A),$$
$$\varepsilon_\pm = \frac{1}{\hbar c}(\tilde{\chi}^{NL}_R \pm i\,\tilde{\chi}^{NL}_A), \qquad (20)$$
$$\delta = \frac{1}{\hbar c}\tilde{\varkappa}^{NL}_R.$$

Squeeezed States of the Field

The preceeding solitions show that the photon statistics of a circularly polarized beam remains unchanged on traversal of the light wave through an isotropic nonlinear medium. However, this does not mean that the state of the field remains unaffected by the nonlinear interaction. The latter, in fact, can cause the field to go over into a squeezed state, having no classical counterpart. In order to prove this, we introduce two Hermitian field operators $\hat{Q}_\sigma$ and $\hat{P}_\sigma$, defined as follows:

$$\hat{Q}_\sigma = \hat{a}_\sigma + \hat{a}_\sigma^+, \qquad \hat{P}_\sigma = -i(\hat{a}_\sigma - \hat{a}_\sigma^+), \qquad (21)$$

with σ standing for $+(-)$ in circular basis, or $x(y)$ in Cartesian basis. The operators (21) satisfy the commutation rule

$$\left[\hat{Q}_\sigma, \hat{P}_{\sigma'}\right] = 2i\delta_{\sigma\sigma'}.$$

The definition of the squeezed state of the electromagnetic field is the following: it is a field state in which the square of the uncertainty in $\hat{Q}_\sigma$ or $\hat{P}_\sigma$ is less than unity [13]:

$$\langle(\Delta\hat{Q}_\sigma)^2\rangle < 1 \quad \text{or} \quad \langle(\Delta\hat{P}_\sigma)^2\rangle < 1, \qquad (22)$$

where $\Delta\hat{Q}_\sigma = \hat{Q}_\sigma - \langle\hat{Q}_\sigma\rangle$.

On normal ordering of the creation and annihilation operators, the definition (22) can be re-written in the form [13, 14]

$$\langle:(\Delta\hat{Q}_\sigma)^2:\rangle < 0 \quad \text{or} \quad \langle:(\Delta\hat{P}_\sigma)^2:\rangle < 0. \qquad (23)$$

To calculate the quantities occurring in the above definition (in the case of propagation in a nonlinear medium) we have to insert the operator solution (19) into the definition (23) and calculate the mean value in the quantum state of the incident beam which we assume as the coherent state $\hat{a}(0)|\alpha\rangle = \alpha|\alpha\rangle$. If one of the normally ordered variances turns out to be negative, the respective field component of the emerging light is in a squeezed state. Such calculations for normally ordered variances of the beam emerging from the medium give

$$\langle:\left[\Delta\hat{Q}_\pm(z)\right]^2:\rangle = \langle:\left[\hat{a}_\pm(z) + \hat{a}_\pm^+(z)\right]^2:\rangle -$$

$$- \langle \hat{a}_\pm(z) + \hat{a}_\pm^+(z) \rangle^2 =$$

$$= 2\,\text{Re}\left\{\alpha_\pm^2 \exp\left[2iz\varphi_\pm + iz\mathcal{E}_\pm + (e^{2iz\mathcal{E}_\pm} - 1)|\alpha_\pm|^2 + \right.\right.$$

$$+ (e^{2iz\delta} - 1)|\alpha_\mp|^2\bigg] - \alpha_\pm^2 \exp\bigg[2iz\varphi_\pm +$$

$$+ 2(e^{iz\mathcal{E}_\pm} - 1)|\alpha_\pm|^2 + 2(e^{iz\delta} - 1)|\alpha_\mp|^2\bigg]\bigg\} +$$

$$+ 2|\alpha_\pm|^2\bigg\{1 - \exp\bigg[2(\cos z\mathcal{E}_\pm - 1)|\alpha_\pm|^2 +$$

$$+ 2(\cos z\delta - 1)|\alpha_\mp|^2\bigg]\bigg\}, \qquad (24)$$

where

$$\alpha_\pm = \frac{\alpha}{\sqrt{2}}(\cos\eta \pm \sin\eta)\,e^{\mp i\theta}$$

η and θ being the ellipticity and azimuth of the elliptically polarized wave.

Of especial interest is the case when the incident beam is polarized circularly: $\eta = \pi/4$, $\theta = 0$. We have $|\alpha_+|^2 = |\alpha|^2$, $|\alpha_-|^2 = 0$ and (24) reduces to

$$\langle :[\Delta\hat{Q}_+(z)]^2: \rangle =$$

$$= 2|\alpha|^2\Big\{\exp\Big[|\alpha|^2(\cos 2z\mathcal{E}_+ - 1)\Big]\cos(z\varphi_+ + z\mathcal{E}_+ +$$

$$+ |\alpha|^2 \sin 2z\mathcal{E}_+) - \exp\Big[2|\alpha|^2(\cos z\mathcal{E}_+ - 1)\Big]\cos(z\varphi_+ + \qquad (25)$$

$$+ 2|\alpha|^2 \sin z\mathcal{E}_+)\Big\} + 2|\alpha|^2\Big\{1 - \exp\Big[2|\alpha|^2(\cos z\mathcal{E}_+ - 1)\Big]\Big\}$$

We gest a similar expression for the other component:

$$\langle :[\Delta\hat{P}_+(z)]^2: \rangle = -2|\alpha|^2\{\ldots\} + 2|\alpha|^2\{\ldots\}. \qquad (26)$$

On putting the initial phase of the field φ_0 in a manner to have $\varphi_+(z) + \varphi_0 = 0$ and on taking a numerical value [37] of $\mathcal{E}_+ z = 1 \times 10^{-6}$, we are in a position to compute the expression (25) numerically, leading to the graphs of Fig. 1,

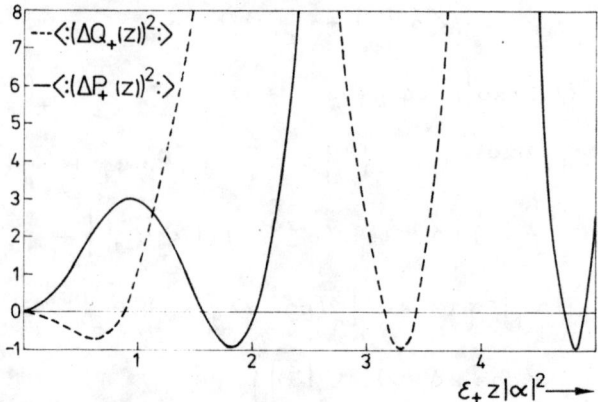

Fig. 1.

where the quantities $\langle :[\Delta \hat{Q}_+(z)]^2: \rangle$ and $\langle :[\Delta \hat{P}_+(z)]^2: \rangle$ are found to be oscillatory in shape depending on $\varepsilon_+ z |\alpha|^2$ and to take positive as well as negative values. In the regions where these values are negative, the respective field component $\hat{Q}_+(z)$ or $\hat{P}_+(z)$ is in a squeezed state. We note that the field can be in squeezed state despite the Poisson photon statistics. Also notable is the magnitude of the squeezing possible to achieve in the process considered above. On our definition of $\hat{Q}_+$ and $\hat{P}_+$ the value permitted for (25) and (26) by quantum mechanics is -1, meaning zero fluctuations of the field. In Fig. 1 the first minimum of (25) amounts to -0.66 whereas its second minimum amounts to as much as -0.97 i.e. 97 per cent of the magnitude permitted by quantum mechanics. As to the first minimum of (26), it amounts to -0.92 also signifying an almost complete elimination of quantum fluctuations.

Finally, it may be worth mentioning that on performing an appropriate interchange of the variables the expressions (25) and (26) become identical with the respective expressions for the isotropic anharmonic oscillator [38].

Conclusions

We have shown previously that the process of light propagation in an isotropic optically active nonlinear medium can be a source of non--classical fields. The photon antibunching effect in this process is insignificant, but can undergo amplification by interference with another beam [39]. In contradistinction to this, the effect of self-squeezing can be almost total and takes place as well for a circularly polarized incident beam, in which case no change in the photon statistics occurs. Hence squeezed states can exist for fields with a Poissonian photon distribution, characteristic for fields in coherent state. Moreover, our results draw a boundary between the phenomenon of photon antibunching and that of squeezing.

We wish to stress once again that we have obtained a strict analytical solution in translation operator form valid for arbitrary polarisation of the incident beam and involving explicitly the nonlinear and multipolar properties of the molecules of the optically active medium [3, 40].

The latest communications [41-44] suggest that the study of quantum optical effects, in particular aqueezed states of light, is no longer a matter of purely academic interest but is fast approaching the stage of experimental testing and applications in telecommunication. In the nearest future some well designed experiments may lead to a dramatic breakthrough in our understanding of the nature of light. Quite recently, theoretical papers [45-56] have appeared analyzing various subtler details of squeezed light and its interaction with matter [57].

References

1. D. F. Walls, Nature, 280, 451 (1979).
2. R. Loudon, Rep. Progress Phys., 43, 913 (1980).
3. S. Kielich, Molekularnaya Nielinieynaya Optika (Izd. Nauka) Moskva 1981.
4. H. Paul, Rev. Mod. Phys., 54, 1061 (1982).
5. J. Peřina, Quantum statistics of linear and nonlinear optical phenomena (Dordrecht, Reidel) 1984.
6. D. F. Walls, Nature, 306, 141 (1983).
7. S. Kielich and R. Tanaś, Proceedings EOC-83, Rydzyna, Poland, 30 May - 4 June, 1983 p. 5.
8. D. Stoler, Phys. Rev. Lett., 33, 1397 (1974).
9. H. P. Yuen, Phys. Rev., A13, 2226 (1976).
10. M. Schubert and W. Vogel, Optics Comm., 36, 164 (1981).
11. G. J. Milburn and D. F. Walls, Optics Comm., 39, 401 (1981).
12. A. Lane, P. Tombesi, H. J. Carmichael and D. F. Walls, Optics Comm., 48, 155 (1983).

13. D. F. Walls and P. Zoller, Phys. Rev. Lett., <u>47</u>, 709 (1981).
14. L. Mandel, Phys. Rev. Lett., <u>49</u>, 136 (1982).
15. H. F. Arnoldus and G. Nienhuis, Optica Acta, <u>30</u>, 1573 (1983).
16. Z. Ficek, R. Tanaś and S. Kielich, J. Opt. Soc. Am., <u>B1</u>, 882, (1984); Phys. Rev., <u>29A</u>, 2004 (1984).
17. Th. Richter, Optica Acta, <u>31</u>, 1045 (1984).
18. H. P. Yuen and J. H. Shapiro, Opt. Lett., <u>4</u>, 334 (1979).
19. J. Peřina, V. Peřina, C. Sibilia and M. Bertolotti, Optics Comm., <u>49</u>, 285 (1984).
20. L. Mandel, Optics Comm., <u>42</u>, 437 (1982).
21. M. Kozierowski and S. Kielich, Phys. Lett., <u>A94</u>, 213 (1983).
22. M. S. Zubairy, M. S. K. Razmi, S. Iqbal and M. Idress, Phys. Lett., <u>98A</u>, 168 (1983).
23. R. Loudon, Optics Comm., <u>49</u>, 67 (1984).
24. P. Chmela, Optica Acta, in press.
25. H. Helm and P. Zoller, Optics Comm., <u>49</u>, 324 (1984).
26. J. Peřina, V. Peřinova and J. Kodousek, Optics Comm., <u>49</u>, 210 (1984).
27. R. Tanaś and S. Kielich, Optica Acta, <u>31</u>, 81 (1984).
28. P. Chmela, R. Horak and J. Peřina, Optica Acta, <u>28</u>, 1209 (1981).
29. K. Wódkiewicz and M. S. Zubairy, Phys. Rev., <u>27A</u>, 2003 (1983).
30. M. Schubert, W. Vogel and D. G. Welsch, Optics Comm., <u>52</u>, 247 (1984).
31. C. M. Caves and B. L. Schumaker, Phys. Rev., <u>31A</u>, 3068 (1985).
32. S. Kielich, Acta Phys. Polonica, <u>30</u>, 851 (1966).
33. S. Kielich, Physica, <u>32</u>, 385 (1966).
34. R. J. Glauber, Quantum Optics and Electronics, (Eds. C. De Witt et. al., Gordon and Breach, New York, 1965) p.63.
35. S. Kielich and R. Tanaś, Izw. Akad. Nauk. Seria Fiz. <u>3</u>, 518 (1984).
36. S. Kielich, R. Tanaś and R. Zawodny (to be published).
37. H. H. Ritze and A. Bandilla, Optics Comm., <u>30</u>, 125 (1979).
38. R. Tanaś, Coherence and Quantum Optics V (Eds. L. Mandel, E. Wolf, Plenum Press, New York) 1984, p. 643.
39. H. H. Ritze, Z. Physik, <u>B39</u>, 353 (1980).
40. A. D. Petrenko and N. I. Zheludev, Optica Acta, <u>31</u>, 1177 (1984).
41. J. H. Shapiro, P. Kumar and M. W. Maeda, J. Opt. Soc. Am., <u>B1</u>, 517 (1984).
42. M. D. Levenson, J. Opt. Soc. Am., <u>B1</u>, 525 (1984).

43. R. E. Slusher, B. Yurke and J. F. Valley, J. Opt. Soc. Am., B1, 525 (1984).
44. P. Kumar and J. H. Shapiro, Phys. Rev., 30A, 1568 (1984).
45. K. Wódkiewicz, Optics Comm., 51, 198 (1984).
46. R. Loudon and T. J. Shepherd, Optica Acta, 31, 1243 (1984).
47. M. J. Collet, D. F. Walls and P. Zoller, Optics Comm., 52, 145 (1984).
48. Z. Ficek, R. Tanaś and S. Kielich, Acta Phys. Polonica, A67, 583 (1985).
49. R. S. Bondurant, P. Kumar, J. H. Shapiro and M. Maeda, Phys. Rev., 30A, 343 (1984).
50. D. Stoler, B. E. A. Saleh and M. C. Teich, Optica Acta, 32, 345 (1985).
51. B. L. Schumaker and C. M. Caves, Phys. Rev., 31A, 3093 (1985).
52. A. Heidmann, J. M. Raimond, S. Reynaud and N. Zagury, Optics Comm., 54, 189 (1985).
53. S. Kielich, M. Kozierowski and R. Tanaś, Optica Acta, 32, ... (1985).
54. S. Y. Kilin, Optics Comm., 53, 409 (1985).
55. B. Yurke, Phys. Rev., A32, 300 and 311 (1985).
56. C. K. Hong and L. Mandel, Phys. Rev. Lett., 54, 323 (1985).
57. G. J. Milburn, Optica Acta, 31, 671 (1984).

G. I. SURDUTOVICH

Institute of Physics
of Semiconductors,
Sibirian Division of USSR
Academy of Sciences,
93 Novosybirsk, USSR

LIGHT PRESSURE AND BISTABILITY PHENOMENA

The pressure of light appears due to recoil of atoms and molecules when photons are scattered on them. This effect manifests itself most explicitely when slow atoms are moving in a non-homogeneous field of the resonant standing electromagnetic wave. Their motion can be described by an equation of the Fokker-Planck type with an effective force depending on the sign of the detuning. The effective dissipative force is caused by the spontaneous transitions which are mixing adiabatical atomic states in the standing wave field. The light pressure affects the absorption coefficient of a standing wave in a gas of resonant atoms with allowed transitions, when the recoil energy aquired by the atom $E_\mu = (\hbar k)^2/2m$ is less than $\hbar \gamma$ (γ is a frequency of spontaneous transitions). In a weak field $dE_0 < \hbar \gamma$ (d is a dipole matrix element of the transition, while E_0 - the amplitude of the field) the two characteristic cases may be identified depending on the parameter $\eta = 4E_\mu \tau /\hbar \cdot (dE_0/\hbar \gamma)$. When η is small the absorption coefficient of a standing wave includes not only the ordinary term associated with saturation and independent on the sign of the detuning, but also a term proportional to $E_\mu \tau$ (where τ is a time of atom-field interaction), which is an odd function of the detuning. When η is large the velocity distribution of the particles is characterised by a narrow dip or peak depending on the sign of the detuning. The width of this narrow peak, which appears for negative detunings, is limited by the diffusion of atoms

-in the velocity space. The dependence of the absorption coefficient of a standing wave on detuning brings about the bistability regime when the radiation transmits through a Fabry-Perrot resonator with the resonant medium in it.

Kazimierz RZĄŻEWSKI

Institute for Theoretical Physics
Polish Academy of Sciences
02-668 Warsaw, al. Lotników 32/46
Poland

SPECTRA GENERATED BY SHORT LASER PULSES

Scattering processes play the central role in many branches of physics. Collisions between microparticles often the main and ofter the only way of studying properties of their interactions. The well developed scattering theory has a certain degree of universality. Such notions as: scattering cross-section, partial waves, phase shifts, resonances are useful regardless the nature of colliding particles. They can be applied to the high energy elementary particles, nuclei, atoms, ions or photons.

Missing from the list of basic concepts of conventional scattering theory is the notion of the spectrum or the energy distribution of scattering particles. The explanation is simple: a particle scattered under certain angle has well defined energy in most cases. The scattering process in most cases involves only two particles and the conservation of energy and momentum determines uniquely the energy of outgoing particles. The only residual width of the energy distribution comes from the velocity distribution in the incident beam and from the finite resolution of the spectrometer used in the detection process.

The only drastic exception to this rule is provided by the resonant scattering of powerful laser light on atoms. Only in this case the density of incoming photons and correlations of their phases (coherence) make the scattering process a collective one.

The first correct results for the spectrum excited by a cw monochromatic laser were found in the late sixties [1]. Several years later they were verified in a series of beautiful experiments [2]. The spectrum, often refered to as the Mollow spectrum, consists of three peaks: one resonant and two satellites displaced by the Rabi frequency Ω_0 from the center.

There are two simple heuristic explanations of the three peak structure. One refers to the dressed states of the composite system consisting of the two-level atom and the highly excited field mode representing the laser light. Hamiltonian of such a system, often called a Jaynes-Cummings [3] model, has a spectrum composed of the infinite ladder of pairs of energy levels. The pairs are displaced by the energy of the single photon while the splitting in each pair is the energy corresponding to the Rabi frequency. The observed fluorescence spectrum arises from the transitions between the pairs of levels.

Another explanation explores a time dependent picture of, so called Bloch equations [4]. Driven by the strong laser signal, atomic dipole moment participates simultaneously in two independent harmonic oscillations: one of them is its natural precession with the frequency of the field and the other is again the Rabi frequency (in fact originally introduced in this context). The spectrum, a quantity quadratic in the emitted field amplitude and hence, quadratic in the atomic dipole moment operators is therefore expected to contain the combination frequencies of the beats.

It is a long way from the heuristic arguments given above to the quantitative theory of the Mollow spectrum but its main three peak structure can be easily understood.

The question of the resonance fluorescence of an atom irradiated by a finite light pulse rather than the CW signal was addressed only recently [5]. The autors treated the case of the square pulse. The basic three-peak structure is still recognisable. The additional structure can then be associated with the, so called, edge effects. They are simply elastically scattered photons from the wavy wings of the spectrum of the square pulse itself.

In this lecture we review the properties of the resonance fluorescence spectrum of the two-level atom driven by a smooth pulse. The problem, complex in general, becomes very simple in the limit of very short pulses. If the lifetime of the transition is much shorter than the duration of pulse the back action of the fuorescence light on the motion of the atom can be neglected. The dynamics of the interacting system can, hence be viewed as going through two stages: first we find the evolution of the two-level atom in the field of the incident pulse. Secondly we compute the properties of the radiation emitted by the given time dependent quantum source.

The first part can be done for arbitrary resonant pulse. Denoting the probability amplitudes of finding the atom in its lower and upper level by $\alpha(t)$ and $\beta(t)$ respectively we can write:

$$\alpha(t) = \cos(\Theta(t)/2) \quad \text{and} \quad \beta(t) = \sin(\Theta(t)/2), \qquad (1)$$

where the running area of the pulse is defined as:

$$\Theta(t) = \Omega_0 \int_{-\infty}^{t} f(\tau)\, d\tau, \qquad (2)$$

with Ω_0 - the Rabi frequency typically corresponding to the peak intensity of the light while $f(t)$ denotes the dimensionless envelope of the pulse. Of course the evolution of all the atomic operators can be easily expressed in terms of the amplitudes $\alpha(t)$ and $\beta(t)$.

The motion of the atomic dipole moment operators is the source of the radiation field. The annihilation operator of the photon field mode of frequency ω is determined by the evolution of the lowering operator $\sigma^- = |0\rangle\langle 1|$:

$$a_\omega \propto \exp(-i\omega t) \int_{-\infty}^{t} \sigma^-(\tau) \exp(i\omega\tau)\, d\tau. \qquad (3)$$

The process studied here is transient in nature. The simplest time independent quantity is the energy spectrum of the emitted field as $t \to \infty$:

$$W(\omega) = \lim_{t \to \infty} \langle 0 | a_\omega^+(t) a_\omega(t) | 0 \rangle. \qquad (4)$$

This quantity gives the total number of photons of frequency ω emitted during the interaction with the pulse. A simple expression for the energy spectrum [6] in terms of the amplitudes $\alpha(t)$, $\beta(t)$ reads:

$$W(\omega) = \left| \int_{-\infty}^{+\infty} \exp\left[-i(\omega - \omega_L)t\right] \beta(t)\, \alpha^*(t)\, dt \right|^2 + \\ + \left| \int_{-\infty}^{+\infty} \exp\left[-i(\omega - \omega_L)t\right] \beta(t)^2\, dt \right|^2. \qquad (5)$$

The above expression yields a finite result for pulses evolving towards the total area of $2\pi n$. Only these pulses leave the atom in its ground state and terminate the production of radiation.

For some pulses the Fourier transform in the formula (5) can be evaluated analytically. The important special case of hyperbolic secant belongs to this class. A Lorentzian pulse is another example. For other pulses the Fourier transforms must be done numerically.

In Fig. 1 we show the resuts for the hyperbolic secant pulse. The dominant feature of the spectrum is its multipeaked structure. The spectrum has $2(n-1)$ zeros for the $2\pi n$ pulse. The result is the spectrum with the number of peaks dynamically determined by the area of the pulse.

No trace of the dressed states forming in the steady state of the CW excitation is visible. The characteristic distance between the peaks as well as their widths is given by the Fourier width of the pulse denoted by Γ_p.

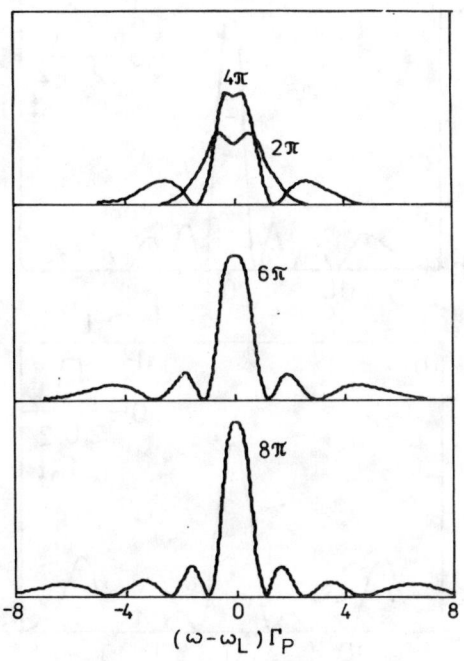

Fig. 1. Spectra of resonance fluorescence produced by the two-level atom driven by the resonant hyperbolic secant pulses of increasing areas 2π, 4π, 6π and 8π

Our initial finding raises more questions than gives the answers. We list some of them:

1. How general is the result?
2. Is it possible to go to the Mollow like limit in a smooth way?
3. How about the time dependence of our spectrum?
4. How does the radiative reaction tend to destroy the multipeak structure?

In the remainder of this review we give brief answers to the above questions. As the first example we consider the family of pulses of the general form:

$$f(t) = \exp\left[-(\Gamma_p t)^C\right].$$

The results for $C = 2$, 6 and 26 is presented in Fig. 2. The pulse in the t domain (see the insets) changes shape from a Gaussian, through one with a well-defined plateau, to a rectangular one. We took the 8π pulses. The energy spectra of the resonance fluorescence change from the pure

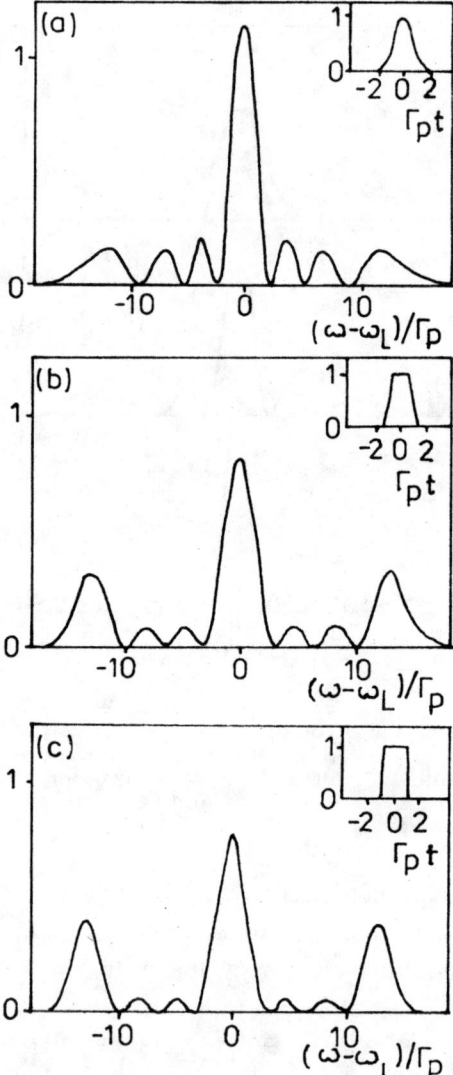

Fig. 2. Energy spectra of resonance fluorescence are plotted for the 8π exponential pulses described in the main text for C = 2, 6 and 26 respectively. The time shape of each pulse is shown in the inset. Their Fourier spectra are shown as dotted lines

multipeak to one with well defined Mollow-type satellites. Their displacement corresponds, of course, to the Rabi frequency at the maximum of the pulse, while their widths are still determined by the width of the pulse. The dotted line represents the Fourier spectrum of the pulse itself. Of course, the Fourier spectrum is single peaked only for the Gaussian pulse. Only in this case do we have pure saturation and coherence type splitting. For C > 2, the Fourier spectrum of the pulse itself is multipeaked and the additional structure in the fluorescence spectrum is partially due to the elastic scattering of the photons from those peaks.

We see from the above example that the pure multipeak splitting occurs for the pulses with simple single peaked Fourier spectrum (hyperbolic secant, Lorentzian [7] or Gaussian).

The additional structure observed in some more complicated cases (pulses described here, more complicated pulses of self-induced transparency [8]) can all be atributed to the elastic scattering. Our example shows also the formation of the dressed states of the atom-plus-field system as the plateau in the pulse imitates the continuous excitation.

So far we have been treating the pulses with Γ_p much larger than the spontaneous width of the transition Γ_s. The distortion of this perfect situation caused by the finite value of Γ_s has been studied in Ref. 7. The multipeaked structure of the spectrum gets smeared with increasing Γ_s but remains very visible for $\Gamma_s \sim 0.1\, \Gamma_p$. The total number of fluorescence photons is also shown to be of the order of Γ_s / Γ_p.

There are, of course, many other sources of smearing of our spectrum in the realistic experimental situation. They are all characterised by their typical frequences. We can say that as long as:

1. the spontaneous width of the transition,
2. the inhomogeneous width,
3. detuning,
4. noise width of the laser,
5. width of the filter used to resolve the spectrum,

are all smaller than the Fourier width of the pulse, the multipeak structure of the spectrum should be experimentally observable.

To shed some light on the problem of the formation of our spectrum it is instructive to study a more conventional notion of the power spectrum, i.e., the rate at which the photons of given frequency are emitted. In the present case of pulse excitation, it is a time dependent quantity.

When discussing a time-dependent power spectrum, one runs into difficulty with the uncertainty relation between time and frequency. This difficulty can be satisfactorily overcome by taking into account a filter necessary to resolve the spectrum [9].

Such a filter is characterized by its width Γ_F, which determines the resolution of the spectrometer and in the same time its reciprocal $1/\Gamma_F$ is the accumulation time of light in the spectrometer. When we have the resolution Γ_F, we cannot know the emission time with the accuracy better than $1/\Gamma_F$.

In the approximation of very short pulses, the time dependent power spectrum is again expressible in terms of certain Fourier transform of the simple atomic motion (1). The details can be found in Ref. 10. Here we simply ilustrate the results.

In Fig. 3, the time dependent spectrum of fluorescence induced by the hyperbolic secant pulse of the area equal to 8π is plotted for a very narrow filter ($\Gamma_F = 0.1\, \Gamma_p$).

The spectrum evolves from its initial single peak form developing more and more peaks as the area of the pulse accumulates. For long times, we get a seven-peak structure which is nearly identical with the energy spectrum of Fig. 1.

The effect of decreasing the accumulation time of the field in the interferometer is illustrated in Fig. 4. With all the parameters as before, now the width of the filter is equal to the width of the pulse. We see the enhancement of the satellite which corresponds to the local, instantaneous Rabi frequency. We get closer to a Mollow-like spectrum.

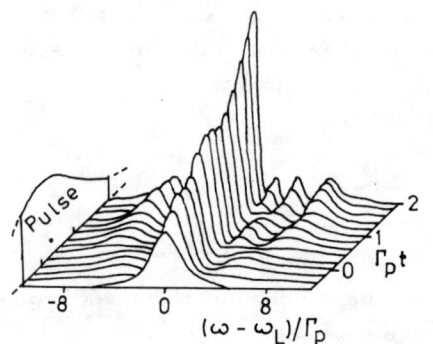

Fig. 3. Physical spectrum of light scattered from a two-level system driven by 8π hyperbolic secant pulse. The seven peak structure develops in stages as the area of the pulse grows. The width of the filter Γ_F is only $0.1\,\Gamma_p$

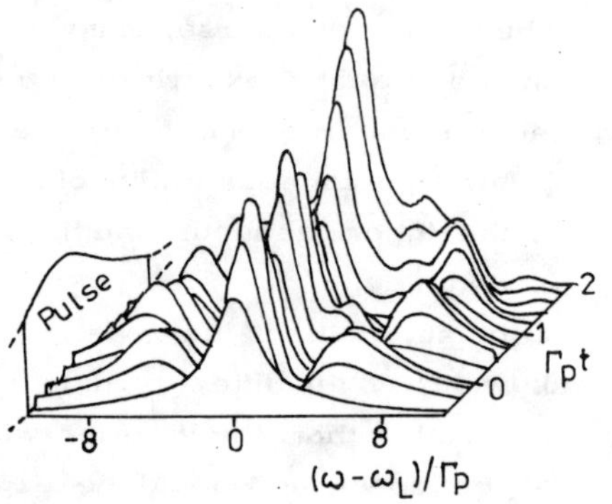

Fig. 4. Same as Fig. 3, but for $\Gamma_F = \Gamma_p$. Shorter accumulation time enhances satellites corresponding to the instantaneous Rabi frequency

The interpretation of this observations is simple. The multipeak structure is a result of the interference of the portions of radiation, emitted by the atom at various stages of the interaction, occuring in the interferometer.

The resonance fluorescence is not the only process in which the short pulses produce the multipeak spectra. A very similar structure can be found in the scattering in which the upper level of the two-level atom driven by the strong and short laser pulse is probed by the transition to some other part of the atomic spectrum. The conventional Autler-Townes two peak spectrum is then replaced by the multi-peaked spectrum with the number of peaks n for the pulse of $2\pi n$ area [11].

Acnowledgements: I am pleased to thank my coworkers and friends with whom the results described in this lecture were obtained. They include: M. Florjańczyk, M. Lewenstein, J. Zakrzewski and J. Haus. I am also indebted to J. H. Eberly, P. Knight, T. Dohnalik and J. Dupont-Roc for interesting discussions and suggestions.

References

1. A. I. Burshtein, Zh. Exp. Teor. Phys., 49, 1362 (1965); M. Newstein, Phys. Rev., 167, 89 (1968); B. R. Mollow, Phys. Rev., 188, 1969 (1969).
2. F. Schuda, C. R. Stroud, Jr., and M. Hercher, J. Phys., B7, L198 (1974); E. Grove, F. Y. Wu, and S. Ezekiel, Phys. Rev., A15, 227 (1977); W. Hartig, W. Rasmussen, R. Schieder, and H. Walther, Z. Phys., A278, 205 (1976).
3. E. T. Jaynes, and F. W. Cummings, Proc I.E.E.E., 51, 89 (1963).
4. For a review of the fundamental properties of the two-level atoms, see L. Allen and J. H. Eberly, Optical Resonance and Two-Level Atoms (Wiley, New York, 1975).
5. X. Y. Huang, R. Tanaś, and J. H. Eberly, Phys. Rev., A26, 892 (1982).
6. K. Rzążewski and M. Florjańczyk, J. Phys., B17, L509 (1984).
7. M. Lewenstein, J. Zakrzewski and K. Rzążewski, JOSA B (in press).
8. M. Florjańczyk, (in preparation).
9. J. H. Eberly and K. Wódkiewicz, JOSA 67, 1252 (1977).
10. M. Florjańczyk, K. Rzążewski and J. Zakrzewski, Phys. Rev., A31, 1558 (1985).
11. K. Rzążewski, Phys. Rev., A28, 2565 (1983); K. Rzążewski, J. Zakrzewski, M. Lewenstein and J. Haus, Phys. Rev., A31, 2995 (1985).

M. BERTOLOTTI

Sezione Fisica
Dipartimento Energetica
Università di Roma "La Sapienza"
Rome, Italy

HISTORICAL PATHS IN LASERS AND QUANTUM OPTICS

This year it is the 25th anniversary of the first laser assembly by Maiman [1]. A number of celebrations are made all around the world, and we wish here briefly retrace the principal steps which brought to its construction.

The concepts and ideas on which the laser is based were developed in a worldwide context and we may start from the introduction of the concept of stimulated emission by Einstein [2], in 1916, remembering the quantum mechanical interpretation of dispersion by Ladenburg [3] and Kramers [4], the first experiments on negative absorption by Ladenburg [5] up to the experiments on population inversion in magnetic resonance and the intoduction of negative temperature by Purcell [6], and optical pumping by Kastler and Brossel [7],[8].

The idea of masers and lasers was then born, and grew independently and through different paths in the USA and in the USSR. The background over which the maser concept grew was the post-war development of radars and microwave electronics. Lasers came out as a natural extension to shorter wavelengths, and the optical background exhisting in Europe was essential to their development.

Excellent recollections from the main laser leaders in USA have recently been published [9], and there is an American Project for the History of Lasers which is hard working to pick up oral witnesses from the most important researchers in the field [10]. A special issue of Optical Acta is in print for celebrating the 25th anniversary.

It is worth mentioning that proposals for using induced emission for producing light sources in a gas discharge, probably for the first time ever, were made by Fabrikant in the Soviet Union in 1939 [11]. Unfortunately the proposal ramained isolated and was not followed by further research, may be because it was still in an embrional form; the usefulness of enclosing the amplifying material in a resonant structure was however appreciated, although no evaluation was performed of the needed population inversion for obtaining a sizeable effect.

Moreover the patent issued to Fabrikant and his students in 1951, was only published, and therefore available to the scientific community, in 1959 [12], after the Schawlow and Townes, Physical Review paper [13]. Fabrikant was interested for some time in gas lasers [14].

The proposal of Schawlow and Townes as detailed as it was aroused the interest of many laboratories, and in June 1960, T. H. Maiman, at Hughes Labs., was the first to obtain stimulated emission in the visible spectrum [1].

By analogy with the term Maser, the word LASER was coined, as an acronym for Light Amplification by Stimulated Emission of Radiation, a word also used by Gould in his early works, and which later gained acceptance over "optical maser" employed by most early workers. Optical maser was Townes-Schawlow terminology. Bell Labs. at first refused to use the Gordon Gould's name laser and in effect discouraged others from using this term, but it won out anyway.

Most of the subsequent early research on solid state lasers was performed, for different reasons, in USA, although European researchers were immediately involved in them, as we will see later.

The first gas laser was developed in USA too, by A. Javan in late 1960 [15], and operated in the near infrared with a helium-neon gas mixture pumped by an electric discharge [16].

Proposals for obtaining, through selective excitation, the inversion of the upper level in a one-gas system were already made in 1959 by J. H. Sanders [17] and A. Javan [18] who both suggested producing the required population inversion by using an electric discharge in a gas.

Sanders did his work while he was at Bell Labs. on leave from the Clarendon Lab., Oxford, England. Coming back to UK he reconsidered, finding a negative result [19], the experiment by Fabrikant and Butayeva [20] in which they claimed light amplification in a discharge of Hg vapor. Lasers action of ionized Hg atoms in a discharge was later found almost contemporarily by Bell in USA [21] and Convert and collab. in Paris [22].

Gas lasers were actively exploited in all the world. The spectroscopic experience of European researchers was of great help for this, and we mention here just a few of the results that were obtained, hoping to mention at least the principal ones.

Amplified spontaneous emission and quasi-cw oscillation in Ar was obtained almost simultaneously by Convert et al. [23] in 1964 in Paris when they were studying different mixtures of Hg and rare gases, Bennett et al. [24] and Bridges [25] in the USA. Laser action in other singly ionized noble gas ions (Kr and Xe) was also reported, almost contemporarily, by Bridges [26] in the USA, and Laures et al. [27] in France.

The previous year in 1963 Legay and Barchewitz at the Laboratoire d'Infrarouge de l'Université, Orsay [28] had observed strong infrared emission from CO_2 when it was mixed with vibrationally excited N_2, and Legay and Legay-Sommaire [29] suggested the possibility of obtaining laser action on rotational-vibrational bands of gases excited by active N_2, and mentioned specifically the CO_2 10.4 μm transition.

At about the same time Patel et al. [30] observed laser action in pure CO_2 at 10.6 μm.

N. Legay-Sommaire, L. Henry and F. Legay improved later the power characteristics of this laser obtaining c.w. operation with powers up to 200 mW [31].

A distinct improvement was made, first in Poland by W. Woliński and then in France by R. Dumanchin and J. Rocca-Serra [32] by introducing the pulsed transverse excitation (TEA) in a CO_2 laser, which allowed to obtain an emitted power density of 1500 W/cm^3, and later by Basov et al. who used electron beams to excite several kinds of lasers [33][34][35] and extended laser action in $C\bar{O}_2$ to high pressures up to 25 atm using an electron-beam-controlled discharge [36].

As early as 1963 before the CO_2 laser was discovered, Basov and Oraevskii [37] suggested that population inversion could be produced thermally by rapid gas expansion. A CO_2 laser based on rapid expansion of a gas flowing through a supersonic nozzle was later discussed in 1966 by Konyukhov and Prokhorov [38]. Similar studies by Hurle and Hertzberg [39] the previous year indicated the feasibility of producing population inversions in Xe in the fast expansion of an arc-heated plasma. A CO_2 laser based on these theoretical predictions was built later by Gerry [40], Konyukhov et al. [41], Dronov et al. [42] and Kuehn and Monson [43].

On the other part of the spectrum at visible and shorter wavelenghts the potential of excimers for laser media was first pointed out by Houtermans at the Physikalisches Institut der Universität, Bern, in 1960 [44], before the first successful operation of the ruby laser by Maiman. Excimer lasers are capable of efficient generation of high-power pulses of radiation at ultra-violet and vacuum ultra-violet wavelengths. The spontaneous radiative decay times of simple excimer systems lie in the range $5 ns - 5 \mu s$, whereas the dissociation time for the lower state may be as short as $10^{-13} s$, if temperature is sufficiently low to avoid thermal population of the lower state, it is consequently possible to invert the population of the bound and dissociative states and so produce laser action. The first laser of this kind was however made to operate with liquid Xe by Basov [34] at $\lambda = 1760$ Å, only in 1970.

Excimer lasers in which the excited molecule contains two different atoms (they should properly be called exciplet lasers) were further studies in USA [45]. However the first time electrically excited xenon or argon were mixed with alogens was in England, by Golde and Thrush [46].

Chemical lasers, first proposed by J. Polanyi [47], were actively considered in many laboratories. Even before their construction in 1965 by Kasper and Pimentel [48], the possibility of making such lasers was discussed in at least two conferences; the 1st Symposium on the Elementary Processes of High-Energy Chemistry, held in Moscow 18-22 March 1963, and the 1st Conference on Chemical Lasers, held in San Diego, USA, 9-11 September 1964. A Supplement (n. 2) of Applied Optics in 1965 was dedicated to Chemical Lasers still before their realization.

Extensive work with electrical initiated chemical lasers was performed in the Soviet Union [49].

The first laser essembled by Kasper and Pimentel [48] used selective vibrational excitation in the chemical reaction

$$H + Cl_2 \rightarrow HCl^* + Cl.$$

The laser worked on emission in a few P-branch transitions of Cl formed in reactions initiated by flash photo-dissociation of chlorine in a $Cl_2 + H_2$ mixture.

The first purely chemical laser was later realized by Cool an Stephens [50] in USA in 1969 using a DF-CO_2 system in which vibrationally excited DF^+ molecules produced by a chemical reaction

$$F + D_2 \rightarrow DF^* + D$$

pumped the upper CO_2 level by intermolecular vibrational energy transfer processes.

The idea of the investigation of photo-dissociation of molecules for the excitation of a laser medium, had already been advanced by S. G. Rautian and I. I. Sobel'man in 1961 [51].

A few months before the attainement of their HCl laser, Kasper and Pimentel succeded in obtaining a laser, in which atomic iodine atoms were excited by flash photolysis of either gaseous CF_3I or CH_3I [52].

In this laser, which subsequently became a very important laser, emission occurs between the two lowest states of the atomic iodine transition $5^2P_{1/2}$ $5^2P_{3/2}$ at a wavelength of 1.315 μm. Excited iodine atoms are produced by UV flash photolysis of various perfluoroalkyl iodides, $C_\Lambda F_{2\Lambda+1} I$, most usually C_3F_7I, through the reaction

$$C_3F_7 \tau + h\nu_{pump} \rightarrow C_3F_7 + \tau^*$$

$$\tau^* \rightarrow \tau + h\nu_{laser}.$$

A iodine laser operated in the Soviet Union at Lebedev Institute a few months after the first Kasper and Pimentel realization [53].

High energy laser pulses, up to 60 J in a single oscillator, were produced shortly thereafter in 1966 by De Maria and Ultee [54].

In 1968, Q-switching and mode locking of the iodine laser was reported by Ferrar [55]. Several researchers described broadening of the gas laser transition by applied magnetic fields [56][57], and by pressure [58]. This broadening of the transition was shown to give control over the laser gain and hence over parasitic oscillation in large volume systems.

Systematic studies of the iodine laser as a high power system were begun in 1970 at the Max-Planck Institute at Garching near Munich by Kompa and Hohla and colleagues [59], and at Lebedev Institute by Basov and Zuev [60]; the major impetus behind the development of this laser being the requirement for a high-power, short-pulse, terawatt laser for laser fusion studies. Other laboratories also began studies of the physics of the iodine laser, including Sandia National Laboratory [61], and Manchester University [62].

By the mid-1970 the feasibility of the iodine laser for short- pulse, high-power operation was clearly demonstrated.

The single-beam, high-power laser, Asterix III, at the Max-Planck Institut für Plasmaphysik (now Quantenoptik) at Garching [63][64], was completed in 1978.

Another important class of lasers which were very much developed in the Soviet Union was the metallic vapor lasers.

Laser action on the 510nm and .578nm lines of copper was first reported by G. Gould and coworkers in 1966 [65]. A furnace was utilized to produce the vapor, and efficiency was low. A major advance was made in the Soviet Union by Isaev [66] who used the waste heat from the gas discharge to provide the power necessary to vaporise the copper present within the plasma tube. Practical efficiences of nearly 1% were reported, with 15 W of average power at repetition rates of 15 KHz. Further improvements were reported by Isaev [67] in 1977, resulting in the attainement of power levels exceeding 40 W for a few minutes. The following year Bokhan [68] studied the basic kinetic processes responsible for the laser operation.

Vapor pressure studies showed that an optimum pressure of 1 torr or lower has to be used [69][70][71]. Increase above this value leads to a decrease in laser power. This optimum metal vapor density was found to depend on a variety of different factors such as buffer species and pressure [69][70], discharge circuit parameters [68][72], interpulse period [68][69][73], and discharge tube diameter [68][73]. A similar dependence of output power on repetition rate [68][69][74], electrical parameters [68][74], tube diameter [68], and inert gas species was found.

Increase in the applied voltage, current density, and rate of rise of current density [68][72][74], improved output; however the large volumes needed for high power operation in the longitudinal discharge devices are incompatible with the above factors.

An alternative design using a transverse discharge does not have the same limitations. It allows use of high field strengths, high metal vapor densities, and low inductance circuitry. It was used for copper vapor lasers by Aleksandrov et al. [75], and resulted in an average laser output power of 75 W at a pulse repetition frequency of 3 KHz in 1980 [76].

Considering now liquid systems, the first unambiguously successful attempt to produce stimulated emission from organic molecules was reported in 1966 by P. P. Sorokin and J. R. Lankard [77]. Similar results were obtained independently, still in the infrared, by M. L. Spaeth and D. P. Bortfeld in the USA [78] and Schafer et al. [79], than at the Physikalisch-Chemisches Institut at Marburg University (Germany).

The possibility of using complex molecules of dye type as active media for tunable lasers had already been pointed out in 1963-64 by B. I. Stepanov, N. A. Borisevich, A. N. Rubinov, and V. A. Tolkachev in Minsk [80].

Later on, B. I. Stepanov, A. N. Rubinov, and V. A. Mostovnikov obtained laser emission in dye solutions independently of western researchers, pumping with a Q-switched ruby laser [81], and in 1967 demonstrated the first flashlamp pumped dye laser [82].

Continuous operation was in principle shown to be possible by B. B. Snavely and F. P. Schafer in 1969 [83], and realized the following year in the USA [84].

In 1973 N. A. Borisevich et al. were the first to built a dye vapor laser [85].

Dye lasers resulted in a very good system to produce picosecond pulses. The first attempt to produce ultrashort pulses with dye lasers involved pumping a dye solution with a mode locked pulse train from a solid state laser and was made almost contemporarily in the USA [86] and in UK [87]. Self-mode locking of flash lamp-pumped dye lasers was first achieved by W. Schmidt and F. P. Schafer in 1968 [88], and later Bradley et al. [89] at the Queen's University of Belfast obtained pulsewidths of only 2 psec.

Measurements of pulsewidths were first made using the two-photon-fluorescence [90]. Later on Bradley et al. [91] developed an ultrafast streak camera capable of picosecond-pulse resolution, which was the first linear detector capable of time-resolving mode-locked dye-laser pulses.

A qualitative model of the pulse generation in a passively mode-locked laser was first presented by Letokhov [92]. The experimental results, mostly obtained by Bradley's group, were later incorporated in a simple rate-equation theory by New at Belfast [93].

A dramatic reduction in the optical pulse width was achieved by using passive cw mode locking techniques, and pulses shorter than a picosecond were obtained in 1974 by Shank and Ippen [94]. More recently using the colliding pulse mode locking, Shank et al. have obtained optical pulses as narrow as 30 fs [95]. The shortest pulse obtained up to now are 8 fs [96].

In Poland experimental research into lasers and their properties started soon after the invention of the laser. First lasers were constructed at the Mickiewicz's University in Poznań and at the Military Academy in Warsaw. Large scale activity in nonlinear optics and quantum optics was developed at the Mickiewicz's University and at the Institute of Physics of Polish Academy of Sciences [97] [98]. Many of the leading figures of the Polish school of quantum optics are represented in these proceedings.

Semiconductors are, in a sense, a very particular kind of lasers.

The possibility of obtaining stimulated emission in a semiconductor was probably considered for the first time although in a very rough way, by John von Neumann in a private communication to John Bardeen in 1953 [99], mostly contemporary to the first speculations about masers, and five years before the Townes and Schawlow paper [13]. Townes [100] remembers a conversation with von Neumann, just after the successful oscillation of his first maser, in which von Neumann was pushing him to investigate stimulated emission in semiconductors.

Extensive research started a few years later (1957) at Lebedev Institute in Moscow, under the direction of Basov [101]. In the meantime, on 22 April 1957, a patent was filed in Japan by Y. Watanabe and J. Nishizawa, in which recombination radiation produced by injection of free carriers in a semiconductor was considered [8].

A few months later, in 1958, P. Aigrain of France, in a speech at Brussels, during the conference Congrès International sur la physique de l'etat solide et ses applications a l'electronique et aux télécommunications [102], showed that population inversion between two localized levels is not required in the case of the emission of two bosons.

At the second conference on Quantum Electronics (Berkeley 1961), attention was still directed towards indirect gap semiconductors such as Ge or Si, and only about one year later it became clear that laser action in Ge or Si was very difficult to achieve, on account of free-carrier absorption, and attention was shifted on direct-gap semiconductors, such as III-V compounds, and notably GaAs. The wrong choice of Ge and Si as first candidates is probably responsible for the achievement of semiconductor lasers after the ruby laser, and not before.

The use of indirect gap semiconductors was also proposed in 1961 by Basov's group [103] in the same year in which Bernard and Duraffourg in Paris discussed in a complete and exhaustive manner the possibility of obtaining stimulated emission in semiconductors by transitions between conduction and valence bands [104].

Ultimately it was the triumph of american technology which led to the operation, almost simultaneously, of GaAs laser diodes in four research groups working at IBM, GE and Lincoln Lab. [105] [106].

Basov's group had already in 1959 proposed a method of generating light by exciting a homogeneous semiconductors with electric field pulses [107]. A new method of electric excitation was considered by the group much later in 1974, which gave light emission in CdS, CdSe, $CdS_x Se_{x-1}$, and ZnSe crystals [108]. This new kind of laser was called <u>semiconductor streamer laser</u> by its inventors [109].

A very significative advance in laser diodes was made by Hayashi [110] and Alferov [111] in 1970, with the realization of double heterostructures which confined the carriers to a very small distance, provided a refractive profile for improving the guiding of the electromagnetic wave, and made possible cw operation at room temperature.

Finally Free-electron lasers (FEL) originate from an early study of Motz in 1951 [112], at the time at Microwave Lab. in Stanford, Ca., who described the use of an undulator for the production of millimetre and submillimetre waves by means of a relativistic beam of electrons. In this same paper Motz states that prior reference to his problem was found in a paper by V. L. Ginsburg [114].

Later on in 1953 he also obtained experimental results [114].

Coming back to UK, at the Engineering Lab., in Oxford, he still continued to contribute to the theory of radiation emitted by electrons in a undulator, and also made in 1956 the first proposal for a traveling-wave maser [115].

In 1976 the first free-electron laser was demonstrated by Madey [116]. This new laser attracted the interest of many laboratories in the world and notably in Italy [117], in the Soviet Union [118], in UK [119], and in France [120] several projects started. The French FEL has been successfully working in 1983.

The few examples that we have considered here show how each kind of laser has been developed through the efforts of many scientists of different countries without any restriction, and how the European contributions have been essential.

For space reasons many contributions have been omitted which nevertheless deserved mention and I apologize for these omissions, hoping to be able to give them space in another occasion.

References

1. T. H. Maiman, Nature, 187, 493 (1960); British Commun. and Electronics, 7, 674 (1960).
2. A. Einstein, Z. Phys., 18, 121 (1917).
3. R. Ladenburg, Z. Phys., 4, 451 (1921); R. Ladenburg and F. Reiche, Naturwiss, 11, 584 (1923).
4. H. A. Kramers, Nature, 113, 673 (1924); H. A. Kramers, Nature, 114, 310 (1924); H. A. Kramers and W. Heisenberg, Z. Phys., 31, 681 (1925).
5. R. Ladenburg, Z. Phys., 48, 15 (1928); R. Ladenburg and H. Kopfermann, Z. Phys., 48, 26 and 51 (1928); and Z. Phys., 65, 167 (1930); R. Ladenburg and S. Levy, Z. Phys., 65, 189 (1930); A. Carst and R. Ladenburg, Z. Phys., 48, 192 (1928).
6. s. for ex. F. Bloch, W. W. Hansen and M. Packard, Phys. Rev., 70, 474 (1946); E. M. Purcell and R. V. Pound, Phys. Rev., 81, 279 (1951).
7. A. Kastler, J. Physique Radium, 11, 255 (1950); J. Brossel and A. Kastler, C. R. Acad. Sci. Paris, 229, 1213 (1949).
8. The reader is referred for ex. to the book, M. Bertolotti: Masers and Lasers: An Historical Approach - Adam Hilger, Bristol 1983 for more informations and bibliography.
9. s. special issue of IEEE J. Quantum Electronics, vol. QE-20, No. 6 June 1984.
10. This study is under the direction of Dr Joan Bromberg. For more informations on this point see f.e. Lasers Focus, Oct. 1984, pp. 58 to 60.
11. V. A. Fabrikant, Thesis (1940) quoted in F. A. Butayeva and V. A. Fabrikant - Investigations in Experimental and Theoretical Physics, A. Memorial to S. G. Landsberg (USSR Acad. of Science Publications, Moscow 1951) pp. 62-70.
12. V. A. Fabrikant, M. M. Vudynskii and F. Butayeva, USSR patent n. 123209 submitted 18 June 1951, and published in 1959.
13. A. L. Schawlow and C. H. Townes, Phys. Rev., 112, 1940 (1958).
14. V. A. Fabrikant, Zhur. Eksp. Teoret. Fiz., 41, 524 (1961) in russian.

15. A. Javan, W. R. Bennett jr., and D. R. Herriott, Phys. Rev. Lett., $\underline{6}$, 106 (1961).
16. Operation in visible was obtained shortly after by A. D. White and J. D. Rigden, PIRE $\underline{50}$, 1697 (1962).
17. J. H. Sanders, Phys. Rev. Lett., $\underline{3}$, 86 (1959).
18. A. Javan, Phys. Rev. Lett., $\underline{3}$, 87 (1959).
19. J. H. Sanders, M. J. Taylor and C. E. Webb, Nature, $\underline{193}$, 767 (1962).
20. F. A. Butayeva and V. A. Fabrikant, Investigations in Experimental and Theoretical Physics - A memorial to G. S. Landsberg, USSR Acad. Sci. Publ. Moscow 1959 in russian pp. 62-70.
21. W. E. Bell, Appl. Phys. Lett., $\underline{4}$, 34 (1964).
22. G. Convert, M. Armand, P. Martinot-Lagarde, C.R., $\underline{258}$, 3259 (1964).
23. G. Convert, M. Armand and P. Martinot-Lagarde, C.R., $\underline{258}$, 4467 (1964).
24. W. R. Bennett jr., J. W. Knutson jr., G. N. Mercer and N. J. L. Detch, Appl. Phys. Lett., $\underline{4}$, 180 (1964).
25. W. B. Bridges, Appl. Phys. Lett., $\underline{4}$, 128 (1964).
26. W. B. Bridges, PIEEE $\underline{52}$, 843 (1964).
27. P. Laures, L. Dana and C. Frapard, C. R. $\underline{259}$, 745 (1964); L. Dana and P. Laures - PIEEE $\underline{53}$, 78 (1965).
28. F. Legay and P. Barchewitz C.R., $\underline{256}$, 5305 (1963); F. Legay and N. Legay-Sommaire, C.R., $\underline{257}$, 2644 (1963).
29. F. Legay and N. Legay-Sommaire, C.R., $\underline{259}$, 99 (1964).
30. C. K. N. Patel, W. L. Faust and R. A. McFarlane - Bull. Am. Phys. Soc., $\underline{9}$, 500 (1964); C. K. N. Patel, Phys. Rev. Lett., $\underline{12}$, 588 (1964); ibidem $\underline{13}$, 617 (1964); C. K. N. Patel, Phys. Rev., $\underline{136A}$, 1187 (1964).
31. N. Legay-Sommaire, L. Henry and F. Legay, C.R., 260, 3339 (1965).
32. R. Dumanchin and J. Rocca-Serra, C.R., $\underline{269}$, 916 (1969); s. also A. J. Beaulieu, APL $\underline{16}$, 504 (1970); W. Wolinski, Polish Patent 56981; Laser Focus, March (1975).
33. N. G. Basov, IEEE J. Quant. Electr. QE-2, 354 (1966).
34. N. G. Basov, V. A. Danilychev, Yu. M. Popov, and D. D. Khodkevich, JETP Lett., $\underline{12}$, 329 (1970).
35. N. G. Basov, E. M. Belenov, V. A. Danilychev, O. M. Kerimov, I. B. Kovsh, A. F. Suchkov, Sov. J. QE, $\underline{1}$, 306 (1971).
36. N. G. Basov, E. M. Belenov, V. A. Danilychev, O. M. Kerimov, I. B. Kovsh, A. F. Suchkov, JETP Lett., $\underline{14}$, 285 (1971); N. G. Basov, E. N. Belenov, V. A. Danilychev, A. P. Suchkov, Sov. Phys. Quant. Electr., $\underline{1}$, 121 (1971) (in russian).

37. N. G. Basov and A. N. Oraevskii, Sov. Phys. JETP, $\underline{17}$, 1171 (1963).
38. V. K. Konyukhov and A. M. Prokhorov, JETP Lett., $\underline{3}$, 286 (1966).
39. I. R. Hurle and A. Hertzberg, Phys. Fluids, $\underline{8}$, 1601 (1965).
40. E. T. Gerry, IEEE Spectrum, $\underline{7}$, 51 (1970); Bull. Am. Phys. Soc., $\underline{15}$, 563 (1970).
41. V. K. Konyukhov, I. V. Matrosov, A. M. Prokhorov, A. T. Shalunov, N. N. Shirokov, JETP Lett., $\underline{12}$, 321 (1970).
42. A. P. Dronov, A. S. D'yakov, E. M. Kudriavtsov, N. N. Sobolev, Sov. Phys. JETP, $\underline{11}$, 353 (1970).
43. D. M. Kuehn, D. J. Monson, APL, $\underline{16}$, 48 (1970).
44. F. G. Houtermans, Helv. Phys. Acta, $\underline{33}$, 933 (1960).
45. J. J. Ewing, L. A. Brau, Phys. Rev., $\underline{A12}$, 129 (1976); C. A. Brau, J. J. Ewing, J. Chem. Phys., $\underline{63}$, 4640 (1976).
46. M. F. Golde, B. A. Trush, Chem. Phys. Lett., $\underline{29}$, 486 (1974); s. also J. E. Valazco and D. W. Setser, J. Chem. Phys., $\underline{62}$, 1990 (1975).
47. T. C. Polanyi, Proc. R. Soc. of Canada, $\underline{54}$, 25 (1960).
48. J. V. V. Kasper and G. C. Pimentel, PRL, $\underline{14}$, 352 (1965).
49. N. G. Basov, E. P. Markin, A. I. Nikitin and A. N. Oraevskii in Proc. Symp. Chemical Laser, St. Louis (1969); O. M. Batovskii, G. K. Vasil'ev, E. F. Makarov and V. L. Tal'rose JETP Lett., $\underline{9}$, 200 (1969); N. G. Basov, L. V. Kulakov, E. P. Markin, A. I. Nikitin and A. N. Oraevskii, JETP Lett., $\underline{9}$, 375 (1969); s. also Int. Symp. Chemical Laser, Moscow, USSR, 2-4 Sept. 1969; V. F. Zharov, V. K. Malinovskii, Yu. S. Neganov, G. M. Chu nak JETP Lett., $\underline{16}$, 154 (1972); V. S. Zuev, V. A. Katulin, V. Yu. Nosach, A. L. Petrov, J. of Soviet Laser Research, $\underline{3}$, 324 (1982).
50. T. A. Cool and R. R. Stephens, APL, $\underline{16}$, 55 (1970); J. Chem. Phys., $\underline{51}$, 5175 (1969); T. A. Cool, R. R. Stephens and T. J. Falk, Int. J. Chem. Kinetics, $\underline{1}$, 495 (1969).
51. S. G. Rautian and I. I. Sobel'man, Sov. Phys. JETP, $\underline{16}$, 1433 (1962) (JETP, $\underline{41}$, 2018 (1961) in russian).
52. J. V. V. Kasper and G. C. Pimentel, APL, $\underline{5}$, 231 (1964).
53. T. L. Andreeva, V. A. Dudkin, V. E. Malyshev, G. V. Mikhailov, V. N. Sorokin and L. A. Novikove, Sov. Phys. JETP, $\underline{22}$, 969 (1966).
54. A. J. De Maria and C. J. Ultee, Appl. Phys. Lett., $\underline{9}$, 67 (1966).
55. C. H. Ferrar, Appl. Phys. Lett., $\underline{12}$, 381 (1968).
56. D. W. Gregg, R. E. Kidder, C. V. Dopler, Appl. Phys. Lett., $\underline{13}$, 297 (1968).

57. P. Gensel, K. Hohla, K. L. Kompa, Appl. Phys. Lett., 18, 48 (1971).
58. W. Full, K. Hohla, Z. Nat., 31a, 569 (1976); V. S. Zuev, V. A. Katulin, V. Yu. Nosach, O. Yu. Nosach, Sov. Phys. JETP, 35, 870 (1972); F. T. Aldridge, Appl. Phys. Lett., 22, 180 (1973).
59. K. Hohla and K. Kompa, Appl. Phys. Lett., 22, 77 (1973).
60. N. G. Basov, V. S. Zuev, Nuov. Cim., 31B, 129 (1976).
61. R. E. Palmer, T. D. Padrick and M. A. Palmer, Opt. Quantum Electron., 11, 61 (1978); E. D. Jones, M. A. Palmer, F. R. Franklin, Opt. Quantum Elec., 8, 231 (1976).
62. H. T. Baker and T. A. King, J. Phys., D9, 2433 (1976); J. Phys., E9, 287 (1976).
63. K. Hohla, High power iodine atom laser in High-power lasers and Applications, Ed. by K. L. Kompa and H. Walther, Springer-Verlag, Berlin, 1978, p. 124.
64. G. Brederlow, R. Brodmann, K. Eidmann, M. Nippus, R. Petsch, S. Witkowski, R. Volk, K. J. Witte, IEEE J. QE 16, 122 (1980).
65. M. Piltch, W. T. Walter, N. Solimene, G. Gould, W. R. Benet, Appl. Phys. Lett., 7, 309 (1965); C. R. Foules, W. T. Silfast, Appl. Phys. Lett., 6, 236 (1965); W. T. Walter, M. Piltch, N. Solimene, G. Gould, Bull. Am. Phys. Soc., 11, 113 (1966); W. T. Walter, N. Solimene, M. Piltch, G. Gould, IEEE J. Quant. Electr. QE 2, 474 (1966); IEEE J. Quant. Electr. QE 4, 355 (1968).
66. A. A. Isaev, M. A. Kazaryan, G. G. Petrash, JETP Lett., 16, 27 (1972).
67. A. A. Isaev, G. Yu. Lemmerman, Sov. J. Quantum Electr., 7, 799 (1977).
68. P. A. Bokhan, V. A. Gerasimov, V. I. Solomonov, V. B. Shcheglov, Sov. J. Quantum Electr., 8, 1220 (1978).
69. B. G. Bricks, T. W. Karras, R. S. Anderson, J. Appl. Phys., 49, 38 (1978).
70. I. Smilanski, G. Erez, A. Kerman, C. A. Levine - Opt. Comm., 30, 70 (1979).
71. P. A. Bokhan, V. D. Burlakov, V. A. Gerasimov, V. I. Solomonov - Sov. J. Quantum Electr., 6, 672 (1976); R. S. Anderson, B. G. Bricks, T. W. Karras, L. W. Springer - IEEE J. Quantum Electr. QE-12, 313 (1976).
72. P. A. Bokhan, V. I. Solomonov, V. B. Shcheglov - Sov. J. Quantum Electr., 7, 1032 (1977).
73. B. G. Bricks, T. W. Karras, R. S. Anderson, R. J. Homsey-BAPS 23, 153 (1978).

74. B. G. Bricks, T. W. Karras, T. E. Buczacki, L. W. Springer, R. S. Anderson - IEEE J. Quantum Electr. QE-11, 57 (1975).
75. I. S. Aleksandrov, Yu. A. Babliko, A. A. Babaev, O. I. Buzhinskii, L. A. Vasil'ev, A. V. Efimov, S. I. Krysanov, G. N. Nikolaev, A. A. Slivitskii, A. V. Sokolov, L. V. Tatarintsev and V. S. Tereshchenkov, Sov. J. Quant. Electr., 5, 1132 (1975); Yu. A. Babelko, L. V. Vasil'ev, V. K. Orlov, A. V. Sokolov and L. V. Tatarintsev - Sov. J. Quant. Electr., 6, 1258 (1976).
76. A. Yu. Artem'ev, Yu. A. Babelko, O. M. Bakhtin, B. L. Borovich, L. A. Vasil'ev, V. E. Gerts, E. P. Nalegach, G. E. Ratnikov, L. V. Tatarintsev and A. N. Ul'yanov - Sov. J. Quant. Electr., 10, 1121 (1980).
77. P. P. Sorokin and J. R. Lankard, IBM J. of Res. and Dev., 10, 162 (1966).
78. M. L. Spaeth and D. P. Bortfield, Appl. Phys. Lett., 9, 179 (1966).
79. F. P. Schäfer, W. Schmidt and J. Volze, Appl. Phys. Lett., 9, 306 (1966).
80. s. scientific reports of the Institute of Physics of Byelorussian Academy of Sciences in 1964-65.
81. B. I. Stepanov, A. N. Rubinov, V. A. Mostovnikov - Sov. Phys. JETP Lett., 5, 117 (1967).
82. B. I. Stepanov, A. N. Rubinov, V. A. Mostovnikov - J. of Applied Spectroscopy, 7, 116 (1967) in russian; s. also B. I. Stepanov and A. N. Rubinov - Sov. Phys. Uspeki, 11, 304 (1968); Pumping with special flash lamps was also obtained by W. Schmidt and F. P. Schäfer - Z. Naturforsch. 22a, 1563 (1967), and P. P. Sorokin and J. P. Lankard - IBM J. Res. Develop., 11, 148 (1967).
83. B. B. Snavely and F. P. Schäfer - Phys. Lett., 28A, 728 (1969).
84. J. B. Marling, D. W. Gregg, S. J. Thomas, IEEE J. Quantum Elec. QE-6, 570 (1970); J. B. Marling, D. W. Gregg, L. Wood, Appl. Phys. Lett., 17, 527 (1979); R. Pappalardo, H. Samelson, A. Lempicki, Appl. Phys. Lett., 16, 267 (1970).
85. N. A. Borisevich, I. I. Kolosha, V. A. Tolkachev, Sov. J. of Appl. Spectroscopy, 19, 1108 (1973) (in russian); The following year a dye vapor laser was assembled by F. P. Schäfer - Opt. Commun., 10, 219 (1974).
86. W. H. Glenn, M. J. Brienza, A. J. De Maria, Appl. Phys. Lett., 12, 54 (1968); B. H. Soffer and J. W. Linn, J. Appl. Phys., 39, 5859 (1968); M. Bass and J. J. Steinfeld, J. Quant. Elec. QE-4, 53 (1968).

87. D. J. Bradley and A. J. F. Durrant, Phys. Lett., 27A, 73 (1968); D. J. Bradley, A. J. F. Durrant, G. M. Gale, M. Moore, P. D. Smith, IEEE J. Quant. Electr. QE-4, 707 (1968).
88. W. Schmidt and F. P. Schäfer, Phys. Lett., 26A, 558 (1968).
89. E. G. Arthurs, D. J. Bradley, B. Liddy, F. O'Neil, A. G. Roddie, W. Sibbett and N. E. Sleat, Proc. of the 10th Inter. Conf. on High Speed Photography (Association Nationale de la Recherche Technique, Paris, 1973).
90. D. J. Bradley, A. J. F. Durrant, F. O'Neil, B. Sutherland, Phys. Lett., 30A, 535 (1969); The two-photons fluorescence technique was introduced by J. A. Giordmaine, P. M. Reutzepis, S. L. Shapiro and K. W. Wecht, Appl. Phys. Lett., 11, 216 (1967); Another non-linear techniques for measuring short pulses was first proposed using second harmonic generation in non-linear crystals by M. Maier, W. Kaiser, J. A. Giordmaine, Phys. Rev. Lett., 17, 1275 (1966); J. A. Armstrong, Appl. Phys. Lett., 10, 16 (1967); H. P. Weber, J. Appl. Phys., 38 (1967).
91. D. J. Bradley, B. Liddy, A.G. Sibbett, W. E. Sleat, Opt. Comm., 3, 426 (1971); D. J. Bradley, B. Liddy, W. E. Sleat, Opt. Comm., 2, 391 (1971) but see also V. V. Korobkin and M. Ya. Schelev in: High Speed Photography, ed. N. R. Nilsson and L. Hogberg, Stockolm, 1968 p. 36.
92. V. S. Letokhov, Sov. Phys. JETP, 27, 746 (1968); P. G. Kryukov, V. S. Letokhov, IEEE J. Quant. Electr. QE-8, 766 (1972).
93. G. H. C. New, Opt. Comm., 6, 188 (1972); IEEE J. Quant. Elec. QE-10, 115 (1974).
94. C. V. Shank and E. P. Ippen, Appl. Phys. Lett., 24, 373 (1974).
95. R. L. Fork, B. J. Green and C. V. Shank, Appl. Phys. Lett., 38, 671 (1981).
96. W. H. Knox, R. L. Fork, M. C. Downer, R. H. Stolen, C. V. Shank, J. A. Valdmanis post-dead line paper at CLEO, Thzz3, May 21-24, 1985, Baltimore.
97. for more details concerning polish activities see for example M. Bertolotti, Opt. Acta (to be published).
98. S. Kielich, Optica Applicate 4, 39 (1974).
99. J. von Neumann, Collected works, vol. 5 (Pergamon, London, 1963).
100. C. N. Townes, IEEE J. Quant. Electr. QE-20, 547 (1984).
101. s. for ex. Yu. M. Popov, Proc. PN Lebedev Physics Institute vol. 31 ed. D. V. Skobel'tsyn, Consultant Bureau, New York, 1968.

102. quoted in P. Aigrain in Quantum Electronics ed. by P. Grivet and N. Bloembergen, Dunod, Paris, 1964 p. 1761; s. also M. Bernard and G. Duraffourg, J. de Phys., 22, 836 (1961) and ref. 104.
103. N. G. Basov, O. N. Krokhin and Yu. M. Popov, JETP, 40, 1879 (1961) in russian.
104. M. G. A. Bernard and G. Duraffourg, Phys. Status Solidi, 1, 669 (1961).
105. R. N. Hall, G. E. Fenner, J. O. Kingsley, T. J. Soltys and R. O. Carlson, Phys. Rev. Lett., 9, 366 (1962); N. I. Nathan, W. P. Dumke, G. Burns, F. H. Dill and G. Lasher, Appl. Phys. Lett., 1, 62 (1962); R. J. Keyes and T. M. Quist, Proc. IRE, 50, 1822 (1962); T. M. Quist, R. H. Rediker, R. J. Keyes, W. E. Krog, B. Lax, A. L. McWhorter and H. J. Zeiger, Appl. Phys. Lett., 1, 91 (1962); N. Holonyak jr. and S. F. Bevacqua, Appl. Phys. Lett., 1, 82 (1962).
106. For an account of the general atmosphere that brought to the construction of the first semiconductor lasers see f.e. H. C. Casey jr. and M. B. Panish - Heterostructure lasers - Part A, Academic Press, New York, 1978, Chp I and reference there in contained; R. N. Hall, IEEE Trans. Electron Devices ED-23, 700 (1976); C. K. N. Patel in ref. 9, p. 561; P. P. Sorokin in ref. 9, p. 585; R. H. Rediker, I. Melngailis, A. Mooradian in ref. 9, p. 602; J. Hecht's interview of R. N. Hall in Lasers of Applications, Feb. 1985, p. 51.
107. N. G. Basov, B. M. Vul, Yu. M. Popov, Sov. Phys. JETP, 10, 416 (1959) (ZEFT, 37, 587 (1959) in russian).
108. N. G. Basov, A. G. Molchanov, A. S. Nasibov, A. Z. Obidin, A. N. Pechenov, Yu. M. Popov, JETP Lett., 19, 336 (1974).
109. N. G. Basov, A. G. Molchanov, A. S. Nasibov, A. Z. Obidin, A. N. Pechenov, Yu. M. Popov, IEEE J. Quant. Elec. QE-10, 794 (1974); N. G. Basov, A. G. Molchanov, A. S. Nasibov, A. Z. Obidin, A. N. Pechenov, Yu. M. Popov, JETP, 43, 912 (1976); N. G. Basov, A. G. Molchanov, A. S. Nasibov, A. Z. Obidin, A. N. Pechenov, Yu. M. Popov, IEEE J. Quantum Elec. QE-13, 699 (1977).
110. I. Hayashi, M. B. Panish, P. W. Foy and S. Sumski, Appl. Phys. Lett., 17, 109 (1970).
111. Zh. I. Alferov, V. M. Andreev, D. Z. Garbuzov, Yu. V. Zhilyaev, E. P. Morozov, E. L. Partnoi and V. G. Trofin, Sov. Phys. Semic., 4, 1573 (1971) (Fiz. Tekh. Poluprovods. 4, 1826 (1970) in russian).

112. H. Motz, J. Appl. Phys., 22, 527 (1951).
113. V. L. Ginsburg, Bull. Acad. Sci. URSS Ser. Phys., 9, (1947) N. 2, 165 (in russian).
114. H. Motz, W. Thon and R. N. Whitehurst, J. Appl. Phys., 24, 827 (1953).
115. H. Motz, J. of Electronics, 2, 571 (1957).
116. R. L. Elias, W. M. Fairbank, J. M. J. Madey, N. A. Schwettman, T. I. Smith, Phys. Rev. Lett., 36, 717 (1976), s. also ref. 8.
117. In Italy there is a large activity in free-electron lasers. In Frascati, near Roma, two groups are working. One supported by the Italian National Institute of Nuclear Physics (INFN) considers a recirculated beam experiment (called LELA) to be performed on the storage ring ADONE (s. R. Barbini, G. Vignola: LELA: A free--electron laser experiment in Adone - Quaderni de "La ricerca scientifica" Roma CNR 1980 p. 81) in collaboration with Naples University and emitting in the visible (R. Barbini et al., J. de Phys., 44, C1-1 (1983); M. Biagini et al., SPIE vol. 453, p. 275 (1983); M. Castellano, N. Cavallo, F. Ceverini, M. R. Masullo, P. Patteri, R. Rinziviollo, S. Solimeno, A. Cutolo, Nuovo Cimento, 81B, 67 (1984); The National Energy Committee (ENEA) supports instead two experiments. The older one makes use of a 20 MeV microtron to obtain IR radiation in the range 10-30 m (U. Bizzarri et al. Nucl. Instr. of Methods, 208, 177 (1983); J. de Phys., 44, C1, 313 (1983)).

A second project, started in 1985, considers the use a 5 MeV microtron for a Cerenkov laser in the range 100 to 300 m (Walsh, Johnson, Dattoli, A. Renieri, Phys. Lett., 53, 779 (1984).

To this experimental activity there is a large theoretical support in Frascati (s. for a general review A. Renieri in Laser Handbook, vol. 4, ed. M. L. Stitch and M. S. Ban), in Milano (s. for ex. R. Bonifacio and F. Casagrande, J. Opt. Soc. America, 2B, 250 (1985); Opt. Comm., 50, 251 (1984); 373 (1984), Roma (s. for ex. C. Sibilia, M. Bertolotti, V. Perinova, A. Luks, Phys. Rev., A28, 329 (1983); F. De Martini and J. M. J. Madey in Physics of Quantum Electronics, ed. by W. Lamb, M. Sargent, M. Scully, N. Y. 1979) and Lecce (I. Boscolo, M. Leo, R. A. Leo, G. Poliani, V. Stagno, Opt. Comm., 36, 337 (1981).
118. Theoretical activity in the Soviet Union is in Moscow, Gorki, Erevan, Tomsk. Experimental activity is in Novosibirsk where a

project is under development s. G. A. Kornynkhin, G. N. Kulipanov, V. N. Litvinenko, N. A. Vinokurov, P. D. Voblyi, Nucl. Instr. Meth., 208, 189 (1983).

119. The UK Free Electron Laser Project represents a consortium of Universities, with the program led by Heriot-Watt and includes Glasgow, Imperial College, Essex University and Daresbury Lab. of the Science and Engineering Research Council. A fully dedicated 100 MeV LINAC at the Kelvin Laboratory is available. N. Marks, G. N. Greaves, M. W. Poole, V. P. Suller, R. P. Walker, Nucl. Instr. Meth., 208, 97 (1983).

120. The French project (D. A. G. Deacon, K. E. Robinson, J. M. J. Madey, C. Bazin, M. Billardon, P. Elleaume, Y. Farge, J. M. Ortega, Y. Petroff and M. F. Velghe, Opt. Comm., 40, 373 (1982)) is only one at present which has given successful operations; M. Billardon, P. Elleaume, J. M. Ortega, M. Bergher, M. Velghe, Y. Petroff, D. A. G. Deacon, J. Madey, K. Robinson, Phys. Rev. Lett., 51, 1652 (1983).

Bruce W. SHORE

Lawrence Livermore
National Laboratory Livermore,
California 94550, USA

CONCLUDING REMARKS

It is a pleasure and an honor to be invited to make some remarks at the close of this very instructive and enjoyable school. I want to do two things in the moments given me: I want to comment very briefly on a few scientific problems that seem particulary interesting to me; and I wish to express thanks to the organizers of this conference. But before beginning I want to acknowledge the financial support of the National Science Foundation of the United States whose travel grant made possible the participation of the American delegation, and I want to acknowledge the personal financial support of the Lawrence Livermore National Laboratory, operated by the University of California under contract to the US Department of Energy under contract No. W-7405-Eng-48.

First let me comment on a very few general topics that I have heard discussed here in this school on quantum optics. By the term "quantum optics" most of us probably mean quantum mechanics as embodied in optical radiation. What could more simply reveal that subject than the study of a single atom interacting with a single photon, a system described by the Jaynes-Cummings model? We have heard how this model, applied to the slightly more general model of a single atom interacting with several photons, leads to Rabi oscillations that are the basis of so much in quantum optics, to decay of these oscillations, and subsequently to revivals. This behavior was, as we have heard, quite as expected; yet theory is now recognized as being quite clear in its predictions. I think that we would all agree that the announcement of experimental detection of these revivals is the most memorable part of this school, and Prof. Walther well deserves the applause with which we greeted his presentation of these remarkable new results.

The notion of a photon as the basic quantized entity of the electromagnetic field is another part of our fundamental theory of matter and radiation. Our ideas and intuitions concerning this entity have changed over the years, as have our ways of thinking of the quantum theory of the electromagnetic field. In the beginning there were integers - the number states - and I think that for many years it was assumed that these states were somehow the unique means of discribing quantum fields. Then came the coherent states, and we began to recognize that we had choices of representations for fields just as we did for particles. The extension of this formalism to squeezed states gives us further opportunity, by broadening our experience with various representations, to distinguish those properties of the field that are intrinsically quantum mechanical from those that merely depend upon our choice of representation or on the source of the light. We see also in this topic a challenging opportunity, only dimly glimpsed at present, to find experimental evidence and subsequently practical application for those new types of elementary fields (i.e. photons).

I have spoken of the atom-field interaction as though an atom absorbed and emitted single photons in distinct separated events. We know, of course, from our familiar Rabi oscillations, that this language is inappropriate for the description of coherent bound state excitation. We have developed a variety of ways to picture that interaction: precesing Bloch vectors, dressed atoms, and so on. We know the relevant parameters; the Rabi frequencies, the detuning frequencies, the relaxation rates. Now we are begining to extend our understanding of this coherent excitation to processes that occur in the ionization continuum; to deal with Rabi oscillations in the continuum. In this enterprise we are still struggling to clarify our basic ideas. We are beginning to understand what we mean by "strong" and "weak" interactions, and how to identify a characteristic localization volume for the interaction. Here too I forsee many opportunities to clarify basic physics.

I have commented that we are revisiting over notion of photons. Radiation must, of course, originate in some source, whether that be a laser or the filtered light from the thermal source. Inevitably the source produces fluctuations, either in phase (as in an idealized laser) or in amplitude (as in a thermal source). We are seeing how it is possible to observe consequences of these fluctuations, as well as the fluctuations in the atomic enviroment. Experiments are now making it possible to observe the consequences of fluctuations and to distinguish between various types of fluctuations. I anticipate the future will bring still further interesting results.

As a theorist I always find it interesting to discover exact soluble models of physical systems, even very idealized systems. Therefore it has been interesting to see at this school that most elementary of soluble models, the harmonic oscillator, still has much to teach us, as we have found in several talks. But it is also clear that the definition of "exactly soluble" now is very broad, for we have seen simple analytic expressions emerge from formidably complicated mathematical expressions. By contemplating such results we gain insight into the expected behavior of systems that are emanate at present, only to numerical description in computer models.

Lastly I must comment on the theme that has recurred persistently throughout this school: the search for quantum chaos. That is a deliberately provacative term. What I mean is more modest. We are beginning to know how the classical dynamics of nonlinear systems leads to extremly irregular behavior, behavior that is commonly termed chaotic. It is natural to ask what the quantum mechanical counterpart is for this behavior. It is very exciting to see the emergence of the body of knowledge on this topic. We are discovering simple models whose behavior can be examined by classical and by quantum theory, and we are learning both of the similarities and the differences. We are learning what questions to ask. It is exciting to anticipate the development of this area of study, the murky boundary between classical and quantum theory.

I have been impressed, during this school, by the number of projects that represent collaborations between scientists from different nations. Such collaborations have been responsible for many of the most exciting results. Therefore I particularly appreciate the opportunity to meet with scientists from many countries, as I see how universal is the science that sustains our interest. I personally welcome the opportunity to renew my acquaintances in Poland and to enlarge my acquaintances with Soviet scientists; to exchange ideas, to learn and to teach. I hope that these contacts, so warm on a personal basis, can in time be communicated to the higher levels of government where, unfortunately, present relations are not so friendly. For this opportunities, and for the opportunity to meet in pleasant surroundings at a place conducive to teaching and to learning (and that, after all is the purpose of a school) I extend to the organizing committee on behalf of all of us, our heartfelt thanks.

LIST OF PRESENTED POSTERS

1. A. BANDILLA, Zentralinstitut für Optik und Spektroskopie, AdW der DDR, 1199 Berlin-Adlershof
   Photon statistics of saturated emission and absorption processes
2. B. BIENIAK and M. GŁÓDŹ, Institute of Physics, Polish Academy of Sciences, Al. Lotników 32/46, 02-668 Warsaw, Poland
   Studies of the time dependent molecular fluorescence from $C(^1\pi_u)$ state of caesium dimer
3. Z. BŁASZCZAK, H. BŁASZCZAK* and P. GAUDEN, Institute of Physics, A. Mickiewicz University, Poznań, Poland
   *Department of Organic Chemistry, School of Medicine, Poznań, Poland
   Laser induced optical birefringence in urea solutions
4. N. N. BOGOLUBOV Jr. Tran QUANG and A. S. SHYMOVSKY, Laboratory of Theoretical Physics, JINR, Dubna, USSR
   Double resonance in a system of atoms
5. W. CHMARA, Department of Physics II, Academy of Technology and Agriculture, Bydgoszcz, Poland
   An operational formula for photon counting statistics
6. Cao LONG VAN and M. TRIPPENBACH, Institute for Theoretical Physics, Polish Academy of Sciences, Al. Lotników 32/46, 02-668 Warsaw, Poland
   Photoelectron spectrum in a model for laser induced autoionisation
7. P. CHMELA, Joint Laboratory of Optics of Czech. Acad. Sci. and Palacky University in Olomouc, Olomouc 77146, Czechoslovakia
   Evolution of light statistics in multiphoton absorption of strong coherent and chaotic radiation
8. A. DROBNIK, A. FRUZIŃSKI and L. WOLF, Institute of Physics, Łódź Technical University, Wólczańska 229, Łódź, Poland
   The recording of thermal change of light reflection coefficient n with holographic interferometry method

9. A. DULCIĆ, Ruder Bosković Institute, University of Zagreb, POB 1016, 41001 Zagreb, Croatia, Yugoslavia
   Diagonal continuum-continuum coupling and ionisation in strong laser fields
10. A. EKERT, W. GAWLIK and J. ZACHOROWSKI, Institute of Physics, Jagiellonian University, Reymonta 4, 30-059 Cracow, Poland
    HFS decoupling in a strong, nonmonochromatic light field
11. W. FALECKI, P. GAWLIK, W. GAWLIK and J. ZACHOROWSKI, Institute of Physics, Jagiellonian University, Reymonta 4, 30-059 Cracow, Poland
    Four-wave mixing and other nonlinear effects in barium
12. Z. FICEK, R. TANAŚ and S. KIELICH, Nonlinear Optics Division, Institute of Physics, A. Mickiewicz University, 60-780 Poznań, Poland
    Spontaneous emission from two nonidentical atoms
13. B. GADOMSKA and W. GADOMSKI, Laboratory of Physics of Dielectrics, Dept. of Chemistry, Warsaw University, Żwirki i Wigury 101, 02-089 Warsaw, Poland
    Bistability in four-wave interaction in crystals
14. W. GAWLIK, Institute of Physics, Jagiellonian University, Reymonta 4, 30-059 Cracow, Poland
    Subnatural laser spectroscopy with nonstationary velocity - selective optical pumping
15. M. GŁÓDŹ and M. KRAIŃSKA-MISZCZAK*, Institute of Physics, Polish Acad. Sci., Al. Lotników 32/46, 02-668 Warsaw, Poland
    *Institute of Experimental Physics, Warsaw University, Hoża 69, 00-681 Warsaw, Poland
    Quantum beat method for the measurment of the HFS A and B constants values below the natural width of the level $6^2D_{3/2}$ in ^{41}K
16. J. GROCHMALICKI, J. MOSTOWSKI and K. RZĄŻEWSKI,* Institute of Physics, Polish Acad. Sci., Al. Lotników 32/46, 02-668 Warsaw, Poland
    Near threshold ionisation of atomic hydrogen
17. K. GNIADEK, Institute of Physics, Warsaw Technical University, Koszykowa 75, Warsaw, Poland
    Dispersion relations for nonlinear TM guided waves
18. W. JASTRZĘBSKI and M. KOLWAS, Institute of Physics, Polish Acad. Sci., Al. Lotników 32/46, 02-668 Warsaw, Poland
    The nonlinear phenomena in optical pumping of Ne

19. W. JAWORSKI, Institute of Physics, Nicolas Copernicus University, Grudziądzka 5, 87-100 Toruń, Poland

   On the microcanonical ensemble for two-level atoms interacting with one mode electromagnetic field

20. A. Ya. KARASIK

   Nonlinear methods of ultrashort pulses formation with the help of optical fibres

21. K. KOLWAS and M. KOLWAS, Institute of Physics, Polish Acad. Sci., Al. Lotników 32/46, 02-668 Warsaw, Poland

   Nonlinear damping of the laser beam on the supersonic sodium flow

22. W. LEOŃSKI, R. TANAŚ and S. KIELICH, Nonlinear Optics Division, A. Mickiewicz University, 60-780 Poznań, Poland

   Double Fano Profile and the strong field photoelectron spectrum

23. M. LEWENSTEIN, J. MOSTOWSKI[*] and M. TRIPPENBACH, Institute for Theoretical Physics, Polish Acad. Sci., Al. Lotników 32/46, 02-668 Warsaw, Poland

   [*]Institute of Physics, Polish Academy of Sciences, Al. Lotników 32/46, 02-668 Warsaw, Poland

   Multichannel decay and continuum-continuum relaxation in above threshold ionisation

24. J. MOSTOWSKI and M. GAJDA, Institute of Physics, Polish Acad. Sci., Al. Lotników 32/46, 02-668 Warsaw

   Equilibrium configurations and temperature effects in the light scattering by a system of many trapped ions

25. R. PARZYŃSKI, Quantum Electronics Laboratory, Institute of Physics, A. Mickiewicz University, Grunwaldzka 6, 60-780 Poznań, Poland

   Light intensity effects on nuclear spin polarisation in resonant two photon ionisation

26. G. A. PASMANIK, Institute of Applied Physics, USSR Academy of Sciences, Gorky, USSR

   Problems of recording space-inhomogeneous quantized fields

27. R. PEPŁOWSKI, Institute of Physics, Nicolas Copernicus University, 87-100 Toruń, Poland

   Codimension two bifurcation in laser system

28. L. POKORA, Institute of Plasma Physics and Laser Microfusion, P.O. Box 49, 00-908 Warsaw, Poland

   Dense plasma as a source of energy for laser pumping

29. D. ROGUŚ and M. LEWENSTEIN, Institute for Theoretical Physics, Polish Acad. Sci., Al. Lotników 32/46, 02-668 Warsaw, Poland
   Resonant ionisation by short laser pulses
30. R. S. ROMANIUK and J. DOROSZ, Institute of Electronics Fundamentals, Warsaw Technical University
   Monomode coherent systems with multicore optical fibres
31. K. RZĄŻEWSKI and R. GROBE[*], Institute for Theoretical Physics, Polish Acad. Sci., Al. Lotników 32/46, 02-668 Warsaw, Poland
   [*]Fachbereich 7, Universität Essen, 4300 Essen, West Germany
   Angular momentum distribution of electrons in above threshold ionisation
32. M. ROMAN, Laboratory of Physicochemistry of Dielectrics, Dept. of Chemistry, Warsaw University, Żwirki i Wigury 101, 02-089 Warsaw, Poland
   Iteration approach to time and spatial resolved stimulated back--scattering of light
33. F. J. SCHÜTTE, K. GERMEY, M. MAREYEN and R. TIEBEL, Pädagogische Hochschule, Potsdam, Section Mathematik/Physik, DDR
   Bifurcations and turbulence in optical systems with several interacting modes
34. A. B. SHVARTSBURG, IZMIRAN, Troitsk, P/O 142092, Moscow Region, USSR
   The statistical phenomena in the nonlinear dynamics of the picosecond pulses in the optical fibres
35. L. SIRKO and K. ROSIŃSKI, Institute of Physics, Polish Acad. Sci., Al. Lotników 32/46, 02-668 Warsaw, Poland
   J-mixing in heavy alkali-atoms excited by nonlinear laser light absorption to intermediate nD_J states
36. K. STEFAŃSKI, Institute of Physics, Nicolas Copernicus University, 87-100 Toruń, Poland
   Semiclassical versus quantum description of a laser bistability region
37. H. STEUDEL, Zentralinstitut für Optik und Spectroskopie, Acad. Sci. of the GDR, Berlin, DDR
   Stimulated Raman scattering with an initial phase shift; the pre--stage of a soliton
38. V. L. VELICHANSKY, P. N. Lebedev Physics Institute, Acad. Sci. USSR, 117924 Moscow, Leninsky Prospekt 53, Moscow, USSR
   Quantum fluctuations and the generation spectrum of injection lasers

39. V. L. VELICHANSKY,
    External cavity diode lasers in high resolution spectroscopy
40. S. VARRO, Central Research Institute of Physics, H-1525 Budapest 114, POB 49, Hungary
    Multiphoton absorption and emission by a free electron in a model laser beam
41. J. SZTUCKI and W. STRĘK, Institute for Low Temperature and Structure Research, Polish Acad. Sci., Wrocław, Poland
    Two-photon transitions in lanthanide systems
42. W. VOGEL and D. G. WELSCH, Friedrich-Schiller-Universität, Sektion Physik, Max-Wien Platz 1, Jena 6900 DDR
    Resonance fluorescence from vibronic systems
43. W. VOGEL and D. G. WELSCH,
    Enhenced squeezing in resonance fluorescence from a regular N-atom system
44. H. VOIGT and A. BANDIŁŁA, Zentralinstitut für Optik und Spektroskopie, AdW der DDR, 1199 Berlin-Adlershof, DDR
    Some remarks on the linewidth of a two-photon laser
45. J. ZAKRZEWSKI, Institute of Physics, Jagiellonian University, Reymonta 4, 30-059 Cracow, Poland
    Pulse shape and chirp influence on multipeak photon and electron spectra
46. J. ZAKRZEWSKI,
    Multiple ionisation of a Hartree atom by intense laser pulses
47. S. ZIELIŃSKA, Institute of Theoretical Physics and Astropyhsics, University of Gdańsk, 80-952 Gdańsk, Poland
    Collision kernels in the eikonal approximation for Lennard-Jones interaction potential

RAYMOND H. FOGLER LIBRARY
DATE DUE

BOOKS ARE SUBJECT TO
RECALL AFTER TWO WEEKS

APR 2 0 1987

AUG 1 7 1987